P9-CSB-121

Stochastic Processes in Information and Dynamical Systems

McGRAW-HILL SERIES IN SYSTEMS SCIENCE

Editorial Consultants: A. V. BALAKRISHNAN, GEORGE DANTZIG, LOTFI ZADEH

BERLEKAMP ALGEBRAIC CODING THEORY
CANON, CULLUM, AND POLAK THEORY OF OPTIMAL CONTROL AND MATHEMATICAL PROGRAMMING
GILL LINEAR SEQUENTIAL CIRCUITS
HARRISON INTRODUCTION TO SWITCHING AND AUTOMATA THEORY
JELINEK PROBABILISTIC INFORMATION THEORY: DISCRETE AND MEMORYLESS MODELS
MANGASARIN NONLINEAR PROGRAMMING
NILSSON LEARNING MACHINES: FOUNDATIONS OF TRAINABLE PATTERN-CLASSIFYING SYSTEMS
PAPOULIS PROBABILITY, RANDOM VARIABLES, AND STOCHASTIC PROCESSES
PAPOULIS SYSTEMS AND TRANSFORMS WITH APPLICATIONS IN OPTICS SIGNAL DESIGN
SAGE AND MELSA ESTIMATION THEORY WITH APPLICATIONS TO COMMUNICATIONS AND CONTROL
SLAGLE ARTIFICIAL INTELLIGENCE: THE HEURISTIC PROGRAMMING APPROACH
VITERBI PRINCIPLES OF COHERENT COMMUNICATION
WEBER ELEMENTS OF DETECTION AND SIGNAL DESIGN
WONG STOCHASTIC PROCESSES IN INFORMATION AND DYNAMICAL SYSTEMS
ZADEH AND DESOER LINEAR SYSTEM THEORY: THE STATE SPACE APPROACH

Stochastic Processes in Information and Dynamical Systems

EUGENE WONG

Department of Electrical Engineering and Computer Sciences
University of California, Berkeley

McGRAW-HILL BOOK COMPANY

NEW YORK SAN FRANCISCO ST. LOUIS DÜSSELDORF JOHANNESBURG
KUALA LUMPUR LONDON MEXICO MONTREAL NEW DELHI PANAMA
RIO DE JANEIRO SINGAPORE SYDNEY TORONTO

This book was set in Monotype Modern 8A, printed on permanent paper, and bound by The Maple Press Company. The designer was Janet Bollow. The editors were Charles R. Wade and Eva Marie Strock. Charles A. Goehring supervised production.

Stochastic Processes in Information and Dynamical Systems

Printed in the United States of America.

Library of Congress catalog card number: 70-139568

1234567890 MAMM 7987654321

07-071568-8

TO JOAN, LINDA, DAVID, AND MICHAEL

Preface

I have written this book to provide a graduate level text in stochastic process for students whose primary interest is its applications. Although stochastic processes have been important in engineering problems for a very long time, recent shifts in emphasis and direction have made a new text seem particularly worthwhile. In the last decade the emphasis has shifted from stationary processes in linear time-invariant systems to stochastic systems in which dynamic structure plays a more profound role. My aim is to provide a high-level, yet readily accessible, treatment of those topics in the theory of continuous-parameter stochastic processes that are important in the analysis of information processing and dynamical systems.

Stochastic process is a subject that can easily become abstract. In dealing with it from an applied point of view, I have found it difficult to decide on the appropriate level of rigor. I intend to provide just enough mathematical machinery so that important results can be stated with precision and clarity; so much of the theory of stochastic processes is inherently simple if the suitable mathematical framework is provided. The price of providing this framework seems worth paying even though my ultimate goal is in applications and not in the mathematics per se.

There are two main topics in the book: second-order properties of stochastic processes (Chapter 3) and stochastic integrals with application to white noise in dynamical systems (Chapters 4 and 6). Each topic provides a convenient core for a one-semester or one-quarter graduate course. I have used this material for several years in two such courses in Berkeley. The level of sophistication required for the first course is considerably lower than that required for the second. For the course centered on second-order properties, a good undergraduate background in probability theory and linear systems analysis is adequate; for the course centered on stochastic integrals, an acquaintance with measure theory is almost necessary. At Berkeley we have required a prerequisite of an undergraduate course on integration. However, a generally high level of mathematical sophistication is probably more important than formal prerequisites.

The supporting material in Chapters 1 and 2 should be used as the instructor sees fit. Much of this material may be inappropriate for one or the other of the two courses and should be omitted.

I have included a fair number of problems and exercises. Solutions have been provided for most of them in order to facilitate self study and to provide a pool of examples and supplementary material for the main text.

I have kept the list of references short and specific. In cases where alternative references exist, I have chosen those which are most familiar to me. The necessary incompleteness of such a list is perhaps compensated by the high degree of relevance that each reference bears to the text. On basic points in probability theory I have relied heavily on the three well-known books by Doob, Loève, and Neveu. On a more elementary level I have found the recent text by Thomasian particularly comprehensive and lucid.

This book could not have been written without direct and indirect assistance from many sources. First, it is obviously a direct result of my teaching and research in stochastic processes. I am grateful to the Department of Electrical Engineering and Computer Sciences of the University of California, Berkeley, for its continuing support. The Army Research Office (Durham), the National Science Foundation, and the Department of Defense through its Joint Services Electronics Program have supported my research activities over the years. The organization of this book and some of its initial writing were done during a very pleasant sabbatical year at the University of Cambridge with the support of a John Simon Guggenheim fellowship. I am indebted to my Cambridge host, Professor J. F. Coales, F. R. S., for his kindness.

Many colleagues and friends have generously provided me with suggestions and criticisms. It is a pleasure to acknowledge the help given me by Pierre Brémaud, Dominque Decavele, Tyrone Duncan, Terence Fine, Larry Shepp, Pravin Varaiya, and Moshe Zakai. I am especially indebted to Bill Root for a painstaking review of a major portion of the manuscript. The number of errors that were found makes me wonder how many still remain, and I dare not emulate my friend Elwyn Berlekamp in offering a cash reward to readers for correction of errors! Nevertheless, I shall be grateful for any report of errors.

Miss Bonnie Bullivant typed the manuscript and its many revisions with both skill and patience.

EUGENE WONG

Contents

1

Elements of Probability Theory

1. EVENTS AND PROBABILITY

The simplest situation in which probability can be considered involves a random experiment with a finite number of outcomes. Let $\Omega = \{\omega_1, \omega_2, \ldots, \omega_N\}$ denote the set of all outcomes of the experiment. In this case a probability p_i can be assigned to each outcome ω_i. The only restrictions that we place on these probabilities are: $p_i \geq 0$ and $\sum_{i=1}^{N} p_i = 1$. Every subset A of Ω in this case also has a well-defined probability which is equal to the sum of the probabilities of the outcomes contained in A.

If the number of outcomes in Ω is countably infinite, the situation is quite similar to the finite case. However, if Ω contains an uncountable number of points, then not more than a countable number of them can have nonzero probabilities, since the number of outcomes with probability $\geq 1/n$ must be less than or equal to n. In the general case, it is necessary to consider probabilities to be defined on subsets of Ω, rather than on points of Ω. A subset A on which a probability is defined is called an **event.** Let $\mathcal{A}$ denote the class of all events. A satisfactory theory of probability demands that the complement of an event should again be an event. Indeed, the occurrence of the complement A^c is just the non-occurrence of A. Similarly, if A and B are two events, then the simultaneous occurrence of both A and B should have a well-defined probability, i.e., the intersection of two events should again be an event. Thus, $\mathcal{A}$ should be closed under complementation and pairwise intersection. This immediately implies that $\mathcal{A}$ is closed under all finite Boolean set operations.[1] A class of sets which is closed under all finite Boolean set operations is called a **Boolean algebra** or simply an **algebra.** An **elementary probability measure** $\mathcal{P}$ is a function defined on an algebra $\mathcal{A}$ such that

$$0 \leq \mathcal{P}(A) \leq 1 \qquad \text{and} \qquad \mathcal{P}(\Omega) = 1 \tag{1.1a}$$

$$\mathcal{P}(A \cup B) = \mathcal{P}(A) + \mathcal{P}(B) \qquad \text{whenever } A \text{ and } B \text{ are disjoint (additivity)} \tag{1.1b}$$

Both (1.1*a*) and (1.1*b*) are clearly natural conditions to be required of a probability measure.

In order that we can consider sequences of events and possible convergence of such sequences, it is necessary that not only finite, but all countable set operations on events again yield events. A class of sets is called a (Boolean) **σ algebra** if it is closed under all countable set operations.[2] It is easy to verify that the intersection of arbitrarily many σ algebras (of subsets of the same Ω) is again a σ algebra (of subsets of Ω). Therefore, given an arbitrary class $\mathcal{C}$ of subsets of Ω, there is a **smallest σ algebra** $\mathcal{A}(\mathcal{C})$ which contains $\mathcal{C}$. This is because we can define $\mathcal{A}(\mathcal{C})$ to be the intersection of all σ algebras containing $\mathcal{C}$, and there is at least one such σ algebra, viz., the collection of all subsets of Ω. We shall say that $\mathcal{C}$ **generates** its minimal σ algebra $\mathcal{A}(\mathcal{C})$. Now, consider a Boolean algebra $\mathcal{B}$ together with an elementary probability measure $\mathcal{P}$. Suppose that in addition to (1.1*a*) and (1.1*b*), $\mathcal{P}$ also satisfies

$$\text{Whenever } \{A_n\} \text{ is a sequence of sets in } \mathcal{B} \text{ such that } A_n \supset A_{n+1} \text{ and } \bigcap_{n=1}^{\infty} A_n = \emptyset \text{ then } \lim_{n\to\infty} \mathcal{P}(A_n) = 0 \text{ (monotone sequential continuity at } \emptyset) \tag{1.1c}$$

[1] Complementation, union, and intersection are the most familiar Boolean set operations. Only complementation and either union or intersection need be defined. All other set operations are then expressible in terms of the two basic operations.

[2] Since all set operations are expressible in terms of complementation and union, to verify that a class is a σ algebra, we only need to verify that it is closed under complementation and countable union.

where $\emptyset$ denotes empty set. Conditions (1.1a) to (1.1c) are equivalent to (1.1a) and the following condition taken together:

Whenever $\{A_n\}$ is a sequence of pairwise disjoint sets in $\mathcal{B}$ such that $\bigcup_{n=1}^{\infty} A_n$ is also in $\mathcal{B}$, then $\mathcal{P}\left(\bigcup_{n=1}^{\infty} A_n\right) = \sum_{n=1}^{\infty} \mathcal{P}(A_n)$

(σ **additivity**) (1.1d)

A set function $\mathcal{P}$ defined on an algebra $\mathcal{B}$ satisfying (1.1a) to (1.1c) is called a **probability measure.** It is a fundamental result of probability theory that a probability measure $\mathcal{P}$ on an algebra $\mathcal{B}$ extends uniquely to a probability measure on the σ algebra generated by $\mathcal{B}$.

Proposition 1.1 (Extension Theorem). Let $\mathcal{B}$ be an algebra and let $\mathcal{A}(\mathcal{B})$ be its generated σ algebra. If $\mathcal{P}$ is a probability measure defined on $\mathcal{B}$, then there is one and only one probability measure $\tilde{\mathcal{P}}$ defined on $\mathcal{A}(\mathcal{B})$ such that the restriction of $\tilde{\mathcal{P}}$ to $\mathcal{B}$ is $\mathcal{P}$ [Neveu, 1965, p. 23].

Thus, we have arrived at the basic concept of a probability space. A **probability space** is a triplet $(\Omega,\mathcal{A},\mathcal{P})$ where Ω is a nonempty set whose elements are usually interpreted as outcomes of a random experiment, $\mathcal{A}$ is a σ algebra of subsets of Ω, and $\mathcal{P}$ is a probability measure defined on $\mathcal{A}$. The set Ω will be called the **basic space,** and its elements are called points. Elements of $\mathcal{A}$ are called **events.**

A subset of an event of zero probability is called a **null set.** Note that a null set need not be an event. A probability space $(\Omega,\mathcal{A},\mathcal{P})$ is said to be **complete** if every null set is an event (necessarily of zero probability). If $(\Omega,\mathcal{A},\mathcal{P})$ is not already complete, $\mathcal{P}$ can be uniquely extended to the σ algebra $\bar{\mathcal{A}}$ generated by $\mathcal{A}$ and its null sets. This procedure is called **completion.** The process of completion is equivalent to the following: For a given probability space $(\Omega,\mathcal{A},\mathcal{P})$, define for every subset A of Ω an outer probability $\mathcal{P}^*(A)$ and an inner probability $\mathcal{P}_*(A)$ by

$$\begin{aligned} \mathcal{P}^*(A) &= \inf_C \{\mathcal{P}(C): C \supset A,\ C \in \mathcal{A}\} \\ \mathcal{P}_*(A) &= \sup_C \{\mathcal{P}(C): C \subset A,\ C \in \mathcal{A}\} \end{aligned} \tag{1.2}$$

Obviously, on $\mathcal{A}$ we have $\mathcal{P}_* = \mathcal{P} = \mathcal{P}^*$. Thus, $\mathcal{P}$ can be uniquely extended to the class of all sets whose inner and outer probabilities are equal. The additional sets for which $\mathcal{P}$ can be so defined are exactly the same as those gotten by completing $(\Omega,\mathcal{A},\mathcal{P})$ [Neveu, 1965, p. 17].

An example might best illustrate the preceding discussions. Set $\Omega = [0,1)$, and let $\mathcal{B}$ be the class consisting of $[0,1)$, $\emptyset$, all semiopen intervals $[a,b)$ with $0 \leq a < b \leq 1$, and all finite unions of disjoint semiopen intervals. The class $\mathcal{B}$ is an algebra, but not a σ algebra. If $A = [a,b)$ is a semiopen interval, we set $\mathcal{P}(A) = b - a$. If $A = \bigcup_{i=1}^{m} A_i$ is a union of disjoint intervals A_i, we set $\mathcal{P}(A) = \sum_{i=1}^{m} \mathcal{P}(A_i)$. Clearly, $\mathcal{P}$ satisfies conditions (1.1a) and (1.1b). We shall show that $\mathcal{P}$ is, in fact, σ additive. Suppose that $A_1, A_2, \ldots$ are disjoint sets in $\mathcal{B}$ such that $A = \bigcup_{i=1}^{\infty} A_i$ is also in $\mathcal{B}$. Then A is a finite union of disjoint semiopen intervals $I_1, \ldots, I_m$, and each $I_k \cap A_i$ is again a finite union of disjoint semiopen intervals. Therefore, to prove that $\mathcal{P}$ is σ additive, it is enough to show that if

$$[a,b) = \bigcup_{i=1}^{\infty} [a_i,b_i) \qquad a \leq a_i \leq b_i \leq a_{i+1} \leq b \qquad i = 1, 2, \ldots$$

then

$$\mathcal{P}([a,b)) = b - a = \sum_{i=1}^{\infty} \mathcal{P}([a_i,b_i)) = \sum_{i=1}^{\infty} (b_i - a_i) \tag{1.3}$$

First, we note that

$$\sum_{i=1}^{n} (b_i - a_i) \leq \sum_{i=1}^{n} (b_i - a_i) + \sum_{i=1}^{n-1} (a_{i+1} - b_i) = b_n - a_1 \leq b - a$$

Hence, $b - a \geq \sum_{i=1}^{\infty} (b_i - a_i)$. Next, we note that for any $\delta > 0$,

$$[a, b - \delta] \subset \bigcup_{i=1}^{\infty} (a_i - \delta/2^i, b_i)$$

The Heine-Borel theorem [see, e.g., Rudin, 1966, p. 36] then states that there is a finite N such that

$$[a, b - \delta] \subset \bigcup_{i=1}^{N} (a_i - \delta/2^i, b_i)$$

It follows that

$$[a,b) \subset [b - \delta, b) \cup \bigcup_{i=1}^{N} [a_i - \delta/2^i, b_i)$$

$$b - a \leq \delta + \sum_{i=1}^{N} (b_i - a_i + \delta/2^i)$$

$$\leq 2\delta + \sum_{i=1}^{\infty} (b_i - a_i)$$

Therefore, for every $\delta > 0$, we have $0 \leq (b - a) - \sum_{i=1}^{\infty} (b_i - a_i) \leq 2\delta$, and (1.3) is proved. Hence, $\mathcal{P}$ is σ additive and is a probability measure. It can be uniquely extended to the σ algebra $\mathcal{A}$ generated by the semiopen intervals. Sets in $\mathcal{A}$ are called Borel sets of $[0,1)$. $\mathcal{P}$ can be further extended by completion to $\mathcal{A}$. Sets in $\bar{\mathcal{A}}$ are called Lebesgue measurable sets of $[0,1)$. We note that $\mathcal{A}$ and $\bar{\mathcal{A}}$ include all intervals in $[0,1)$ and not just semiopen ones. In particular, a point x in $[0,1)$ can be considered to be a degenerate interval $[x,x]$ and is in $\mathcal{A}$.

Let Ω be a basic space, and let $\mathcal{A}$ be a σ algebra of subsets of Ω. The pair $(\Omega,\mathcal{A})$ is called a **measurable space** and is sometimes referred to as a **preprobability space.** A nonnegative σ additive set function μ defined on $\mathcal{A}$ is called a **measure.** Thus, a probability measure is a measure satisfying $\mathcal{P}(\Omega) = 1$. More generally, μ is said to be a **finite measure** if $\mu(\Omega) < \infty$. Even more generally, if Ω is a countable union of sets $A_1, A_2, \ldots$ in $\mathcal{A}$ such that $\mu(A_i)$ is finite for each i, then μ is said to be a **σ-finite measure.** For example, let $\Omega = R$ be the real line, and let $\mathcal{A}$ be the σ algebra generated by the class of all intervals, and for any finite interval A define $\mu(A) =$ length of A. Then, μ can be extended to a unique σ-finite measure on $(\Omega,\mathcal{A})$, and upon completion it is just the Lebesgue measure of the real line.

2. MEASURES ON FINITE-DIMENSIONAL SPACES

Let R denote the real line $(-\infty,\infty)$, and let R^n denote the collection of all ordered n-tuples $\mathbf{x} = (x_1, x_2, \ldots, x_n)$, $x_i \in R$. A subset of R^n of the form $\{\mathbf{x}: x_i \in A_i, i = 1, 2, \ldots, n\}$, where $A_1, \ldots, A_n$ are intervals, will be called a **rectangle**. We denote by $\mathcal{R}^n$ the smallest σ algebra of subsets of R^n containing all rectangles. The σ algebra $\mathcal{R}^n$ is called the n-dimensional **Borel σ algebra,** and sets in $\mathcal{R}^n$ are called **Borel sets.**

We note that $\mathcal{R}^n$ can be generated in many ways. $\mathcal{R}^n$ is the smallest σ algebra containing every set of the form $\{\mathbf{x}: -\infty < x_i < a\}$ where i is one of the integers $1, 2, \ldots, n$ and a is a real number. Thus, the class of sets generating $\mathcal{R}^n$ can be smaller than the class of all rectangles. Of course, the class of sets generating $\mathcal{R}^n$ can also be larger. For example, $\mathcal{R}^n$ is also the smallest σ algebra containing every set of the form $\{\mathbf{x}: x_i \in A_i, i = 1, \ldots, n\}$ where A_i are one-dimensional Borel sets. We note that not every set in $\mathcal{R}^n$ can be written in this way, for example, $\{(x_1,x_2): x_1 + x_2 = 2\}$. A set of the form $\{\mathbf{x}: x_i \in A_i, i = 1, \ldots, n\}$ will be called a **product** and will be denoted by $\prod_{i=1}^{n} A_i$.

The pair $(R^n,\mathcal{R}^n)$ is obviously a measurable space. Measures defined on $(R^n,\mathcal{R}^n)$ are called **Borel measures,** and probability measures defined on $(R^n,\mathcal{R}^n)$ are called **Borel probability measures.** Let μ be a finite Borel measure defined on $(R^n,\mathcal{R}^n)$, and define a function M on R^n by

$$M(a_1, a_2, \ldots, a_n) = \mu(\{\mathbf{x}, -\infty < x_i < a_i, i = 1, 2, \ldots, n\}) \tag{2.1}$$

The function M satisfies the following rather obvious conditions:

$$M \text{ is nonnegative and bounded by } \mu(R^n) \tag{2.2a}$$

$$\text{For each } i \quad \lim_{a_i \to -\infty} M(\mathbf{a}) = 0 \tag{2.2b}$$

$$M(\mathbf{a}) \xrightarrow[\mathbf{a} \to (\infty, \infty, \ldots, \infty)]{} \mu(R^n) \tag{2.2c}$$

Condition (2.2*a*) is a simple property of measures, and conditions (2.2*b*) and (2.2*c*) follow from monotone sequential continuity of μ.

Let A be any rectangle of the form

$$A = \{\mathbf{x}: a_i \le x_i < b_i,\ i = 1, \ldots, n\} = \bigcap_{i=1}^{n} \{\mathbf{x}: a_i \le x_i < b_i\}$$

Denoting $A_i = \{\mathbf{x}: a_i \le x_i < b_i\}$, we have (because of additivity)

$$\mu\left(\bigcap_{i=1}^{n-1} A_i \cap (-\infty, b_n)\right) = \mu\left(\bigcap_{i=1}^{n-1} A_i \cap (-\infty, a_n)\right) + \mu\left(\bigcap_{i=1}^{n} A_i\right)$$

Hence,

$$\mu(A) = \mu\left(\bigcap_{i=1}^{n-1} A_i \cap (-\infty, b_n)\right) - \mu\left(\bigcap_{i=1}^{n-1} A_i \cap (-\infty, a_n)\right)$$

Continuing in this way, we find

$$\mu(A) = \sum_{\mathbf{x} \in S(A)} (-1)^{k(\mathbf{x})} M(\mathbf{x})$$

where M is defined by (2.1), $S(A)$ denotes the collection of 2^n vertices $\{\mathbf{x}: x_i = a_i, b_i\}$ of A, and $k(\mathbf{x})$ is the number of a's in $\mathbf{x}$. It follows that M satisfies two additional conditions:

$$\sum_{\mathbf{x} \in S(A)} (-1)^{k(\mathbf{x})} M(\mathbf{x}) \ge 0 \qquad \text{for all rectangles } A \text{ of the form}$$

$$A = \bigcap_{i=1}^{n} \{\mathbf{x}: a_i \le x_i < b_i\} \tag{2.3}$$

$$\text{For each } i \lim_{a_i \uparrow b_i} \sum_{\mathbf{x} \in S(A)} (-1)^{k(\mathbf{x})} M(\mathbf{x}) = 0 \tag{2.4}$$

Condition (2.3) might be referred to as the **monotonicity condition.** In one dimension, it reduces to simply

$$M(b) - M(a) \ge 0 \qquad b \ge a$$

Condition (2.4) is a continuity condition. In one dimension, it reduces to left continuity

$$\lim_{a \uparrow b} M(a) = M(b)$$

A function satisfying (2.2) to (2.4) will be called a **distribution function.** Equation (2.1) defines a distribution function in terms of a finite Borel measure.

Conversely, every distribution function defines a finite Borel measure. To verify this, let C be the class of all rectangles of the form

$$A = \bigcap_{i=1}^{n} \{\mathbf{x}: a_i \leq x_i < b_i\} \tag{2.5}$$

Let $\mathcal{B}(C)$ be the smallest algebra (but not σ algebra) containing C. Then, every set in $\mathcal{B}(C)$ is a finite union of disjoint rectangles of the form (2.5). Given a distribution function M, we can define a nonnegative set function μ on C by

$$\mu(A) = \sum_{\mathbf{x}\in S(A)} (-1)^{k(\mathbf{x})} M(\mathbf{x})$$

and extend μ to $\mathcal{B}(C)$ by additivity. Monotone sequential continuity of μ follows from conditions (2.4) and (2.2). Hence, μ can be extended to a σ-additive set function on $\mathcal{A}(C)$, which is just $\mathcal{R}^n$. If M satisfies

$$M(\mathbf{a}) \xrightarrow[\mathbf{a}\to(\infty,\infty,\ldots,\infty)]{} 1 \tag{2.6}$$

μ is a probability measure. Distribution functions satisfying (2.6) are called **probability distribution functions** for obvious reasons.

We have established a one-to-one relationship between distribution functions and finite Borel measures. Distribution functions are useful characterizations of finite Borel measures, because being point functions rather than set functions, they are easier to specify. However, when they are used as integrators, distribution functions usually have to be interpreted as measures (cf. Sec. 1.5).

3. MEASURABLE FUNCTIONS AND RANDOM VARIABLES

Suppose that $(\Omega_1,\mathcal{A}_1)$ and $(\Omega_2,\mathcal{A}_2)$ are two measurable spaces, and f is a function with domain Ω_1 and range in Ω_2. The function f is said to be a **measurable function** or a **measurable mapping** of $(\Omega_1,\mathcal{A}_1)$ into $(\Omega_2,\mathcal{A}_2)$ if for every set A in $\mathcal{A}_2$, the set

$$f^{-1}(A) = \{\omega: f(\omega) \in A\} \tag{3.1}$$

is in $\mathcal{A}_1$. The set $f^{-1}(A)$ is called the **inverse image** of A.

If f is a measurable function $(R^n,\mathcal{R}^n) \to (R^m,\mathcal{R}^m)$, then f is called a **Borel function.** All continuous functions are Borel functions. The pointwise limit of a sequence of Borel functions is again a Borel function. Indeed, the class of Borel functions is the smallest class of functions, which contains all continuous functions and is closed under pointwise limit.

If $(\Omega,\mathcal{A})$ is a measurable space, and X is a measurable mapping of $(\Omega,\mathcal{A})$ into $(R,\mathcal{R})$, then X is usually referred to as a real-valued **$\mathcal{A}$-measurable** function. If a probability measure $\mathcal{P}$ on $(\Omega,\mathcal{A})$ is defined, then X is called a **real random variable** defined on the probability space $(\Omega,\mathcal{A},\mathcal{P})$. We note that if X is $\mathcal{A}$ measurable, and $\mathcal{A}'$ is a σ algebra containing $\mathcal{A}$, then X is also $\mathcal{A}'$ measurable.

Because $\mathcal{R}$ is generated by the collection of half lines $\{x\colon -\infty < x < a\}$, for X to be a real random variable, it is sufficient for every set for the form $\{\omega\colon X(\omega) < a\}$ to be an event. The function P_X defined by

$$P_X(a) = \mathcal{P}(\{\omega\colon X(\omega) < a\}) \tag{3.2}$$

is called the **probability distribution function** of X. The relationship

$$\mathbf{P}_X(A) = \mathcal{P}(\{\omega\colon X(\omega) \in A\}) \tag{3.3}$$

defines a probability measure $\mathbf{P}_X$ on $(R,\mathcal{R})$. Our discussion in Sec. 2 indicated that the distribution function P_X and the Borel measure $\mathbf{P}_X$ are mutually uniquely determined.

Suppose that $X_1, X_2, \ldots, X_n$ are n real random variables and $\mathbf{X} = (X_1, X_2, \ldots, X_n)$. Then, $\mathbf{X}$ is a measurable function mapping $(\Omega,\mathcal{A})$ into $(R^n,\mathcal{R}^n)$. The function

$$\begin{aligned} P_X(\mathbf{a}) &= P_X(a_1, \ldots, a_n) \\ &= \mathcal{P}(\{\omega\colon X_i(\omega) < a_i,\ i = 1, \ldots, n\}) \qquad \mathbf{a} \in R^n \end{aligned} \tag{3.4}$$

is called the **joint probability distribution function** of $\mathbf{X}$. At the same time, the relationship

$$\mathbf{P}_X(A) = \mathcal{P}(\{\omega\colon \mathbf{X}(\omega) \in A\}) \qquad A \in \mathcal{R}^n \tag{3.5}$$

defines a Borel probability measure.

A real random variable X is said to be **discrete** (with probability 1) if there exists a countable set $S = \{x_i\}$ such that

$$\sum_{x_i \in S} \mathcal{P}(\{\omega\colon X(\omega) = x_i\}) = 1 \tag{3.6}$$

If X is discrete, then its distribution function P_X is a function which is constant except for jumps at x_i, $i = 1, 2, \ldots$, the size of the jump at x_i being $\mathcal{P}(\{\omega\colon X(\omega) = x_i\})$. The probability measure $\mathbf{P}_X$ is concentrated on the points in S. For an arbitrary Borel set A, we have

$$\mathbf{P}_X(A) = \sum_{x_i \in A \cap S} \mathcal{P}(\{\omega\colon X(\omega) = x_i\}) \tag{3.7}$$

Let $\mathbf{P}$ be a probability measure defined on $(R^n, \mathcal{R}^n)$. It is said to be **singular** (with respect to the Lebesgue measure) if there exists a set S in $\mathcal{R}^n$ such that $\mathbf{P}(S) = 1$ and the Lebesgue measure of S is zero. On the other hand, $\mathbf{P}$ is said to be **absolutely continuous** (with respect to the Lebesgue measure) if Lebesgue measure $(A) = 0$ implies $\mathbf{P}(A) = 0$. It is clear that if $X_1, X_2, \ldots, X_n$ are discrete random variables, then $\mathbf{P}_X$ is singular. However, $\mathbf{P}_X$ being singular does not imply that $X_1, X_2, \ldots, X_n$ are necessarily discrete, since the set S having simultaneous zero Lebesgue measure and $\mathbf{P}_X$ measure 1 need not be countable.

If $X_1, \ldots, X_n$ are such that $\mathbf{P}_X$ is absolutely continuous, then there exists a nonnegative Borel function $p_X(\mathbf{x})$, $\mathbf{x} \in R^n$ such that

$$\mathbf{P}_X(A) = \int_A p_X(\mathbf{x})\, d\mathbf{x} \qquad A \in \mathcal{R}^n \tag{3.8}$$

The function p_X is called the **probability density function** for the random variables $X_1, \ldots, X_n$. Representation (3.8) results from an application of the Radon-Nikodym theorem [Loève, 1963, p. 132]. In terms of the distribution function, (3.8) takes on the more familiar form

$$P_X(a_1, a_2, \ldots, a_n) = \int_{-\infty}^{a_1} \cdots \int_{-\infty}^{a_n} p_X(x_1, x_2, \ldots, x_n)\, dx_1\, dx_2 \cdots dx_n \tag{3.9}$$

In general, $\mathbf{P}_X$ is neither absolutely continuous nor singular, but we can always write

$$\mathbf{P}_X = \alpha \mathbf{P}_X^{(1)} + (1 - \alpha)\mathbf{P}_X^{(2)} \tag{3.10}$$

where $0 \leq \alpha \leq 1$, and $\mathbf{P}_X^{(1)}$ and $\mathbf{P}_X^{(2)}$ are, respectively, absolutely continuous and singular. The decomposition (3.10) is known as the Lebesgue decomposition [Neveu, 1965, p. 108].

Given a probability distribution function $P(\mathbf{x})$, $\mathbf{x} \in R^n$, a set of random variables $X_1, X_2, \ldots, X_n$ having P as their joint distribution function is called a **realization** of P. Every probability distribution function, i.e., every nonnegative function satisfying (2.3) to (2.6), has at least one realization. Indeed, it has numerous realizations. One standard realization can be constructed as follows: An n-dimensional probability distribution function P defines a probability measure $\mathbf{P}$ on $(R^n, \mathcal{R}^n)$. Let $\mathbf{X}$: $R^n \to R^n$ be the **coordinate function** defined by

$$X_j(\mathbf{x}) = x_j \tag{3.11}$$

Then $\mathbf{X} = (X_1, \ldots, X_n)$ as random variables on $(R^n, \mathcal{R}^n, \mathbf{P})$ will have P as the joint distribution function because

$$\begin{aligned} &\mathbf{P}(\{\mathbf{x}: X_1(\mathbf{x}) < a_1, X_2(\mathbf{x}) < a_2, \ldots, X_n(\mathbf{x}) < a_n\}) \\ &= \mathbf{P}(\{\mathbf{x}: x_i < a_i, i = 1, 2, \ldots, n\}) = P(a_1, a_2, \ldots, a_n) \end{aligned} \tag{3.12}$$

If $\mathbf{X} = (X_1, X_2, \ldots, X_n)$ are real random variables defined on a probability space $(\Omega, \mathcal{A}, \mathcal{P})$, and $f: R^n \to R^m$ is a Borel function, then

$$\mathbf{Y} = (Y_1, Y_2, \ldots, Y_m) = f(\mathbf{X}) \tag{3.13}$$

are again random variables defined on $(\Omega, \mathcal{A}, \mathcal{P})$. If we denote the Borel probability measure of $\mathbf{X}$ by $\mathbf{P}_X$, then for $A \in \mathcal{R}^m$,

$$\begin{aligned} \mathbf{P}_Y(A) &= \mathcal{P}(\{\omega: \mathbf{Y}(\omega) \in A\}) \\ &= \mathcal{P}(\{\omega: f(\mathbf{X}(\omega)) \in A\}) \\ &= \mathbf{P}_X(\{\mathbf{x}: f(\mathbf{x}) \in A\}) \\ &= \mathbf{P}_X(f^{-1}(A)) \end{aligned} \tag{3.14}$$

Equation (3.14) expresses the transformation rule of Borel probability measures (equivalently, distribution functions). We note that (3.14) would be somewhat awkward to state in terms of distribution functions directly, without introducing the Borel probability measures.

Suppose that $f: R^n \to R^n$ is a one-to-one and invertible mapping such that both f and f^{-1} have continuous partial derivatives with respect to the coordinates. Further, suppose that $\mathbf{X} = (X_1, X_2, \ldots, X_n)$ are random variables with a density function p_X. Then $\mathbf{Y} = f(\mathbf{X})$ also has a density function, which is given by [see Thomasian, 1969, pp. 362–363]

$$p_Y(\mathbf{y}) = p_X(f^{-1}(\mathbf{y}))|J(\mathbf{y})| \tag{3.15}$$

where J is the matrix with elements $\partial f_i^{-1}(\mathbf{y})/\partial y_j$, and $|J(\mathbf{y})|$ stands for the absolute value of the determinant. As an example of (3.15), suppose that (X_1, X_2) has a joint density function

$$p_X(x_1, x_2) = \frac{1}{2\pi} e^{-\frac{1}{2}(x_1^2 + x_2^2)}$$

and $\begin{bmatrix} Y_1 \\ Y_2 \end{bmatrix} = \mathbf{A} \begin{bmatrix} X_1 \\ X_2 \end{bmatrix}$, where $\mathbf{A}$ is a constant nonsingular matrix. Denoting by $\mathbf{x}'$ the transpose of $\mathbf{x}$, we can write $(x_1^2 + x_2^2) = \mathbf{x}'\mathbf{x}$. Hence,

$$\begin{aligned} p_X(f^{-1}(\mathbf{y})) &= p_X(\mathbf{A}^{-1}\mathbf{y}) \\ &= \frac{1}{2\pi} e^{-\frac{1}{2}\mathbf{y}'(\mathbf{AA}')^{-1}\mathbf{y}} \end{aligned}$$

and

$$p_Y(\mathbf{y}) = \frac{1}{2\pi|\mathbf{A}|} e^{-\frac{1}{2}\mathbf{y}'(\mathbf{AA}')^{-1}\mathbf{y}}$$

Other examples can be found in the exercise section.

Finally, we note that the notation

$$\mathcal{P}(\{\omega: \mathbf{X}(\omega) \in A\})$$

is unduly clumsy. We shall use the simpler, but less exact, notation

$$\mathcal{P}(\mathbf{X} \in A)$$

instead.

4. SEQUENCES OF EVENTS AND RANDOM VARIABLES

For a given probability space $(\Omega, \mathcal{A}, \mathcal{P})$, let $\{A_n, n = 1, 2, \ldots\}$ be a sequence of events. The sequence $\{A_n\}$ is said to be **increasing** if $A_{n+1} \supset A_n$ for every n and **decreasing** if $A_{n+1} \subset A_n$ for every n. A sequence which is either increasing or decreasing is said to be **monotone.** The limit $\lim_{n\to\infty} A_n$ is defined for every monotone sequence as $\bigcup_{n=1}^{\infty} A_n$ or $\bigcap_{n=1}^{\infty} A_n$ according as $\{A_n\}$ is increasing or decreasing. For a general sequence $\{A_n\}$ (not necessarily monotone), we define superior and inferior limits as follows:

$$\limsup_{n\to\infty} A_n = \bigcap_{n=1}^{\infty} \bigcup_{k\geq n} A_k \tag{4.1}$$

$$\liminf_{n\to\infty} A_n = \bigcup_{n=1}^{\infty} \bigcap_{k\geq n} A_k \tag{4.2}$$

The superior limit is the set of all points which occur in an infinite number of A_n's, while the inferior limit is the set of all points which occur in all but a finite number of A_n's. Hence, $\limsup A_n \supset \liminf A_n$. If the superior limit and the inferior limit coincide, then we say that $\{A_n\}$ is a convergent sequence, and we set

$$\lim A_n = \limsup A_n = \liminf A_n \tag{4.3}$$

We note that lim sup, lim inf, and lim all involve only countable set operations. Hence all such sequential limits of events are again events.

Suppose that $\{A_n\}$ is a convergent sequence of events, and A is its limit. Then A is an event, and the following proposition relates $\mathcal{P}(A)$ to $\mathcal{P}(A_n)$.

Proposition 4.1. Every probability measure $\mathcal{P}$ is **sequentially continuous** in the sense that if $\{A_n\}$ is a convergent sequence of events, then

$$\lim_{n\to\infty} \mathcal{P}(A_n) = \mathcal{P}(\lim_{n\to\infty} A_n) \tag{4.4}$$

Proof: First, if $\{A_n\}$ decreases to the empty set $\emptyset$, then (4.4) is simply (1.1*c*) (sequential monotone continuity at $\emptyset$). If $\{A_n\}$ is a decreasing sequence with a nonempty limit A, then

$$\mathcal{P}(A_n) = \mathcal{P}(A_n - A) + \mathcal{P}(A)$$

and $\{A_n - A\}$ decreases to $\emptyset$. If $\{A_n\}$ is increasing with limit A, then

$$\mathcal{P}(A_n) = \mathcal{P}(A) - \mathcal{P}(A - A_n)$$

and $\{A - A_n\}$ decreases to $\emptyset$. In either case, (4.4) follows once again from (1.1*c*).

Now suppose $\{A_n\}$ is convergent, but not necessarily monotone. Set $B_n = \bigcap_{k \geq n} A_k$ and $C_n = \bigcup_{k \geq n} A_k$. Then, $B_n \subset A_n \subset C_n$, and $\{B_n\}$ and $\{C_n\}$ are monotone sequences. Therefore,

$$\begin{aligned} \mathcal{P}(\lim B_n) &= \lim \mathcal{P}(B_n) \leq \lim \mathcal{P}(A_n) \\ \mathcal{P}(\lim C_n) &= \lim \mathcal{P}(C_n) \geq \lim \mathcal{P}(A_n) \end{aligned}$$

Since by definition $\lim A_n = \lim B_n = \lim C_n$, we have

$$\mathcal{P}(\lim A_n) \leq \lim \mathcal{P}(A_n) \leq \mathcal{P}(\lim A_n)$$

so that (4.4) is proved. ▮

The following rather obvious consequence of Proposition 4.1 is known as the **Borel-Cantelli lemma.** For such an obvious result, it is amazingly powerful and is a standard tool in proving properties which are true with probability 1.

Proposition 4.2. For an arbitrary sequence of events $\{A_n\}$, $\sum_{n=1}^{\infty} \mathcal{P}(A_n) < \infty$ implies $\mathcal{P}(\limsup A_n) = 0$.

Proof: From Proposition 4.1 we have

$$\begin{aligned} \mathcal{P}(\limsup A_n) &= \mathcal{P}\left(\lim_{n\to\infty} \bigcup_{k\geq n} A_k\right) \\ &= \lim_{n\to\infty} \mathcal{P}\left(\bigcup_{k\geq n} A_k\right) \leq \lim_{n\to\infty} \sum_{k\geq n}^{\infty} \mathcal{P}(A_n) \end{aligned}$$

Because $\sum_{n=1}^{\infty} \mathcal{P}(A_n) < \infty$,

$$\lim_{n\to\infty} \sum_{k\geq n}^{\infty} \mathcal{P}(A_k) = 0$$

which proves the proposition. ▮

Before proceeding to the consideration of sequences of random variables, we first recall the definitions for infimum and supremum. Let S be a set of numbers in $[-\infty, \infty]$. A lower bound b (upper bound) of S is defined by the property $b \leq x (b \geq x)$ for all x in S. The infimum of S, denoted by inf S, is the greatest lower bound of S, and the supremum of S, denoted by sup S, is the least upper bound of S. Suppose that $\{X_n\}$ is a sequence of random variables defined on a common probability space $(\Omega, \mathcal{A}, \mathcal{P})$. If for every $\omega \in \Omega$ and every n, $X_{n+1}(\omega) \geq X_n(\omega)$, then $\{X_n\}$ is said to be an **increasing** sequence. The sequence $\{X_n\}$ is said to be **decreasing** if $\{-X_n\}$ is increasing. If a sequence $\{X_n\}$ is either increasing or decreasing, then it is said to be **monotone.**

If $\{X_n\}$ is a monotone sequence, then we define its limit as

$$\lim_{n \to \infty} X_n = \sup_n X_n \quad \text{or} \quad \lim_{n \to \infty} X_n = \inf_n X_n$$

according as $\{X_n\}$ is increasing or decreasing. We note that such limits may assume values $+\infty$ and $-\infty$. If $\{X_n\}$ is an arbitrary sequence, not necessarily monotone, we define inferior and superior limits as

$$\liminf_{n \to \infty} X_n = \lim_{n \to \infty} \inf_{k \geq n} X_k \tag{4.5}$$

and

$$\limsup_{n \to \infty} X_n = \lim_{n \to \infty} \sup_{k \geq n} X_k \tag{4.6}$$

If the inferior and superior limits of $\{X_n\}$ agree, we say that the sequence $\{X_n\}$ converges and set

$$\lim_{n \to \infty} X_n = \liminf_{n \to \infty} X_n = \limsup_{n \to \infty} X_n \tag{4.7}$$

Again, we note that even though X_n is finitely valued for each n, the limit $\lim_{n \to \infty} X_n$ may assume values $\pm\infty$.

If $\{X_n\}$ is a sequence of random variables, then both lim inf X_n and lim sup X_n are random variables *provided that they are finite at every* $\omega \in \Omega$. This is because

$$\begin{aligned} \{\omega: \liminf X_n(\omega) < a\} &= \{\omega: \sup_n \inf_{k \geq n} X_n(\omega) < a\} \\ &= \bigcap_n \bigcup_{k \geq n} \{\omega: X_k(\omega) < a\} \end{aligned} \tag{4.8}$$

and

$$\begin{aligned} \{\omega: \limsup X_n(\omega) < a\} &= \{\omega: \inf_n \sup_{k \geq n} X_n(\omega) < a\} \\ &= \bigcup_n \bigcap_{k \geq n} \{\omega: X_k(\omega) < a\} \end{aligned} \tag{4.9}$$

so that every set of the form

$$\{\omega: \liminf X_n(\omega) < a\} \quad \text{or} \quad \{\omega: \limsup X_n(\omega) < a\}$$

is an event. Indeed, if we had defined random variables to be *extended* real-valued functions, as is often done, then inferior and superior limits of a sequence of random variables are always random variables. The following is an immediate consequence of (4.8) and (4.9).

Proposition 4.3. Let $\{X_n\}$ be a sequence of random variables converging to a limit X. Suppose that $X(\omega)$ is finite at every $\omega \in \Omega$, then X is a random variable.

Let A be an event. Define the indicator function I_A as follows:

$$I_A(\omega) = \begin{cases} 1 & \omega \in A \\ 0 & \omega \in \Omega - A \end{cases} \tag{4.10}$$

It is obvious that I_A is a random variable. Suppose that X is a function that can be written as

$$X = \sum_{\nu=1}^{n} \alpha_\nu I_{A_\nu}$$

where $\alpha_1, \alpha_2, \ldots, \alpha_n$, are real numbers, and $A_1, A_2, \ldots, A_n$, are events. Then, X is called a **simple** random variable.

Proposition 4.4. Every random variable is the limit of a sequence of simple random variables. Every nonnegative random variable is the limit of an increasing sequence of nonnegative simple random variables.

Proof: We define $\{X_n\}$ as follows:

$$X_n(\omega) = \begin{cases} -2^n & \text{if } X(\omega) < -2^n \\ k/2^n & \text{if } X(\omega) \in [k/2^n, k+1/2^n) \\ & \qquad -2^{2n} \le k \le 2^{2n} - 1 \\ 2^n & \text{if } X(\omega) \ge 2^n \end{cases} \tag{4.11}$$

For a fixed ω and for $n \ge \log_2 |X(\omega)|$, we have

$$\sup_{k \ge n} |X(\omega) - X_k(\omega)| \le 2^{-n} \xrightarrow[n\to\infty]{} 0$$

so that $\{X_n\}$ converges to X at every ω. If X is nonnegative, then $\{X_n\}$ as defined by (4.11) is an increasing sequence of nonnegative random variables. ▌

5. EXPECTATION OF RANDOM VARIABLES

The expectation of a random variable X can be defined as the Stieltjes integral

$$EX = \int_{-\infty}^{\infty} x\, dP(x) \tag{5.1}$$

where $P(\cdot)$ is the probability distribution function for X. Provided that at least one of the two integrals $\int_0^{\infty} x\, dP(x)$ and $-\int_{-\infty}^{0} x\, dP(x)$ is less than ∞, the integral in (5.1) is well defined. As a definition for the expectation, (5.1) is less than satisfactory on two counts. First, the definition is not elementary in that it depends on the definition of a Stieltjes integral. Secondly, it obscures the fact that EX is really the integral of X on Ω with respect to the probability measure $\mathcal{P}$. In other words,

$$EX = \int_{\Omega} X(\omega)\, \mathcal{P}(d\omega) \tag{5.2}$$

where the integral needs to be defined. For these reasons, we shall give a definition for EX by defining the integral in (5.2) for a random variable X.

First, we define EX when X is a simple random variable. By definition, this means that X has the form

$$X = \sum_{i=1}^{n} x_i I_{A_i} \tag{5.3}$$

where A_i are events and x_i are real constants. For such X we define

$$EX = \sum_{i=1}^{n} x_i \mathcal{P}(A_i) \tag{5.4}$$

Thus defined, EX satisfies the following properties:

Additivity:	$E(X + Y) = EX + EY$	(5.5a)
Homogeneity:	$E(cX) = cEX$	(5.5b)
Order preservation:	$X \geq Y \Rightarrow EX \geq EY$	(5.5c)

In addition, EX also satisfies the following important property.

Lemma. Let $\{X_n\}$ be a *monotone* sequence of simple random variables converging to a *simple random variable* X. Then

$$\lim_{n \to \infty} EX_n = EX$$

Proof: If $\{X_n\}$ is decreasing, then $\{X_n - X\}$ decreases to zero. If $\{X_n\}$ is increasing, then $\{X - X_n\}$ decreases to zero. Since expectation is additive, we only need to prove that if $\{X_n\}$ decreases to zero, then

$$\lim_{n\to\infty} EX_n = 0$$

To do this, we note that for every $\epsilon > 0$,

$$\begin{aligned} 0 \le EX_n &\le (\max X_n)\mathcal{P}(X_n > \epsilon) + \epsilon \\ &\le (\max X_1)\mathcal{P}(X_n > \epsilon) + \epsilon \\ &\xrightarrow[n\to\infty]{} \epsilon \end{aligned}$$

which completes the proof of the lemma. ∎

Next, let X be a nonnegative random variable, and let $\{X_n\}$ be an increasing sequence of nonnegative simple random variables converging to X. Since $\{EX_n\}$ is a nondecreasing sequence of nonnegative numbers, $\lim_{n\to\infty} EX_n = \sup_n EX_n$ always exists, but may be infinite. We shall show that two such sequences $\{X_n\}$, $\{Y_n\}$, both converging to X, have the same limit $\lim_{n\to\infty} EX_n = \lim_{n\to\infty} EY_n$. Hence, we can unambiguously define

$$EX = \lim_{n\to\infty} EX_n \tag{5.6}$$

for nonnegative random variables. For a general X, we write

$$X = X_+ - X_- \tag{5.7}$$

where X_+ and X_- are both nonnegative and define

$$EX = EX_+ - EX_- \tag{5.8}$$

provided that the right-hand side is not of the form $\infty - \infty$. It is easy to verify that, thus defined, EX satisfies properties (5.5). The only thing that remains to be shown is the uniqueness of the right-hand limit in (5.6).

Proposition 5.1. Let $\{X_n\}$ and $\{Y_n\}$ be two increasing sequences of nonnegative simple random variables converging to the same limit X. Then

$$\lim_{n\to\infty} EX_n = \lim_{n\to\infty} EY_n$$

Proof: Let p be a fixed integer and set

$$Z_n = \min(X_n, Y_p)$$

It is clear that $\{Z_n\}$ is an increasing sequence of simple random variables. Since $X_n \xrightarrow[n\to\infty]{} X \ge Y_p$, we have

$$\lim_{n\to\infty} Z_n = Y_p$$

Using the lemma, we get

$$\lim_{n\to\infty} EX_n \geq \lim_{n\to\infty} EZ_n = EY_p$$

Hence, $\lim_{n\to\infty} EX_n \geq \lim_{p\to\infty} EY_p$. Reversing the roles of $\{X_n\}$ and $\{Y_n\}$ yields $\lim_{n\to\infty} EY_n \geq \lim_{n\to\infty} EX_n$ and proves the theorem. ∎

The basic properties (5.5) of EX are precisely those of an integral. When we want to emphasize the nature of EX as an integral, we shall write

$$EX = \int_\Omega X(\omega)\mathcal{P}(d\omega)$$

We shall also use the abbreviation $\int_\Omega X\,d\mathcal{P}$. A random variable X is said to be **integrable** if $E|X| < \infty$. The following results on sequences of integrable random variables are counterparts of standard results on sequences of integrable functions in integration theory. Proofs will be omitted [Loève, 1963, pp. 124–125].

Proposition 5.2 (Monotone Convergence). Let $\{X_n\}$ be an increasing sequence of nonnegative random variables converging to X. Then,

$$EX = \lim_{n\to\infty} EX_n \tag{5.9}$$

Proposition 5.3 (Fatou's Lemma). Let $\{X_n\}$ be a sequence of random variables. Suppose that there exists an integrable random variable X such that $X_n(\omega) \geq X(\omega)$ for all n and ω. Then,

$$\liminf_{n\to\infty} EX_n \geq E \liminf_{n\to\infty} X_n \tag{5.10}$$

Proposition 5.4 (Dominated Convergence). Let $\{X_n\}$ be a sequence of random variables converging to X. Suppose that there exists an integrable random variable Y such that $|X_n(\omega)| \leq Y(\omega)$ for all n and ω. Then,

$$\lim_{n\to\infty} EX_n = EX \tag{5.11}$$

Suppose that $\mathbf{X} = (X_1, \ldots, X_n)$ are random variables defined on $(\Omega,\mathcal{A},\mathcal{P})$ with a joint distribution function P_X. Let f be a nonnegative Borel function. Then $f(\mathbf{X}(\omega))$, $\omega \in \Omega$, is a nonnegative random variable defined on $(\Omega,\mathcal{A},\mathcal{P})$, and $f(\mathbf{x})$, $\mathbf{x} \in R^n$, is a random variable defined on

$(R^n, \mathcal{R}^n, \mathbf{P}_X)$. Both of the integrals $\int_\Omega f(\mathbf{X}(\omega))\, \mathcal{P}(d\omega)$ and $\int_{R^n} f(\mathbf{x})\, \mathbf{P}_X(d\mathbf{x})$ are well defined, and we have

$$\int_\Omega f(x(\omega))\, \mathcal{P}(d\omega) = \int_{R^n} f(\mathbf{x})\, \mathbf{P}_X(d\mathbf{x}) \tag{5.12}$$

We leave the proof of (5.12) to Exercise 1.3. If f is not nonnegative, then we write $f = f^+ - f^-$ as before, and (5.12) still holds provided that one of the pair $Ef^+(\mathbf{X})$ and $Ef^-(\mathbf{X})$ is finite. The integral $\int_{R^n} f(\mathbf{x})\, \mathbf{P}_X(d\mathbf{x})$ is called a **Lebesgue-Stieltjes integral** and is often written as $\int_{R^n} f(\mathbf{x})\, dP_X(\mathbf{x})$ to emphasize the role of P_X as a distribution function. However, for the definition of the integral, it is the role of $\mathbf{P}_X$ as a measure which is crucial.

If $\mathbf{X} = (X_1, \ldots, X_n)$ are real random variables, the function $F_X(\mathbf{u})$, $\mathbf{u} \in R^n$, defined by

$$F_X(\mathbf{u}) = E \exp\left(i \sum_{k=1}^n u_k X_k\right) = E \cos \sum_{k=1}^n u_k X_k + iE \sin \sum_{k=1}^n u_k X_k \tag{5.13}$$

is called the **characteristic function** for $\mathbf{X}$. Expressed in terms of $\mathbf{P}_X$, the characteristic function is given by

$$F_X(\mathbf{u}) = \int_{R^n} \exp\left(i \sum_{k=1}^n u_k x_k\right) \mathbf{P}_X(d\mathbf{x}) \tag{5.14}$$

Every distribution function is uniquely determined by its corresponding characteristic function [compare with Eqs. (3.5.21) and (3.5.32)]. This is easy to see when P_X is absolutely continuous with density p_X. In that case,

$$F_X(\mathbf{u}) = \int_{R^n} \exp\left(i \sum_{k=1}^n u_k x_k\right) p_X(\mathbf{x})\, d\mathbf{x} \tag{5.15}$$

and p_X can be obtained from F_X by the inversion formula of the Fourier integral, viz.,

$$p_X(\mathbf{x}) = \frac{1}{(2\pi)^n} \int_{R^n} \exp\left(-i \sum_{k=1}^n u_k x_k\right) F_X(\mathbf{u})\, d\mathbf{u} \tag{5.16}$$

6. CONVERGENCE CONCEPTS

Thus far, when we speak of the convergence of a sequence of random variables, we have been referring to pointwise convergence at every point in Ω. For this type of convergence, the probability measure $\mathcal{P}$ plays no role. In probability theory, convergence concepts which depend on $\mathcal{P}$ are of greater interest. In this section, we shall define some of these convergence concepts and explore their interrelation.

Definition. A sequence of random variables $\{X_n\}$ is said to **converge almost surely** to X if there exists a set A such that $\mathcal{P}(A) = 0$, and for every $\omega \notin A$

$$\lim_{n\to\infty} |X_n(\omega) - X(\omega)| = 0 \tag{6.1}$$

We abbreviate the words "almost sure" and "almost surely" to a.s., and we adopt the equivalent notations lim a.s. $X_n = X$ and $X_n \xrightarrow[n\to\infty]{\text{a.s.}} X$ to denote the a.s. convergence of $\{X_n\}$ to X. We observe that as we have defined it, the limit a.s. convergent sequence of random variables is always finite except on a set of probability zero. Furthermore, two functions which are both a.s. limits of the same sequence are equal except on a set of probability zero. With these considerations, we see that we can always take the limit X of an a.s. convergent sequence of random variables $\{X_n\}$ to be a random variable, i.e., it is finite and measurable.

We note that the convergence theorems for expectation, Propositions 5.2 to 5.4, remain valid if in the statements of these theorems "convergence at every point" is replaced by "convergence almost surely." This follows simply from the fact that if $\mathcal{P}(A) = 0$, then $\int_A X\,d\mathcal{P} = 0$. Thus, if $X_n \xrightarrow[n\to\infty]{\text{a.s.}} X$, and we denote by A the set on which convergence does not take place, then $\int_{\Omega-A} X_n\,d\mathcal{P} \to \int_{\Omega-A} X\,d\mathcal{P}$ is the same as $\int_\Omega X_n\,d\mathcal{P} \to \int_\Omega X\,d\mathcal{P}$.

Given a sequence $\{X_n\}$ which may or may not be almost surely convergent, it is difficult to apply the definition of a.s. convergence to it, because we have no candidate for the limit X. For this reason, the concept of **mutual convergence** is useful. We say that a sequence $\{X_n\}$ converges mutually a.s. if $\sup_{m\geq n} |X_m - X_n| \xrightarrow[n\to\infty]{\text{a.s.}} 0$. By virtue of the Cauchy criterion for the convergence of a sequence of real numbers, for each ω, we have $\sup_{m\geq n} |X_m(\omega) - X_n(\omega)| \xrightarrow[n\to\infty]{} 0$, if and only if the sequence of real numbers $\{X_n(\omega)\}$ converges. Therefore, mutual a.s. convergence and a.s. convergence are equivalent, and we have the following.

Proposition 6.1. A sequence of random variables $\{X_n\}$ converges a.s. if and only if it converges mutually a.s.

Definition. A sequence of random variables $\{X_n\}$ is said to **converge mutually in probability** if for every $\epsilon > 0$,

$$\sup_{m \geq n} \mathcal{P}(|X_m - X_n| \geq \epsilon) \xrightarrow[n \to \infty]{} 0 \tag{6.2}$$

A sequence $\{X_n\}$ is said to **converge in probability,** if there exists X such that for every $\epsilon > 0$,

$$\mathcal{P}(|X_n - X| \geq \epsilon) \xrightarrow[n \to \infty]{} 0 \tag{6.3}$$

We use the notations lim in p. $X_n = X$ or $X_n \xrightarrow[n \to \infty]{\text{in p.}} X$ to denote that $\{X_n\}$ converges in probability to X.

Proposition 6.2. Let $\{X_n\}$ be a sequence of random variables:
(*a*) If $\{X_n\}$ converges a.s., then it converges in probability to the same limit.
(*b*) $\{X_n\}$ converges in probability if and only if it converges mutually in probability.
(*c*) If $\{X_n\}$ converges in probability, then there is a subsequence converging a.s. to the same limit.

Remark: The only part of Proposition 6.2 that is easy to prove is (*a*). If $X_n \xrightarrow[n \to \infty]{\text{a.s.}} X$, then

$$\mathcal{P}(\sup_{m \geq n} |X_m - X| \geq \epsilon) \xrightarrow[n \to \infty]{} 0 \tag{6.4}$$

for every $\epsilon > 0$, which certainly implies that

$$\mathcal{P}(|X_n - X| \geq \epsilon) \xrightarrow[n \to \infty]{} 0 \tag{6.5}$$

for every $\epsilon > 0$. The proof for parts (*b*) and (*c*) will be omitted.

There is a third type of convergence which is important to us. We define it as follows.

Definition. A sequence of random variables $\{X_n\}$ is said to **converge in νth mean** ($\nu > 0$) to X if

$$E|X_n - X|^\nu \xrightarrow[n \to \infty]{} 0 \tag{6.6}$$

We use the notation $\lim_{n \to \infty} \nu.\text{m. } X_n = X$ or $X_n \xrightarrow[n \to \infty]{\nu.\text{m.}} X$.

Remark: The case $\nu = 2$ is of particular importance and is known as convergence in **quadratic mean.** We abbreviate quadratic mean by q.m.

Proposition 6.3.

(*a*) If $\{X_n\}$ converges in νth mean, then it converges in probability to the same limit.

(*b*) $\{X_n\}$ converges in νth mean if and only if

$$\sup_{m\geq n} E|X_m - X_n|^\nu \xrightarrow[n\to\infty]{} 0 \tag{6.7}$$

Proof:

(*a*) We make use of the following rather obvious inequality known as the **Markov inequality:**

$$\begin{aligned} E|Z|^\nu = \int_\Omega |Z|^\nu \, d\mathcal{P} &\geq \int_{|Z|\geq\epsilon} |Z|^\nu \, d\mathcal{P} \\ &\geq \epsilon^\nu \mathcal{P}(|Z| \geq \epsilon) \end{aligned} \tag{6.8}$$

Therefore, if $\{X_n\}$ converges in νth mean to X, then

$$\mathcal{P}(|X_n - X| \geq \epsilon) \leq \frac{1}{\epsilon^\nu} E|X_n - X|^\nu \xrightarrow[n\to\infty]{} 0$$

which proves that $X_n \xrightarrow[n\to\infty]{\text{in p.}} X$.

(*b*) We first suppose that $\{X_n\}$ converges in νth mean to X. Then,

$$\sup_{m\geq n} E|X_m - X_n|^\nu \leq 2^\nu\{E|X_n - X|^\nu + \sup_{m\geq n} E|X_m - X|^\nu\} \xrightarrow[n\to\infty]{} 0$$

Conversely, suppose that (6.7) holds. Then, by virtue of the Markov inequality (6.8), $\{X_n\}$ also converges mutually in probability. It follows from Proposition 6.2 that there is a subsequence $\{X'_n\}$ converging almost surely, and we denote the limit by X. Using Fatou's lemma, we find

$$E|X_n - X|^\nu \leq \liminf_{m\to\infty} E|X_n - X'_m|^\nu$$

Since $\{X'_n\}$ is a subsequence of $\{X_n\}$, we have

$$\lim_{n\to\infty} \liminf_{m\to\infty} E|X_n - X'_m| = \lim_{m,n\to\infty} E|X_m - X_n|^\nu$$

which is zero, because $\{X_n\}$ converges mutually in ν.m. Hence,

$$E|X_n - X|^\nu \xrightarrow[n\to\infty]{} 0$$

and $\{X_n\}$ converges in ν.m. to X. ∎

If we are given the pairwise joint distributions of a sequence of random variables $\{X_n\}$, we can determine whether $\{X_n\}$ converges in probability and whether it converges in νth mean. This is simply because if we know the joint distribution P_{mn} of X_m and X_n, we can compute $E|X_m - X_n|^\nu$ and $\mathcal{P}(|X_m - X_n| \geq \epsilon)$. Thus, mutual convergence in ν.m.

and in probability can be determined, hence, also convergence. On the other hand, to determine whether $\{X_n\}$ converges a.s. generally requires that we know the joint distribution of all finite subsets of random variables from the sequence $\{X_n\}$. There are, however, sufficient conditions on pairwise distributions which ensure a.s. convergence. We state one of the simplest and most useful of such conditions as follows.

Proposition 6.4. Suppose that for every $\epsilon > 0$,

$$\sum_n \sup_{m \geq n} \mathcal{P}(|X_m - X_n| \geq \epsilon) < \infty \tag{6.9}$$

Then $\{X_n\}$ converges almost surely.

Proof: Since (6.9) implies $\sup_{m \geq n} \mathcal{P}(|X_m - X_n| \geq \epsilon) \xrightarrow[n \to \infty]{} 0$, the sequence $\{X_n\}$ converges in probability. Let X denote the limit and define

$$A_n^\epsilon = \{\omega: |X(\omega) - X_n(\omega)| \geq 2\epsilon\}$$

Since $A_n^\epsilon \subset \{\omega: \max(|X(\omega) - X_m(\omega)|, |X_m(\omega) - X_n(\omega)|) \geq \epsilon\}$, we have

$$\mathcal{P}(A_n^\epsilon) \leq \mathcal{P}(|X - X_m| \geq \epsilon) + \mathcal{P}(|X_m - X_n| \geq \epsilon)$$

By letting $m \to \infty$, we get

$$\mathcal{P}(A_n^\epsilon) \leq \sup_{m \geq n} \mathcal{P}(|X_m - X_n| \geq \epsilon)$$

It follows from (6.9) that $\sum_n \mathcal{P}(A_n^\epsilon) < \infty$, and it follows from the Borel-Cantelli lemma (Proposition 4.2) that

$$\mathcal{P}(\limsup_n A_n^\epsilon) = 0$$

so that for every $\epsilon > 0$, $|X_n - X| \geq 2\epsilon$ for, at most, a finite number of values of n with probability 1. If we take $A = \bigcup_{k \geq 1} \limsup_n A_n^{1/k}$, then $\mathcal{P}(A) = 0$ and $\omega \notin A$ implies that

$$\lim |X_n(\omega) - X(\omega)| = 0$$

proving the theorem. ∎

The following example illustrates some of the ideas introduced in this section. Suppose $\{X_n\}$ is a sequence of random variables such that

$$\mathcal{P}[(X_m - X_n) < a] = \frac{1}{\sqrt{2\pi\sigma_{mn}^2}} \int_{-\infty}^{a} \exp\left(-\frac{x^2}{2\sigma_{mn}^2}\right) dx \tag{6.10}$$

and $\sigma_{mn}^2 = |m - n|/mn$. We want to investigate the convergence of $\{X_n\}$. First, we compute $E|X_m - X_n|^2$ and find

$$E|X_m - X_n|^2 = \frac{1}{\sqrt{2\pi\sigma_{mn}^2}} \int_{-\infty}^{\infty} x^2 \exp\left(-\frac{x^2}{2\sigma_{mn}^2}\right) dx$$
$$= \sigma_{mn}^2 = \frac{|m - n|}{mn} \tag{6.11}$$

Therefore,

$$\lim_{n\to\infty} \sup_{m\geq n} E|X_m - X_n|^2 = \lim_{n\to\infty} \frac{1}{n} = 0$$

and we have shown that $\{X_n\}$ converges in quadratic mean. It follows that it also converges in probability.

To prove a.s. convergence is more difficult. First, we note the formula

$$\int_{-\infty}^{\infty} x^{2k+2} \exp(-\tfrac{1}{2}x^2)\, dx = -\int_{-\infty}^{\infty} x^{2k+1} \frac{d}{dx} [\exp(-\tfrac{1}{2}x^2)]\, dx$$
$$= (2k + 1) \int_{-\infty}^{\infty} x^{2k} \exp(-\tfrac{1}{2}x^2)\, dx$$

Therefore,

$$E|X_m - X_n|^4 = 3\left(\frac{|m - n|}{mn}\right)^2$$
$$E|X_m - X_n|^6 = 15\left(\frac{|m - n|}{mn}\right)^3 \tag{6.12}$$
$$\cdots\cdots\cdots\cdots\cdots\cdots$$

Using the Markov inequality, we have

$$\sup_{m\geq n} \mathcal{P}(|X_m - X_n| \geq \epsilon) \leq \sup_{m\geq n} \frac{E|X_m - X_n|^4}{\epsilon^4} = \frac{3}{n^2\epsilon^4}$$

Therefore, for every $\epsilon > 0$,

$$\sum_n \sup_{m\geq n} \mathcal{P}(|X_m - X_n| \geq \epsilon) \leq \frac{3}{\epsilon^4} \sum_n \frac{1}{n^2} < \infty$$

and it follows from Proposition 6.4 that $\{X_n\}$ converges almost surely.

As the final topic in this section, we consider the distribution function of the limit of a convergent sequence of random variables.

Proposition 6.5. Let $\{X_n\}$ converge in probability to X. Let $P_n(\cdot)$ denote the distribution function of X_n, and let $P(\cdot)$ denote the distribution function of X. Then, at every continuity point x of $P(\cdot)$,

$$\lim_{n\to\infty} P_n(x) = P(x) \tag{6.13}$$

Proof: We write

$$P(x - \epsilon) = \mathcal{P}(X < x - \epsilon, X_n < x) + \mathcal{P}(X < x - \epsilon, X_n \geq x)$$
$$P_n(x) = \mathcal{P}(X < x - \epsilon, X_n < x) + \mathcal{P}(X \geq x - \epsilon, X_n < x)$$

Therefore,

$$P(x - \epsilon) \leq P_n(x) + \mathcal{P}(|X - X_n| > \epsilon)$$

Similary, we find

$$P_n(x) \leq P(x + \epsilon) + \mathcal{P}(|X - X_n| > \epsilon)$$

Because $\{X_n\}$ converges to X in probability, we have

$$P(x - \epsilon) \leq \lim_{n \to \infty} P_n(x) \leq P(x + \epsilon)$$

Because $P(\cdot)$ is continuous at x, we get (6.13) by letting $\epsilon \to 0$. ▮

Remark: We note that a distribution function is completely determined by its values at points of continuity. Because of this, $P(\cdot)$ is determined everywhere by the sequence $\{P_n(\cdot)\}$. The proof that a distribution function $P(\cdot)$ is completely determined by its values at points of continuity can be outlined as follows: Because $P(\cdot)$ is nondecreasing and bounded by 1, the number of jumps of size $1/n$ or greater must be less than n. Hence, the points of discontinuity are, at most, countable. This means that points of continuity are dense in $(-\infty, \infty)$, that is, every $x \epsilon (-\infty, \infty)$ is the limit of a sequence of continuity points. Because $P(\cdot)$ is left continuous at every x, it is completely determined by its values at points of continuity.

7. INDEPENDENCE AND CONDITIONAL EXPECTATION

Two events A and B are said to be **independent** if

$$\mathcal{P}(A \cap B) = \mathcal{P}(A)\mathcal{P}(B) \tag{7.1}$$

N events $A_1, A_2, \ldots, A_N$ are said to be independent if for any subset $\{k_1, \ldots, k_r\}$ of $\{1, 2, \ldots, N\}$,

$$\mathcal{P}\left(\bigcap_{i=1}^{r} A_{k_i}\right) = \prod_{i=1}^{r} \mathcal{P}(A_{k_i}) \tag{7.2}$$

More generally, we say that a family (countable or not) of events is a family of independent events if every finite subfamily is a set of independent events. If A and B are events such that $\mathcal{P}(B) > 0$, then we define the conditional probability of A given B by

$$\mathcal{P}(A|B) = \frac{\mathcal{P}(A \cap B)}{\mathcal{P}(B)} \tag{7.3}$$

If A and B are independent, then clearly

$$\mathcal{P}(A|B) = \mathcal{P}(A) \tag{7.4}$$

We say that the random variables $X_1, \ldots, X_N$, are independent if for arbitrary constants $x_1, \ldots, x_N$, the events $\{X_1 < x_1\}, \ldots, \{X_N < x_N\}$ are independent. If we denote the distribution function of X_i by $P_i(\cdot)$ and the joint distribution function of $X_1, \ldots, X_N$ by $P(\cdot)$, then $X_1, \ldots, X_N$ are independent if and only if

$$P(x_1, \ldots, x_N) = \prod_{i=1}^{N} P_i(x_i) \tag{7.5}$$

We leave the proof of this fact as an exercise. More generally, we define an infinite and possibly uncountable collection of random variables to be independent if every finite subset is independent.

Let $X_1, X_2, \ldots, X_N$ and Y be random variables. We define the conditional distribution function of Y relative to $X_1, \ldots, X_N$ as follows:

1. If $x_1, x_2, \ldots, x_N$ are such that $\mathcal{P}(X_1 = x_1, \ldots, X_N = x_N) > 0$, we set

$$P(y|x_1, \ldots, x_N) = \mathcal{P}(Y < y|X_1 = x_1, \ldots, X_N = x_N) \tag{7.6}$$

2. If $X_1, X_2, \ldots, X_N$ and Y have a joint density function p, we set

$$P(y|x_1, \ldots, x_N) = \frac{\int_{-\infty}^{y} p(x_1, x_2, \ldots, x_N, \eta)\, d\eta}{\int_{-\infty}^{\infty} p(x_1, x_2, \ldots, x_N, \eta)\, d\eta} \tag{7.7}$$

for all $(x_1, \ldots, x_N)$ such that the denominator is positive.

3. If neither of the above conditions is satisfied, $P(y|x_1, \ldots, x_N)$ remains undefined for the present.

If the numerator and the denominator in (7.7) are continuous at $(x_1, \ldots, x_N)$ then it is not hard to see that

$$P(y|x_1, \ldots, x_N) = \lim_{\substack{\epsilon_i \downarrow 0 \\ i=1,2,\ldots,N}} \mathcal{P}(Y < y|x_i \le X_i < x_i + \epsilon_i,\ i = 1, \ldots, N) \tag{7.8}$$

The function defined by

$$p(y|x_1, \ldots, x_N) = \frac{p(x_1, \ldots, x_N, y)}{\int_{-\infty}^{\infty} p(x_1, \ldots, x_N, \eta)\, d\eta} \tag{7.9}$$

is called the conditional density function of Y relative to $X_1, \ldots, X_N$, and it has the interpretation

$$\begin{aligned} p(y|x_1, \ldots, x_N)\, dy \\ = \mathcal{P}(y \le Y < y + dy | X_1 = x_1, \ldots, X_N = x_N) \end{aligned} \tag{7.10}$$

Conditional distributions involve rather cumbersome notations, and their use is limited to situations involving only a finite number of random variables. In more advanced applications of probability theory, they are all but abandoned in favor of the simpler yet more powerful concept of conditional expectation.

Before defining the conditional expectation, we shall try to get an intuitive picture of what it is. Suppose that $\mathcal{A}$ is a σ algebra of subsets of some basic space Ω. We call A an **atom** of $\mathcal{A}$ if $A \in \mathcal{A}$, and no subset of A belongs to $\mathcal{A}$ other than A itself and the empty set $\emptyset$. Atoms are a kind of irreducible units of a σ algebra. It is obvious that two distinct atoms must be disjoint. Suppose that there is a *countable* collection of sets $B_1, B_2, \ldots$, which generates $\mathcal{A}$, that is, $\mathcal{A}$ is the minimal σ algebra containing all the B_j's. Then $\mathcal{A}$ is said to be **separable.** Atoms of a separable σ algebra have some important properties that we summarize as follows.

Proposition 7.1. Let $\mathcal{A}$ be a separable σ algebra generated by a countable collection of sets $B_1, B_2, \ldots$.

(*a*) The collection of all atoms of $\mathcal{A}$ can be indexed by a subset of the real line. We denote the index set by $T (T \in R)$ and denote the collection of atoms by $\{A_t, t \in T\}$.

(*b*) Every set B in $\mathcal{A}$ is a union of atoms.

(*c*) $\Omega = \bigcup_{t \in T} A_t$.

Proof: Take the generating sets $B_1, B_2, \ldots$ and define $\tilde{B}_i$ to be B_i or its complement. Now, the atoms of $\mathcal{A}$ can be constructed by forming the countable intersections $\bigcap_i \tilde{B}_i$, because for any atom A, the intersection $A \cap B_i$ must be either A or empty. Since each atom, thus constructed, is indexed by a sequence (possibly infinite) of binary numbers, the index set T can be taken to be a subset of the real line. Each ω belongs to either B_i or its complement for each i. Hence, each ω belongs to one and only one A_t so that $\Omega = \bigcup_{t \in T} A_t$. Each set B in $\mathcal{A}$ is the union of all intersections $\bigcap_i \tilde{B}_i$ such that $B \cap \tilde{B}_i$ is nonempty for every i. Hence, B is a union of atoms. ▮

On an atom of $\mathcal{A}$, an $\mathcal{A}$-measurable function X can only assume a single value. Let A be an atom, and let x be a real number, then $\{\omega: X(\omega) = x\}$ must be in $\mathcal{A}$ and so must $A \cap \{\omega: X(\omega) = x\}$. By definition, A being an atom implies

$$A \cap \{\omega: X(\omega) = x\} = \emptyset \quad \text{or} \quad A$$

Now suppose that $\mathcal{A}_1$ and $\mathcal{A}_2$ are two separable σ algebras such that $\mathcal{A}_1 \supset \mathcal{A}_2$. Since every atom of $\mathcal{A}_2$ is a set in $\mathcal{A}_1$, every atom of $\mathcal{A}_2$ is a union of atoms of $\mathcal{A}_1$. Therefore, the atoms of $\mathcal{A}_2$ are bigger than the atoms of $\mathcal{A}_1$, and the collection of atoms of $\mathcal{A}_2$ gives a coarser partition of Ω than the corresponding collection from $\mathcal{A}_1$. A function measurable with respect to $\mathcal{A}_2$ is necessarily measurable with respect to $\mathcal{A}_1$, but not conversely, because an $\mathcal{A}_1$-measurable function may take on more than one value on an atom of $\mathcal{A}_2$.

Let X be a random variable defined on a probability space $(\Omega,\mathcal{A},\mathcal{P})$. We can write X as the difference $X^+ - X^-$ of a pair of nonnegative random variables X^+ and X^-. We assume that at least one of the pair EX^+ and EX^- is finite so that EX is well defined. Let $\mathcal{A}'$ be a sub-σ algebra of $\mathcal{A}$. Roughly speaking, the **conditional expectation** of X with respect to $\mathcal{A}'$ is an $\mathcal{A}'$-measurable random variable obtained by averaging X on the atoms of $\mathcal{A}'$. The precise definition is as follows.

Definition. $E^{\mathcal{A}'}X$ is uniquely defined up to sets in $\mathcal{A}'$ of probability zero by the following two conditions:

(*a*) $E^{\mathcal{A}'}X$ is measurable with respect to $\mathcal{A}'$.

(*b*) Let I_A denote the indicator function of A. Then

$$EI_AE^{\mathcal{A}'}X = EI_AX \qquad \text{for all } A \text{ in } \mathcal{A}' \tag{7.11}$$

The existence of $E^{\mathcal{A}'}X$ as a possibly extended real-valued function is guaranteed by the Radon-Nikodym theorem, one version of which we state as follows.

Proposition 7.2. Let $(\Omega,\mathcal{A},\mathcal{P})$ be a probability space, and let μ be a nonnegative σ-additive set function on $(\Omega,\mathcal{A})$, that is, μ is a measure. Suppose that for every A in $\mathcal{A}$ such that $\mathcal{P}(A) = 0$, we also have $\mu(A) = 0$. Then, there exists a nonnegative, $\mathcal{A}$-measurable, extended real-valued function φ such that for every $A \in \mathcal{A}$,

$$\mu(A) = \int_A \varphi \, d\mathcal{P} \tag{7.12}$$

Furthermore, φ is unique up to sets of $\mathcal{P}$-measure zero.

Remark: The function φ is called the **Radon-Nikodym derivative** of μ with respect to $\mathcal{P}$ and is sometimes denoted by $d\mu/d\mathcal{P}$.

We now apply the Radon-Nikodym theorem to the problem of defining conditional expectation. Suppose that X is a random variable on $(\Omega,\mathcal{A},\mathcal{P})$, and $\mathcal{A}'$ is a sub-σ algebra of $\mathcal{A}$. We write $X = X^+ - X^-$ as usual and assume that at least one of the pair EX^+ and EX^- is finite. Now, define measures μ^+ and μ^- on $(\Omega,\mathcal{A}')$ by

$$\begin{aligned} \mu^+(B) &= EI_BX^+ \\ \mu^-(B) &= EI_BX^- \end{aligned} \qquad B \in \mathcal{A}' \tag{7.13}$$

If $B \in \mathcal{B}$ and $\mathcal{P}(B) = 0$, then

$$\mu^+(B) = \int_B X^+ \, d\mathcal{P} = 0$$

$$\mu^-(B) = \int_B X^- \, d\mathcal{P} = 0$$

Therefore, there exist $\mathcal{A}'$-measurable functions φ^+ and φ^- such that for all B in $\mathcal{A}'$,

$$\mu^{\pm}(B) = \int_B \varphi^{\pm} \, d\mathcal{P} \tag{7.14}$$

If we set

$$\begin{aligned} E^{\mathcal{A}'}X &= E^{\mathcal{A}'}X^+ - E^{\mathcal{A}'}X^- \\ &= \varphi^+ - \varphi^- \end{aligned} \tag{7.15}$$

then $E^{\mathcal{A}'}X$ will have the defining properties, and uniqueness follows from the uniqueness of φ^+ and φ^-.

If EX^+ and EX^- are both finite, that is, if $E|X|$ is finite, then $E^{\mathcal{A}'}X$ can always be taken to be finite valued so that it is a random variable as we have defined it. If not, $E^{\mathcal{A}'}X$ may have to assume values of $\pm\infty$ and is a random variable only in an extended sense. Of course, if we had defined random variables to be extended-valued functions to begin with, this difficulty would be avoided. This is done by many authors. However, this approach also has its disadvantages. For example, extended-valued functions cannot always be added, because the sum may involve $\infty - \infty$. We shall continue to define random variables as real-valued functions. When the need arises, we shall make free use of extended-valued measurable functions. While they are not random variables in our sense, the difference is seldom important.

Roughly speaking, a conditional expectation has (almost surely) all the properties of an expectation. We make this precise by the following proposition.

Proposition 7.3. If $X = c$ a.s. then $E^{\mathcal{A}'}X = c$ a.s., and if $X \geq Y$ a.s., then $E^{\mathcal{A}'}X \geq E^{\mathcal{A}'}Y$ a.s. Furthermore $E^{\mathcal{A}'}$ is a linear operation, that is,

$$E^{\mathcal{A}'}(\alpha X + \beta Y) = \alpha E^{\mathcal{A}'}X + \beta E^{\mathcal{A}'}Y \qquad \text{a.s.} \tag{7.16}$$

Proof: Everything follows directly from the definition of $E^{\mathcal{A}'}$.

Conditional expectation also has the convergence property of expectation. The results corresponding to Propositions 5.2 to 5.4 and a result concerning convergence in ν.m. are stated in the following proposition, but proof will be omitted.

Proposition 7.4 (Convergence Properties).

(*a*) If $X_n \xrightarrow[n\to\infty]{\nu.\text{m.}} X$ and $\nu \geq 1$, then

$$E^{\mathcal{A}'}X_n \xrightarrow[n\to\infty]{\nu.\text{m.}} E^{\mathcal{A}'}X \tag{7.17}$$

(*b*) If $\{X_n\}$ is a monotone sequence of integrable random variables converging to X, then $\{E^{\mathcal{A}'}X_n\}$ is almost surely monotone and

$$E^{\mathcal{A}'}X_n \xrightarrow[n\to\infty]{\text{a.s.}} E^{\mathcal{A}'}X \tag{7.18}$$

(*c*) Let $\{X_n\}$ be a sequence of integrable random variables. Suppose that there exists an integrable random variable X such that $X_n \geq X$ for all n, then

$$\liminf E^{\mathcal{A}'}X_n \geq E^{\mathcal{A}'}\liminf X_n \qquad \text{a.s.} \tag{7.19}$$

(*d*) Suppose that $X_n \xrightarrow[n\to\infty]{\text{a.s.}} X$ and $|X_n| \leq Y$, for some integrable Y. Then

$$E^{\mathcal{A}'}X_n \xrightarrow[n\to\infty]{\text{a.s.}} E^{\mathcal{A}'}X \tag{7.20}$$

Conditional expectation is really a smoothing operation. Roughly speaking, $E^{\mathcal{A}'}X$ is obtained by averaging X on the atoms of $\mathcal{A}'$. Proposition 7.5 below summarizes some of the properties related to the fact that $E^{\mathcal{A}'}X$ is a smoothed version of X.

Proposition 7.5 (Smoothing Properties).

(*a*) If B is an atom of $\mathcal{A}'$ and $\mathcal{P}(B) > 0$, then the value of $E^{\mathcal{A}'}X$ on B is given by

$$(E^{\mathcal{A}'}X)_B = \frac{1}{\mathcal{P}(B)} \int_B X \, d\mathcal{P} \tag{7.21}$$

(b) If every event in $\mathcal{A}'$ is independent of every event of the form $\{\omega: X(\omega) < x\}$, then

$$E^{\mathcal{A}'}X = EX \qquad \text{a.s.} \tag{7.22}$$

In particular, if we denote $\mathcal{A}_\emptyset = \{\Omega,\emptyset\}$, then $E^{\mathcal{A}_\emptyset}X = EX$ a.s.
(c) If Y is $\mathcal{A}'$ measurable, then

$$E^{\mathcal{A}'}YX = YE^{\mathcal{A}'}X \qquad \text{a.s.} \tag{7.23}$$

In particular, $E^{\mathcal{A}'}Y = Y$ almost surely, and for any random variable X, $E^{\mathcal{A}}X = X$ a.s.
(d) If $\mathcal{A}_1 \supset \mathcal{A}_2$, then

$$E^{\mathcal{A}_2}E^{\mathcal{A}_1}X = E^{\mathcal{A}_2}X \qquad \text{a.s.} \tag{7.24}$$

Proof:
(a) If B is an atom of $\mathcal{A}'$, then $E^{\mathcal{A}'}X$ must be constant on B, because $E^{\mathcal{A}'}X$ is $\mathcal{A}'$ measurable. By definition,

$$EI_BE^{\mathcal{A}'}X = EI_BX$$

Since $E^{\mathcal{A}'}X$ is constant on B, we also have

$$EI_BE^{\mathcal{A}'}X = (EI_B)(E^{\mathcal{A}'}X)$$

Hence, $(E^{\mathcal{A}'}X)_B = (1/EI_B)EI_BX$, which proves (a).
(b) Let $B \in \mathcal{A}'$. Then I_B and X are independent. Therefore,

$$EI_BE^{\mathcal{A}'}X = EI_BX = (EX)(EI_B)$$

On the other hand, EX is $\mathcal{A}'$ measurable and

$$E(I_BEX) = (EX)(EI_B)$$

Hence, $E^{\mathcal{A}'}X = EX$, a.s. by virtue of the uniqueness of $E^{\mathcal{A}'}X$.
(c) If B is an event in $\mathcal{A}'$, and $Y = I_B$, then for every $A \in \mathcal{A}'$,

$$EI_AE^{\mathcal{A}'}YX = EI_AI_BX = EI_{A\cap B}X$$
$$EI_AYE^{\mathcal{A}'}X = EI_AI_BE^{\mathcal{A}'}X$$
$$= EI_{A\cap B}X$$

so that (7.23) follows. By linearity (7.23) must also be true when Y is a linear combination of indicator functions (that is, Y is a simple function). For the rest, we note that by virtue of linearity, we only need to prove (7.23) for the case when both X and Y are nonnegative. Since Y is nonnegative, we can find an increasing sequence of nonnegative simple functions $\{Y_n\}$ converging to Y. Because (7.23) is true for simple functions,

$$Y_nE^{\mathcal{A}'}X = E^{\mathcal{A}'}XY_n$$

From Proposition 5.2 we have for every $A \in \mathcal{A}'$,

$$
\begin{aligned}
EI_A Y_n E^{\mathcal{A}'} X &\xrightarrow[n\to\infty]{} EI_A Y E^{\mathcal{A}'} X \\
EI_A E^{\mathcal{A}'} X Y_n &= EI_A X Y_n \xrightarrow[n\to\infty]{} EI_A XY \\
&= EI_A E^{\mathcal{A}'} XY
\end{aligned}
$$

whence (7.23) follows.
(*d*) Suppose that $\mathcal{A}_1 \supset \mathcal{A}_2$ and $A \in \mathcal{A}_2$. Then I_A is both $\mathcal{A}_2$ measurable and $\mathcal{A}_1$ measurable. Now for every $A \in \mathcal{A}_2$,

$$
\begin{aligned}
EI_A E^{\mathcal{A}_2} E^{\mathcal{A}_1} X &= EI_A E^{\mathcal{A}_1} X \\
&= EI_A X \\
&= EI_A E^{\mathcal{A}_2} X
\end{aligned}
$$

which proves (7.24). ▮

We define two σ algebras $\mathcal{A}_1$ and $\mathcal{A}_2$ to be independent if, whenever $B_1 \in \mathcal{A}_1$ and $B_2 \in \mathcal{A}_2$,

$$EI_{B_1} I_{B_2} = EI_{B_1} EI_{B_2} \tag{7.25}$$

Conditional independence can be defined in a similar way as follows: Two σ algebras $\mathcal{A}_1$ and $\mathcal{A}_2$ are said to be **conditionally independent** given $\mathcal{A}'$ if, whenever $B_1 \in \mathcal{A}_1$ and $B_2 \in \mathcal{A}_2$, we have

$$E^{\mathcal{A}'} I_{B_1} I_{B_2} = E^{\mathcal{A}'} I_{B_1} E^{\mathcal{A}'} I_{B_2} \tag{7.26}$$

Conditional independence plays a vital role in the theory of Markov processes.

Suppose that X and Y are two random variables, and $E|Y| < \infty$. Let $\mathcal{A}_X$ denote the smallest σ algebra with respect to which X is measurable. Now, we want to show that $E^{\mathcal{A}_X} Y$ can always be expressed as a Borel function of X, that is, there exists a real-valued Borel function $f(x)$, $x \in R$, such that

$$(E^{\mathcal{A}_X} Y)(\omega) = f(X(\omega)) \qquad \text{a.s.} \tag{7.27}$$

The notation $E^{\mathcal{A}_X} Y$ is unnecessarily cumbersome, and one usually writes $E(Y|X)$ instead.

To show (7.27), we begin by recalling the probability measure $\mathbf{P}_X$ on the σ algebra of Borel sets $\mathcal{R}$ defined by

$$\mathbf{P}_X(B) = \mathcal{P}(\{\omega: X(\omega) \in B\}) \tag{7.28}$$

Let $I_B(x)$, $x \in R$, and $B \in \mathcal{R}$, be the indicator function

$$I_B(x) = \begin{cases} 1 & \text{if } x \in B \\ 0 & \text{if } x \notin B \end{cases} \tag{7.29}$$

It is clear that $EI_B(X)Y = 0$ whenever $\mathcal{P}(X \in B) = \mathbf{P}_X(B) = 0$. Therefore, by writing $Y = Y^+ - Y^-$ as the difference of two nonnegative functions and applying the Radon-Nikodym theorem, we find that there exists a Borel function f such that

$$\begin{aligned} EI_B(X)Y &= \int_B f(x)\, \mathbf{P}_X(dx) \qquad B \in \mathcal{R} \\ &= \int_{\{\omega:\, X(\omega) \in B\}} f(X(\omega))\, \mathcal{P}(d\omega) \end{aligned} \tag{7.30}$$

By the definition of $E^{\mathcal{A}_X} Y$, we also have

$$EI_B(X)Y = \int_{\{\omega:\, X(\omega) \in B\}} (E^{\mathcal{A}_X} Y)(\omega)\, \mathcal{P}(d\omega) \tag{7.31}$$

Because $\mathcal{A}_X$ is generated by X, every event A in $\mathcal{A}_X$ is of the form $A = \{\omega: X(\omega) \in B\}$ for some $B \in \mathcal{R}$ (see Exercise 16). A comparison of (7.30) and (7.31) now yields (7.27).

Exactly the same procedure can now be used to show that if $\mathcal{A}_X$ is the smallest σ algebra with respect to which $\mathbf{X} = (X_1, X_2, \ldots, X_n)$ are all measurable, then

$$(E^{\mathcal{A}_X} Y)(\omega) = f(\mathbf{X}(\omega)) \qquad \text{a.s.} \tag{7.32}$$

where Y satisfies $E|Y| < \infty$, and $f: R^n \to R$ is a Borel function. Again, we shall often write $E(Y|\mathbf{X})$ instead of $E^{\mathcal{A}_X} Y$.

Earlier in this section we defined the conditional distribution function $P(y|x_1, \ldots, x_n)$ [see (7.6) and (7.7)] and interpreted it as

$$P(y|x_1, \ldots, x_N) = \mathcal{P}(Y < y | X_\nu = x_\nu,\ \nu = 1, \ldots, n) \tag{7.33}$$

However, $P(y|x_1, \ldots, x_n)$ was defined only if either of two conditions is satisfied. We can now remove this restriction. Let $I_{\{Y<y\}}$ denote the indicator function

$$I_{\{Y<y\}}(\omega) = \begin{cases} 1 & \text{if } Y(\omega) < y \\ 0 & \text{otherwise} \end{cases} \tag{7.34}$$

According to (7.32) there exists a Borel function, say $g_y(\mathbf{x})$, $\mathbf{x} \in R^n$, such that

$$E(I_{\{Y<y\}}|\mathbf{X}) = g_y(\mathbf{X}) \tag{7.35}$$

We now define

$$P(y|x_1, \ldots, x_n) = g_y(x_1, \ldots, x_n) \tag{7.36}$$

By exactly the same procedure, we can define the joint distribution function

$$\begin{aligned} &P(y_1, \ldots, y_m|x_1, \ldots, x_n) \\ &\quad = \mathcal{P}(Y_j < y_j,\ j = 1, \ldots, m | X_k = x_k,\ k = 1, \ldots, n) \end{aligned} \tag{7.37}$$

It is not hard to verify that (7.36) reduces to (7.6) and (7.7) under the corresponding conditions.

EXERCISES

1. Let $\Omega = [0, \infty)$.

(*a*) Let C_1 be the class of all intervals of the form $[0,a)$. Show that C_1 is not an algebra.

(*b*) Let C_2 be the class of all unions of a finite number of intervals of the form $[a,b)$. Show that C_2 is an algebra, but not a σ algebra.

(*c*) Show that C_2 is the smallest algebra which contains C_1.

(*d*) Show that the σ algebra generated by C_2 contains all intervals in $[0, \infty)$, closed or open at either end.

2. Suppose that $\Omega = [0, \infty)$, and C is the class of all intervals of the form $[0,a)$, $0 < a \leq \infty$. Let $P(x)$, $0 \leq x < \infty$, be a left-continuous nondecreasing function such that $P(0) = 0$ and $\lim_{x \to \infty} P(x) = 1$. We define $\mathcal{P}$ on C by

$$\mathcal{P}([0,a)) = P(a)$$

(*a*) Show that $\mathcal{A}(C)$ includes all intervals in $[0, \infty)$.

(*b*) How must $\mathcal{P}((a,b))$, $\mathcal{P}([a,b))$, and $\mathcal{P}([a,b])$ be defined in terms of P in order for $\mathcal{P}$ to be σ additive?

3. Let $\mathbf{X} = (X_1, \ldots, X_n)$ be random variables defined on a probability space $(\Omega,\mathcal{A},\mathcal{P})$. We can uniquely define a probability measure $\mathbf{P}_X$ on $(R^n,\mathcal{R}^n)$ by

$$\mathbf{P}_X(\{\mathbf{x}: \alpha_\nu \leq x_\nu < \beta_\nu, \nu = 1, \ldots, n\}) = \mathcal{P}(\{\omega: \alpha_\nu \leq X_\nu(\omega) < \beta_\nu, \nu = 1, \ldots, n\})$$

Let $f: R^n \to R$ be a nonnegative Borel function.

(*a*) Show that for arbitrary a and b

$$\mathbf{P}_X(\{\mathbf{x}: a \leq f(\mathbf{x}) < b\}) = \mathcal{P}(\{\omega: a \leq f(\mathbf{X}(\omega)) < b\})$$

(*b*) Both $\int_{R^n} f(\mathbf{x})\, \mathbf{P}_X(d\mathbf{x})$ and $\int_\Omega f(\mathbf{X}(\omega))\mathcal{P}(d\omega)$ are well defined by the procedure given in Sec. 1.5. Show that they are equal.

We note that $\int_{R^n} f(\mathbf{x})\, \mathbf{P}_X(d\mathbf{x})$ so defined is called a Lebesgue-Stieltjes integral and is sometimes written as $\int_{R^n} f(\mathbf{x})\, dP(\mathbf{x})$. By writing $f = f^+ - f^-$, we can extend the definition to functions which are not necessarily nonnegative.

4. Let $\{A_n, n = 1, 2, \ldots\}$ be a sequence of sets. We can show that a point ω belongs to $\bigcap_{n=1}^{\infty} \bigcup_{k \geq n} A_k$ if and only if it belongs to an infinite number of A_n's as follows.

Suppose that ω belongs to an infinite number of A_n's, then for every n, $\omega \in \bigcup_{k \geq n} A_k$. Therefore, $\omega \in \bigcap_{n=1}^{\infty} \bigcup_{k \geq n} A_k$. On the other hand, if ω belongs to only a finite number of A_n's, then there is some n_0 such that $\omega \notin \bigcup_{k \geq n_0} A_k$. Since $\bigcup_{k \geq n_0} A_k$ contains $\bigcap_{n=1}^{\infty} \bigcup_{k \geq n} A_k$, this proves that ω cannot belong to $\bigcap_{n=1}^{\infty} \bigcup_{k \geq n} A_k$ if ω belongs to only a finite number of A_n's. Show that $\omega \in \bigcup_{n=1}^{\infty} \bigcap_{k \geq n} A_k$ if and only if ω belongs to all but a finite number of A_n's.

5. Suppose that $X_1, \ldots, X_n$ are n random variables with a joint density function p_X. Let $Y_1, \ldots, Y_n$ be defined by $Y_i = f_i(X_1, \ldots, X_n)$. Suppose that $\mathbf{f}$ has a differentiable inverse $\mathbf{g}$ so that $X_i = g_i(Y_1, \ldots, Y_n)$. Show that the joint density function of $Y_1, \ldots, Y_n$ is given by

$$p_Y(y_1, \ldots, y_n) = p_X(g_1(\mathbf{y}), \ldots, g_n(\mathbf{y}))|J(\mathbf{y})|$$

where $|J|$ denotes the absolute value of the determinant of $J(\mathbf{y}) = \partial g_i/\partial y_j$. Suggestion: Consider the incremental "rectangle" in R^n with sides $dy_1, dy_2, \ldots, dy_n$ and located at the point $\mathbf{y}$. Under the transformation $\mathbf{g}$ this rectangle is mapped approximately into a "parallelepiped" located at $\mathbf{g}(\mathbf{y})$ with sides $J(\mathbf{y})\,d\mathbf{y}$, the volume of which is $|J(\mathbf{y})|\,dy_1, dy_2, \ldots, dy_n$ [see, for example, Birkhoff and MacLane, 1953, pp. 307–310]. The desired result now follows from the interpretation of probability density function as probability per unit volume [see also, Thomasian, 1969, pp. 362–363].

6. Suppose that X_1, X_2, X_3 are random variables with a joint density

$$p_X(x_1,x_2,x_3) = \frac{1}{(2\pi)^{\frac{3}{2}}} \exp\left[-\tfrac{1}{2}(x_1^2 + x_2^2 + x_3^2)\right]$$

Find the density function for $Y = \sqrt{X_1^2 + X_2^2 + X_3^2}$.
Hint: Introduce random variable Θ and Φ so that

$$X_1 = Y \cos\Theta$$
$$X_2 = Y \sin\Theta \cos\Phi$$
$$X_3 = Y \sin\Theta \sin\Phi$$

7. Suppose that $X_1, \ldots, X_n$ have a joint density function

$$p_X(x_1, \ldots, x_n) = \frac{1}{(2\pi)^{n/2}} \exp\left[-\tfrac{1}{2}(x_1^2 + \cdots + x_n^2)\right]$$

Let $Y_k = \sum_{j=1}^{k} X_j$, $k = 1, \ldots, n$. Find the joint density p_Y for $Y_1, \ldots, Y_n$.

8. Prove that

$$E\frac{|X|}{1+|X|} - \frac{\epsilon}{1+\epsilon} \leq \mathcal{P}(|X| \geq \epsilon) \leq \frac{1+\epsilon}{\epsilon} E\frac{|X|}{1+|X|}$$

9. Show that $X_n \xrightarrow[n\to\infty]{\text{in p.}} X$ if and only if

$$\lim_{n\to\infty} E\left(\frac{|X_n - X|}{1 + |X_n - X|}\right) = 0$$

10. Suppose that $\{X_n\}$ is a sequence of random variables such that

$$\mathcal{P}(X_n - X_m < x) = \frac{1}{\pi\sigma_{mn}} \int_{-\infty}^{x} \frac{1}{1 + \xi^2/\sigma_{mn}^2}\, d\xi$$

with $\sigma_{mn}^2 = |m - n|/mn$. Does $\{X_n\}$ converge in probability?
Hint: Compute $E[|X_n - X_m|/(1 + |X_n - X_m|)]$.

11. Starting from the Schwarz inequality $|EXY|^2 \le EX^2EY^2$, prove that $X_n \xrightarrow[n\to\infty]{\text{q.m.}} X$ implies that

$$EX_n \xrightarrow[n\to\infty]{} EX$$

and

$$EX_n^2 \xrightarrow[n\to\infty]{} EX^2$$

Hint: For the second part verify that

$$(\sqrt{EX_n^2} - \sqrt{EX^2})^2 \le E(X_n - X)^2$$

12. If $\{X_n\}$ converges in q.m. to X and each X_n has a density function $p_n(x) = (1/\sqrt{2\pi\sigma_n^2}) \exp[-\frac{1}{2}(x - \mu_n)^2/\sigma_n^2]$, prove that X has a density

$$p(x) = (1/\sqrt{2\pi\sigma^2}) \exp[-\tfrac{1}{2}(x - \mu)^2/\sigma^2]$$

provided that $E(X - EX)^2 > 0$.

13. Prove that $X_1, \ldots, X_n$ are mutually independent if and only if

$$\mathcal{P}(X_\nu < x_\nu, \nu = 1, \ldots, n) = \prod_{\nu=1}^{n} \mathcal{P}(X_\nu < x_\nu)$$

14. Let X_1 and X_2 be two random variables such that $EX_1 = 0$ and $EX_2 = 0$.

(*a*) Suppose that we can find a linear combination $Y = X_1 + \alpha X_2$ which is independent of X_2. Show that $E(X_1|X_2) = -\alpha X_2$.

(*b*) If X_1 and X_2 have a joint density function

$$p(x_1,x_2) = \frac{1}{2\pi\sqrt{1 - \rho^2}} \exp\left[-\frac{1}{2(1 - \rho^2)}(x_1^2 - 2\rho x_1x_2 + x_2^2)\right]$$

find $E(X_1|X_2)$.

15. Suppose that X_1 and X_2 have a joint density given by

$$p(x_1,x_2) = \frac{1}{2\pi} \exp\left[-\tfrac{1}{2}(x_1^2 + x_2^2)\right]$$

Let $Y = \sqrt{X_1^2 + X_2^2}$. Find $E(X_1|Y)$.
Hint: Introduce random variable Φ so that $X_1 = Y \cos \Phi$ and $X_2 = Y \sin \Phi$.

16. Let $\mathcal{A}_X$ denote the smallest σ algebra with respect to which the random variables $\mathbf{X} = (X_1, X_2, \ldots, X_n)$ are all measurable. Let $X^{-1}(B)$ denote $\{\omega: \mathbf{X}(\omega) \in B\}$. Show that $\mathcal{A}_X = \{X^{-1}(B),\ B \in \mathcal{R}^n\}$. In other words, $\mathcal{A}_X$ is the collection of all inverse images of Borel sets.

2

Stochastic Processes

1. DEFINITION AND PRELIMINARY CONSIDERATIONS

A **stochastic process** $\{X_t, t \in T\}$ is a family of random variables, indexed by a real parameter t and defined on a common probability space $(\Omega,\mathcal{A},\mathcal{P})$. Unless otherwise specified, the parameter set T will always be taken to be an interval. By definition, for each t, X_t is an $\mathcal{A}$-measurable function. For each ω, $\{X_t(\omega), t \in T\}$ is a function defined on T and is called a **sample function** of the process.

If $T_n = \{t_1, \ldots, t_n\}$ is a finite set from T, we denote by P_{T_n} the joint distribution function of $\{X_{t_1}, \ldots, X_{t_m}\}$. The collection $\{P_{T_n}\}$ as T_n ranges over all finite sets in T is called the family of **finite-dimensional distributions** of the process $\{X_t, t \in T\}$. Loosely speaking, problems which can be answered directly in terms of the finite-dimensional distributions involve no mathematical difficulties of a measure-theoretical nature. The more elementary applications of stochastic processes are problems of this type. Let $\{X_t, t \in T\}$ be a stochastic process defined on $(\Omega,\mathcal{A},\mathcal{P})$. Let $\mathcal{B}_X$ and $\mathcal{A}_X$ be, respectively, the smallest algebra and the smallest σ algebra with respect to which X_t is measurable for every $t \in T$. The difference between $\mathcal{B}_X$ and $\mathcal{A}_X$ is that $\mathcal{B}_X$ is only closed under all finite set operations while $\mathcal{A}_X$ is closed under all countable set operations. Now, if all finite-dimensional distributions are known, then the probability of every event in $\mathcal{B}_X$ is uniquely and directly determined, because every set in $\mathcal{B}_X$ involves only a finite number of X_t's. In other words, $\mathcal{P}$ is uniquely determined on $\mathcal{B}_X$. Now, since $\mathcal{P}$ is a probability measure, it must be σ additive. Hence, by the extension theorem (Proposition 1.1.1), we can extend $\mathcal{P}$ from $\mathcal{B}_X$ to the smallest σ algebra containing $\mathcal{B}_X$, and that is just $\mathcal{A}_X$. Therefore, the restriction of $\mathcal{P}$ on $\mathcal{A}_X$ is uniquely determined by the set of all finite-dimensional distributions.

We illustrate the previous remarks by an example. Suppose that $\{X_t, 0 \le t \le 1\}$ is a stochastic process. The event

$$\left\{\omega: X_t(\omega) \ge 0, t = \frac{1}{k}, k = 1, 2, \ldots, n\right\} \tag{1.1}$$

is a set in $\mathcal{B}_X$. The events

$$\left\{\omega: X_t(\omega) \ge 0, t = \frac{1}{k}, k = 1, 2, \ldots\right\} \tag{1.2}$$

and

$$\left\{\omega: X_t(\omega) \ge 0, t = \frac{l}{k}, l = 0, 1, \ldots, k; k = 1, 2, \ldots\right\} \tag{1.3}$$

are sets in $\mathcal{A}_X$. However, the set

$$\{\omega: X_t(\omega) \ge 0, 0 \le t \le 1\} = \bigcap_{t \in [0,1]} \{\omega: X_t(\omega) \ge 0\} \tag{1.4}$$

may not even be in $\mathcal{A}$, that is, it may not be an event, because it involves uncountable set operations on events. Whether it is an event or not depends on the precise nature of the probability space and the process. We shall consider these questions in greater detail in Sec. 2.

In practice, one seldom begins with a given probability space and a given family of random variables defined on it. Instead one often starts with a proposed collection of finite-dimensional distributions $\{P_{T_n}$, all finite T_n in $T\}$, which is usually obtained by a combination of observations and hypotheses. The question then arises as to whether we can always find a stochastic process having these distributions. We shall answer this question as clearly as we can, because it is a source of some confusion.

First, the collection of finite-dimensional distributions must be **compatible** in the following sense: If T_n and T_m are two ordered finite sets from T such that T_n contains T_m, then P_{T_m} must be equal to P_{T_n} with the appropriate variables set to ∞. For example,

$$P_{t_1}(x_1) = P_{t_1,t_2}(x_1, \infty) \tag{1.5}$$

Given a compatible family of finite-dimensional distributions $\{P_{T_n}$, all finite T_n in $T\}$, we can always find a probability space $(\Omega, \mathcal{A}, \mathcal{P})$ and a family of random variables $\{X_t, t \in T\}$ having the given finite-dimensional distributions. The proof is by construction. Let $\Omega = R^T = \{$the set of all real-valued functions defined on $T\}$. Let $X_t(\omega) =$ the value of ω at t. Let $\mathcal{B}_X$ and $\mathcal{A}_X$ be, respectively, the smallest Boolean algebra and the smallest σ algebra with respect to which every X_t is measurable. Now, every set in $\mathcal{B}_X$ is of the form

$$\{\omega: (X_{t_1}(\omega), X_{t_2}(\omega), \ldots, X_{t_n}(\omega)) \in B\} \tag{1.6}$$

where B is an n-dimensional Borel set. Given a compatible family of finite-dimensional distributions $\{P_{T_n}\}$, we set

$$\mathcal{P}(\{\omega: (X_{t_1}(\omega), X_{t_2}(\omega), \ldots, X_{t_n}(\omega)) \in B\}) = \int_B dP_{t_1,t_2,\ldots,t_n}(x_1, \ldots, x_n) \tag{1.7}$$

This defines an elementary probability measure $\mathcal{P}$ on $(\Omega, \mathcal{B}_X)$. Now, it can be shown that $\mathcal{P}$, so defined, is not only finitely additive, but also σ additive. This means that $\mathcal{P}$ is a probability measure and can be uniquely extended to $\mathcal{A}_X$. To show that $\mathcal{P}$ is σ additive, and not merely finitely additive, is fairly difficult [see Neveu, 1965, pp. 82–83], and the proof will be omitted. To summarize, we take $\Omega = R^T$,

$$X_t(\omega) = \text{value of } \omega \text{ at } t \tag{1.8}$$

and $\mathcal{A}_X$ to be the minimal σ algebra generated by $\{X_t, t \in T\}$. The probability measure $\mathcal{P}$ is defined by (1.7). So defined, the process $\{X_t, t \in T\}$ has the prescribed finite-dimensional distributions. We note that $\{X_t(\omega), \omega \in R^T, t \in T\}$ as defined by (1.8) is called the **coordinate function,** because $X_t(\omega)$ is the tth coordinate of ω. Sets of the form (1.6) are called **cylinder sets.**

In the construction that we have just given, the basic space Ω was taken to be R^T, and the σ algebra was taken to be $\mathcal{A}_X$, the minimal σ algebra containing all cylinder sets. In a sense, R^T is too big and $\mathcal{A}_X$ is rather small. For example, sets of the form

$$\{\omega: a \leq X_t(\omega) \leq b \text{ for all } t \in T\}$$

are not in $\mathcal{A}_X$, if T is uncountable. Sometimes, it may be convenient to work with a different Ω. Suppose that we are given a compatible family of finite-dimensional distributions $\{P_{T_n}$, all finite T_n in $T\}$, and that we are also given a basic space Ω together with a family of functions $\{X_t(\omega), \omega \in \Omega, t \in T\}$. Let $\mathcal{A}_X$ be defined as before. The question is, "can we always find a probability measure $\mathcal{P}$ defined on $(\Omega,\mathcal{A}_X)$ so that $\{X_t, t \in T\}$ has the prescribed finite-dimensional distribution?" The answer is an immediate "no." In order for this to be possible, $\{P_{T_n}\}$ must satisfy other conditions, in addition to the compatibility condition. First of all, it may happen that for two different Borel sets A and B,

$$\{\omega\colon (X_{t_1}(\omega), X_{t_2}(\omega), \ldots, X_{t_n}(\omega)) \in A\} = \{\omega\colon (X_{t_1}(\omega), X_{t_2}(\omega), \ldots, X_{t_n}(\omega)) \in B\} \tag{1.9}$$

Since the same set in $\mathcal{A}_X$ can have only one probability value, we must require

$$\int_A dP_{t_1,\ldots,t_n} = \int_B dP_{t_1,\ldots,t_n} \tag{1.10}$$

whenever A and B satisfy (1.9). This consistency condition is sufficient to ensure that (1.7) defines an elementary probability $\mathcal{P}$ on $(\Omega,\mathcal{B}_X)$. However, $\mathcal{P}$ need not be σ additive (or equivalently, monotone-sequential continuous at $\emptyset$). If $\mathcal{P}$ is not σ additive, it means that we cannot define a probability measure on $(\Omega,\mathcal{A}_X)$ so that $\{X_t, t \in T\}$ has the prescribed finite-dimensional distributions. We illustrate this discussion by an example. Let $T = [0,1]$, and take $\Omega = C[0,1] = \{$all continuous functions on $[0,1]\}$. Take $X_t(\omega)$ to be the coordinate function as before. Suppose that $\{P_{T_n}\}$ is given by

$$P_{t_1,t_2,\ldots,t_n}(x_1, \ldots, x_n) = \prod_{\nu=1}^{n} \int_{-\infty}^{x_\nu} \frac{1}{\sqrt{2\pi}} \exp\left(-\tfrac{1}{2}z^2\right) dz \tag{1.11}$$

In order to be consistent with (1.11), we must have

$$\mathcal{P}(X_t > \epsilon, X_s < -\epsilon) = \left[\int_\epsilon^\infty \frac{1}{\sqrt{2\pi}} \exp\left(-\tfrac{1}{2}z^2\right) dz\right]^2 \tag{1.12}$$

Because $\Omega = C[0,1]$, the sets

$$A_n = \{\omega\colon X_t(\omega) > \epsilon, X_{t+1/n}(\omega) < -\epsilon\}$$

must converge to $\emptyset$ for every $\epsilon > 0$. However,

$$\mathcal{P}(X_t > \epsilon, X_{t+1/n} < -\epsilon) = \left[\int_\epsilon^\infty \frac{1}{\sqrt{2\pi}} \exp\left(-\tfrac{1}{2}z^2\right) dz\right]^2 \underset{n\to\infty}{\not\to} 0$$

Hence, $\mathcal{P}$ defined by (1.7) and (1.11) cannot be sequentially continuous at $\emptyset$ and cannot be extended to $\mathcal{A}_X$. Intuitively, the reason is clear. Since $\Omega = C[0,1]$, any probability measure on $(\Omega,\mathcal{A}_X)$ gives $\mathcal{P}(\Omega) = 1$, which means that with probability 1, every sample function is continuous.

Clearly, not every compatible family of finite-dimensional distributions is consistent with this fact. In particular, (1.11) implies that $\{X_t, t \in [0,1]\}$ is a family of independent and identically distributed random variables, and the independence between X_t and X_s, no matter how close t and s are, is incompatible with continuous sample functions. Later, we shall develop conditions on the finite-dimensional distributions which guarantee continuous sample functions.

2. SEPARABILITY AND MEASURABILITY

The definition of a stochastic process X_t requires it to be an $\mathcal{A}$-measurable ω function for each t, but places no condition on it as a function of t. When T is an interval, questions of an analytical nature concerning sample functions of the process usually involve an uncountable number of events and random variables and are not always answerable without additional assumptions. For example, the set $\{\omega\colon X_t(\omega) \geq 0 \text{ for all } t \in T\} = \bigcap_{t\in T} \{\omega\colon X_t(\omega) \geq 0\}$ may not be an event since it involves an *uncountable* intersection of events. Similarly, $Y = \sup_{t\in T} X_t$ may not be a random variable, because sets of the form

$$\{\omega\colon Y(\omega) \leq y\} = \bigcap_{t\in T} \{\omega\colon X_t(\omega) \leq y\} \tag{2.1}$$

may not be events. The same comment applies to $\inf_{t\in T} X_t$ and to

$$\limsup_{u\to t} X_u = \lim_{n\to\infty} \sup_{|u-t|<1/n} X_u \tag{2.2}$$

Hence, even when $\lim_{u\to t} X_u$ exists, it may not be a random variable. Further, an integral of the form

$$Z(\omega) = \int_a^b X_t(\omega)\, dt$$

may not be well defined for almost all ω, and may not be a random variable even when it is well defined. These difficulties motivated the introduction of the concepts of a separable process and a measurable process [Doob, 1953, pp. 50–71].

Definition. A process $\{X_t, t \in T\}$ is said to be **separable** if there exist a countable set $S \subset T$ and a fixed null event Λ such that for any closed set $K \subset [-\infty, \infty]$ and any open interval I, the two sets

$$\{\omega\colon X_t(\omega) \in K, t \in I \cap T\} \qquad \{\omega\colon X_t(\omega) \in K, t \in I \cap S\}$$

differ by a subset of Λ.

The countable set S is called the **separant** or **separating set.** If $\{X_t,\ t \in T\}$ is separable, then every set of the form $\{\omega\colon X_t(\omega) \in K,\ t \in I \cap T\}$ differs from an event by at most a null set and can be rendered an event by completing the probability space on which the process is defined. If the process $\{X_t,\ t \in T\}$ is separable and $\omega \notin \Lambda$, then for any open interval I, $X_t(\omega) \leq a$, $t \in I \cap S$ implies $X_t(\omega) \leq a$ for all $t \in I \cap T$. Thus,

$$\sup_{t \in I \cap S} X_t(\omega) \geq \sup_{t \in I \cap T} X_t(\omega)$$

Since $T \supset S$, the opposite inequality holds also. Hence for any open interval I

$$\sup_{t \in I \cap S} X_t(\omega) = \sup_{t \in I \cap T} X_t(\omega) \tag{2.3}$$

for all $\omega \notin \Lambda$. Because S is countable, $\sup_{t \in I \cap S} X_t$ is a random variable, thus $\sup_{t \in I \cap T} X_t$ is equal almost everywhere to a random variable. If the probability space is complete, then $\sup_{t \in I \cap T} X_t$ is itself a random variable. Similar results hold for $\inf_{t \in I \cap T} X_t$ and, hence, also for

$$\limsup_{s \to t} X_s = \lim_{n \to \infty} \sup_{|s-t|<1/n} X_s$$

and

$$\liminf_{s \to t} X_s = \lim_{n \to \infty} \inf_{|s-t|<1/n} X_s$$

Thus, $\lim_{s \to t} X_s$, when it exists, is equal almost everywhere to a random variable and can be made a random variable by completing the underlying probability space. Henceforth, the underlying probability space will always be assumed to be complete.

Given a probability space $(\Omega, \mathcal{A}, \mathcal{P})$ and a process $\{X_t,\ t \in T\}$ defined on it, then $\{X_t,\ t \in T\}$ is either separable or not, and there is nothing one can do about it. In practice, the situation is never this rigid. Instead, one is usually free to choose the way in which $\{X_t, t \in T\}$ is to be defined, as long as some specified finite-dimensional distributions are satisfied. Thus, the following result is an exceedingly important one.

Proposition 2.1. For every stochastic process $\{X_t,\ t \in T\}$ there exists a process $\{\tilde{X}_t, t \in T\}$, defined on the same probability space, such that

(*a*) $\{\tilde{X}_t, t \in T\}$ is separable
(*b*) $\mathcal{P}(X_t = \tilde{X}_t) = 1$ for each $t \in T$

Remarks:

(*a*) Although the set $\{\omega: X_t(\omega) \neq \tilde{X}_t(\omega)\}$ is a null event for each t, the set

$$\{\omega: X_t(\omega) \neq \tilde{X}_t(\omega) \text{ for at least one } t \text{ in } T\} = \bigcup_{t\in T} \{\omega: X_t \neq \tilde{X}_t(\omega)\} \tag{2.4}$$

need not be an event and need not have zero probability even if it is an event. If it is a null event, then $\{X_t, t \in T\}$ is itself a separable process.

(*b*) Obviously, $\{X_t, t \in T\}$ and $\{\tilde{X}_t, t \in T\}$ have the same finite-dimensional distributions.

(*c*) It may be necessary for $\tilde{X}_t$ to assume values $\pm\infty$.

A proof of Proposition 2.1 will be omitted. Suffice it to note that the standard proof is by construction, and the separating set S that results is, in general, quite complicated. The situation improves if a continuity condition is satisfied by the finite-dimensional distributions. A process $\{X_t, t \in T\}$ is said to be **continuous in probability at** t if

$$\mathcal{P}(|X_s - X_t| \geq \epsilon) \xrightarrow[s\to t]{} 0 \tag{2.5}$$

for every $\epsilon > 0$. If $\{X_t, t \in T\}$ is continuous in probability at every point in T, then we shall say simply $\{X_t, t \in T\}$ is continuous in probability. We note that continuity in probability is verifiable in terms of two-dimensional distributions.

Proposition 2.2. Let $\{X_t, t \in T\}$ be a separable process which is continuous in probability. Then every countable set dense in T is a separating set.

Thus, whenever a process $\{X_t, t \in T\}$ is both separable and continuous in probability, the probability of an event involving an uncountable number of X_t can be computed by choosing a sequence of partitions of T in the usual manner. For example, suppose $\{X_t, 0 \leq t \leq 1\}$ is both separable and continuous in probability, and we wish to compute $\mathcal{P}(X_t \geq 0, 0 \leq t \leq 1)$. First, we note that we can take S to be the set of all dyadic rationals, that is,

$$S = \left\{\frac{k}{2^n}, 0 \leq k \leq 2^n, n = 0, 1, \ldots\right\}$$

Hence,

$$\begin{aligned}\mathcal{P}(X_t \geq 0, 0 \leq t \leq 1) &= \mathcal{P}(X_{k/2^n} \geq 0, 0 \leq k \leq 2^n, n = 0, 1, \ldots) \\ &= \mathcal{P}\left(\bigcap_{n=0}^{\infty} \{\omega: X_{k/2^n}(\omega) \geq 0, 0 \leq k \leq 2^n\}\right)\end{aligned}$$

Because $A_n = \{\omega: X_{k/2^n}(\omega) \geq 0,\ 0 \leq k \leq 2^n\}$ is a decreasing sequence in n, and because every probability measure is sequentially continuous, we have

$$\mathcal{P}(X_t \geq 0,\ 0 \leq t \leq 1) = \lim_{n\to\infty} \mathcal{P}(\{\omega: X_{k/2^n}(\omega) \geq 0,\ 0 \leq k \leq 2^n\}) \tag{2.6}$$

As an example of a nonseparable process, consider the following. Let $\Omega = [0,1]$, $\mathcal{A}$ be the σ algebra of Lebesgue measurable sets, and let $\mathcal{P}$ be the Lebesgue measure. Consider a process $\{X_t,\ t \in [0,1]\}$ defined on $(\Omega,\mathcal{A},\mathcal{P})$ by

$$X_t(\omega) = \begin{cases} 1 & \text{if } \omega = t \\ 0 & \text{otherwise} \end{cases} \tag{2.7}$$

The process $\{X_t,\ t \in [0,1]\}$ is nonseparable because for any set $S \subset [0,1]$,

$$\{\omega: X_t(\omega) = 0 \text{ for all } t \in S\} = [0,1] - S$$

Therefore,

$$\mathcal{P}(\{\omega: X_t(\omega) = 0 \text{ for all } t \text{ in } [0,1]\}) = 0$$

while if S is any countable set, then

$$\mathcal{P}(\{\omega: X_t(\omega) = 0 \text{ for all } t \text{ in } S\}) = 1$$

Now, let $\{\tilde{X}_t,\ t \in [0,1]\}$ be defined by

$$\tilde{X}_t(\omega) = 0 \qquad \text{for all } t \text{ and } \omega \tag{2.8}$$

The process $\{\tilde{X}_t,\ t \in [0,1]\}$ is clearly separable. Indeed, for every closed set K,

$$\begin{aligned}\{\omega: \tilde{X}_t(\omega) \in K \text{ for all } t \in [0,1]\} &= \{\omega: X_0(\omega) \in K\} \\ &= \begin{cases} [0,1] & \text{if } K \ni 0 \\ \emptyset & \text{if } K \not\ni 0 \end{cases}\end{aligned}$$

For each t,

$$\{\omega: X_t(\omega) = \tilde{X}_t(\omega)\} = [0,1] - \{t\}$$

which is an event with probability 1.

It is often desirable to be able to define integrals of the form $\int_a^b X_t(\omega)\, dt$. If the integral is to be interpreted as a Lebesgue integral of sample functions, Lebesgue integrability of almost all sample functions is clearly a necessity. Even if almost all sample functions of $\{X_t,\ t \in T\}$ are Lebesgue integrable, the resulting integral $\int_a^b X_t(\omega)\, dt$ still may not be a random variable. What is needed is that $X_t(\omega)$ defines a (t,ω) function measurable with respect to $\mathcal{L} \otimes \mathcal{A}$, where $\mathcal{L}$ denotes the σ algebra of Lebesgue measurable sets in T.† We recall that $\mathcal{L}$ is the smallest σ algebra containing all intervals and completed with respect to the measure which assigns lengths to intervals.

Definition. A process $\{X_t, t \in T\}$ with a Lebesgue measurable parameter set T is said to be a **measurable process** if $X_t(\omega)$ is a (t,ω) function measurable with respect to $\mathcal{L} \otimes \mathcal{A}$, where $\mathcal{L}$ is the σ algebra of Lebesgue measurable sets in T, and $\mathcal{A}$ is the σ algebra of events in the defining probability space, i.e., for every $x \in (-\infty, \infty)$,

$$\{(t,\omega): X_t(\omega) \leq x\} \in \mathcal{L} \otimes \mathcal{A}$$

We note that whereas separability imposes no restriction on the finite-dimensional distributions, in general, measurability does. For example, it can be shown that if T is an interval, and if $\{X_t, t \in T\}$ represents a collection of independent and identically distributed random variables, then it cannot be measurable unless the distribution of X_t is concentrated at a point (that is, X_t is a.s. a constant). The following proposition gives a sufficient condition on the finite-dimensional distribution for a measurable process to exist and also summarizes some preceding results [Doob, 1953, pp. 61–62].

Proposition 2.3. Let $\{X_t, t \in T\}$ be a process continuous in probability, and let T be an interval. Then, there exists a process $\{\tilde{X}_t, t \in T\}$ defined on the same probability space such that

(a) $\mathcal{P}(X_t = \tilde{X}_t) = 1$ for each $t \in T$
(b) $\{\tilde{X}_t, t \in T\}$ is separable
(c) $\{\tilde{X}_t, t \in T\}$ is measurable
(d) Any countable set dense in T is a separating set for $\{\tilde{X}_t, t \in T\}$

We call the process $\{\tilde{X}_t, t \in T\}$ in Proposition 2.3 a separable and measurable modification of $\{X_t, t \in T\}$. Whenever a process $\{X_t, t \in T\}$ is continuous in probability, we can always assume that it has already been replaced by a separable and measurable modification. Doing so is almost never incompatible with any important assumption.

Finally, we note that if A is a Lebesgue measurable set of the real line, and if

$$\int_A E|X_t|\, dt < \infty$$

then almost all sample functions are Lebesgue integrable on A, and $\int_A X_t\, dt$ defines a random variable. This follows directly from Fubini's theorem.

† Let $\mathcal{A}_1$ and $\mathcal{A}_2$ be, respectively, σ algebras of subsets of Ω_1 and Ω_2. $\mathcal{A}_1 \otimes \mathcal{A}_2$ denotes the smallest σ algebra which includes all sets of the form $A_1 \times A_2$, $A_1 \in \mathcal{A}_1$, $A_2 \in \mathcal{A}_2$.

3. GAUSSIAN PROCESSES AND BROWNIAN MOTION

Many random phenomena in physical problems are well approximated by stochastic processes that are called Gaussian processes. This is fortunate, because Gaussian processes have distributions which enjoy great analytical simplicity. Brownian motion, or Wiener process as it is sometimes called, is a specific kind of Gaussian process and plays a vital role in the modern theory of stochastic processes. Generalizing on some of the properties of Brownian motion has led to the theory of diffusion processes and sample-continuous martingales. Brownian motion is also the key to a proper understanding of "white noise," a widely used model for noise phenomena. Through this intervening role of Brownian motion, systems with white-noise disturbances can be studied using results of diffusion theory and martingale theory. This theme will be developed in considerable detail in Chaps. 4 and 6.

Let Z be a random variable such that $EZ^2 < \infty$. Let $\mu = EZ$ and $\sigma^2 = E(Z - \mu)^2$. The random variable Z is said to be Gaussian either if $\sigma^2 = 0$, in which case Z is equal to μ with probability 1, or if

$$\mathcal{P}(Z < a) = \int_{-\infty}^{a} \frac{1}{\sqrt{2\pi\sigma^2}} \exp\left[-\frac{1}{2}\frac{(z-\mu)^2}{\sigma^2}\right] dz \tag{3.1}$$

In other words, Z has a density function

$$p_Z(z) = \frac{1}{\sqrt{2\pi\sigma^2}} \exp\left[-\frac{1}{2}\frac{(z-\mu)^2}{\sigma^2}\right] \tag{3.2}$$

whenever $\sigma^2 > 0$. We can compute the characteristic function and find

$$\begin{aligned} F(u) &= Ee^{iuZ} = \int_{-\infty}^{\infty} \frac{1}{\sqrt{2\pi\sigma^2}} \exp\left[-\frac{1}{2\sigma^2}(z-\mu)^2\right] e^{iuz}\, dz \\ &= e^{iu\mu - \frac{1}{2}u^2\sigma^2} \end{aligned} \tag{3.3}$$

Since the distribution function is uniquely determined by the characteristic function, a necessary and sufficient condition for Z to be Gaussian is that Z should satisfy

$$Ee^{iuZ} = e^{iuEZ - \frac{1}{2}E(Z-EZ)^2} \tag{3.4}$$

We note that this condition is valid even for the case $E(Z - EZ)^2 = 0$.

A stochastic process $\{X_t, t \in T\}$ is said to be a **Gaussian process** if every finite linear combination of the form

$$Z = \sum_{i=1}^{N} \alpha_i X_{t_i} \tag{3.5}$$

is a Gaussian random variable. For $\{X_t, t \in T\}$ to be a Gaussian process, it is clearly necessary that for each t, X_t be a Gaussian random variable.

But this is not enough. A necessary and sufficient condition is given by the following.

Proposition 3.1. A process $\{X_t, t \in T\}$ is Gaussian if and only if

(a) $EX_t^2 < \infty$ for each $t \in T$

(b) For every finite collection $(t_1, \ldots, t_N) \subset T$

$$E \exp\left(i \sum_{k=1}^{N} u_k X_{t_k}\right) = \exp\left[i \sum_{k=1}^{N} u_k \mu(t_k) - \frac{1}{2} \sum_{k,l=1}^{N} u_k u_l R(t_k,t_l)\right] \tag{3.6}$$

where

$$\mu(t) = EX_t \tag{3.7}$$

and

$$R(t,s) = E[X_t - \mu(t)][X_s - \mu(s)] \tag{3.8}$$

Proof: Suppose that $\{X_t, t \in T\}$ is a Gaussian process. Then by definition, $Z = \sum_{k=1}^{N} u_k X_{t_k}$ is Gaussian. Therefore,

$$Ee^{iZ} = e^{iEZ} e^{-\frac{1}{2}E(Z-EZ)^2}$$

By direct computation we find

$$EZ = \sum_{k,l=1}^{N} u_k \mu(t_k)$$

and

$$E(Z - EZ)^2 = \sum_{k,l=1}^{N} u_k u_l R(t_k,t_l)$$

Conversely, suppose conditions (a) and (b) are satisfied. Let $Z = \sum_{k=1}^{N} \alpha_k X_{t_k}$ be an arbitrary finite linear combination. Then, using (3.6) we find

$$\begin{aligned} Ee^{iuZ} &= E \exp\left(i \sum_{k=1}^{N} u\alpha_k X_{t_k}\right) \\ &= \exp\left[iu \sum_{k=1}^{N} \alpha_k \mu(t_k) - \tfrac{1}{2}u^2 \sum_{k,l} \alpha_k \alpha_l R(t_k,t_l)\right] \\ &= e^{iuEZ} e^{-\frac{1}{2}u^2 E(Z-EZ)^2} \end{aligned}$$

Therefore, Z is Gaussian. Since this is true for every linear combination, $\{X_t, t \in T\}$ is a Gaussian process by definition. ∎

Remark: Suppose that $\{X_n\}$ is a sequence of Gaussian random variables converging to a random variable X. Then (see Chap. 1, Exercise 10)

$$\mu_n = EX_n \xrightarrow[n\to\infty]{} \mu = EX$$

and

$$\sigma_n^2 = E(X_n - EX_n)^2 \xrightarrow[n\to\infty]{} \sigma^2 = E(X - EX)^2$$

Since X_n has a density function $p_n(\cdot)$ given by

$$p_n(x) = \frac{1}{\sqrt{2\pi\sigma_n^2}} \exp\left[-\frac{1}{2\sigma_n^2}(x - \mu_n)^2\right]$$

and the distribution functions of X_n converge to that of X, the density function $p(\cdot)$ of X must be given by

$$p(x) = \frac{1}{\sqrt{2\pi\sigma^2}} \exp\left[-\frac{1}{2\sigma^2}(x - \mu)^2\right]$$

In other words, the limit of a q.m. convergent sequence of Gaussian random variables is always a Gaussian random variable. Therefore, if $\{X_t, t \in T\}$ is a Gaussian process, then not only is every sum of the form

$$\sum_{k=1}^{N} \alpha_k X_{t_k}$$

a Gaussian random variable, but so is the limit of every q.m. convergent sequence of such sums. This makes precise the often stated result that a random variable obtained by a linear operation on a Gaussian process is always Gaussian. Linear operation is taken to mean the q.m. limit of a sequence of finite linear combinations.

The function $\mu(t)$ is called the **mean function** of $\{X_t, t \in T\}$ or simply the **mean.** The function $R(t,s)$ defined by (3.8) is called the **covariance function.** Covariance functions will be considered in some detail in Chap. 3. One important property of covariance functions is that for any finite collection $(t_1, t_2, \ldots, t_n) \subset T$, the matrix $\mathbf{R}$ formed by setting $R_{ij} = R(t_i,t_j)$ is nonnegative definite.[1] This is simply because for any complex constants $a_1, a_2, \ldots, a_n$, we have

$$\sum_{i,j=1}^{n} a_i\bar{a}_j R_{ij} = \sum_{i,j}^{n} E\{a_i[X_{t_i} - \mu(t_i)]\}\{\bar{a}_j[X_{t_j} - \mu(t_j)]\}$$

$$= E\left|\sum_{i=1}^{n} a_i[X_{t_i} - \mu(t_i)]\right|^2 \geq 0 \tag{3.9}$$

[1] The term positive semidefinite is more conventional than nonnegative definite. We have adopted the latter for a closer correspondence with the terminology usually associated with covariance functions.

It is apparent from (3.6) that the characteristic function F for every finite collection $X_{t_1}, \ldots, X_{t_N}$ is completely determined by the mean $\mu(t)$ and the covariance function $R(t,s)$, viz.,

$$F(u_1, \ldots, u_N) = E \exp\left(i \sum_{k=1}^{N} u_k X_{t_k}\right) = \exp\left[i \sum_{k=1}^{N} u_k \mu(t_k) - \frac{1}{2} \sum_{k,l=1}^{N} u_k u_l R(t_k,t_l)\right] \quad (3.10)$$

If the matrix $\mathbf{R} = [R(t_k,t_l)]$ is positive definite and not merely nonnegative definite, then the inversion formula for Fourier integral in R^n can be used to obtain the probability density function for $X_{t_1}, \ldots, X_{t_N}$. Specifically, we have

$$p(x_1,t_1; x_2,t_2; \ldots; x_N,t_N) = \frac{1}{(2\pi)^N} \int_{-\infty}^{\infty} \cdots \int_{-\infty}^{\infty} F(u_1, \ldots, u_N) \exp\left(-i \sum_{k=1}^{N} u_k x_k\right) du_1 \cdots du_N = \frac{1}{(2\pi)^{N/2}|\mathbf{R}|^{\frac{1}{2}}} e^{-\frac{1}{2}(\mathbf{x}-\boldsymbol{\mu})'\mathbf{R}^{-1}(\mathbf{x}-\boldsymbol{\mu})} \quad (3.11)$$

where $\mathbf{R}^{-1}$ denotes the inverse of $\mathbf{R}$, $\mathbf{x}$ denotes the column vector with components $x_1, \ldots, x_N$, $\boldsymbol{\mu}$ denotes the column vector with components $\mu(t_1), \ldots, \mu(t_N)$, $|\mathbf{R}|$ denotes the determinant of $\mathbf{R}$, and prime denotes transpose. When the matrix $\mathbf{R}$ is not positive definite, then (3.11) fails, and $X_{t_1}, \ldots, X_{t_N}$ do not have a joint density function. However, the joint distribution function can still be obtained from the characteristic function given by (3.6). In particular, if A is a rectangle such that the distribution function is continuous on its boundary, then we have

$$\mathcal{P}((X_{t_1}, \ldots, X_{t_N}) \epsilon A) = \int_{R^N} F(u_1, \ldots, u_N) \psi_A(u_1, \ldots, u_N)\, du_1 \cdots du_N \quad (3.12)$$

where

$$\psi_A(u_1, \ldots, u_N) = \int_A \exp\left(-i \sum_{k=1}^{N} u_k x_k\right) dx_1 \cdots dx_k$$

Finally, the distribution function $\mathcal{P}(X_{t_i} < x_i,\ i = 1, \ldots, N)$ can be obtained by using (3.12) and taking limits.

These considerations show that all finite-dimensional distributions of a Gaussian process are completely determined once we specify its mean $\mu(t)$ and covariance function $R(t,s)$. This is indeed the simplest way of specifying the finite-dimensional distributions of a Gaussian process.

As an example, consider the **Brownian motion** process $\{X_t, t \geq 0\}$ defined by

$$\{X_t, t \geq 0\} \text{ is a Gaussian process} \tag{3.13a}$$

$$EX_t = 0 \qquad EX_tX_s = \min(t,s) \tag{3.13b}$$

If $0 < t_1 < t_2 < \cdots < t_N$, the matrix $\mathbf{R} = [\min(t_k,t_l)]$ is positive definite. Thus, the density function can be written down immediately by the use of (3.11). After a little rearrangement, we find

$$p(x_1,t_1; x_2,t_2; \ldots ; x_N,t_N) = \prod_{\nu=1}^{N} \frac{1}{\sqrt{2\pi(t_\nu - t_{\nu-1})}} \exp\left[-\frac{1}{2}\frac{(x_\nu - x_{\nu-1})^2}{t_\nu - t_{\nu-1}}\right] \tag{3.14}$$

where $t_0 = 0 = x_0$. Equation (3.14) shows that $\{X_{t_1}, X_{t_2} - X_{t_1}, \ldots, X_{t_N} - X_{t_{N-1}}\}$ must be a collection of independent random variables for every increasing $(t_1, \ldots, t_N)$. Any process having this property is called a **process with independent increments.** Thus, a Brownian motion process is a process with independent increments.

A Brownian motion is also a Markov process, which is defined as follows.

Definition. A process $\{X_t, t \in T\}$ is said to be a **Markov process** if for any increasing collection $t_1, t_2, \ldots, t_n$ in T,

$$\mathcal{P}(X_{t_n} \leq x_n | X_{t_\nu} = x_\nu, \nu = 1, \ldots, n-1) = \mathcal{P}(X_{t_n} \leq x_n | X_{t_{n-1}} = x_{n-1}) \tag{3.15}$$

In other words, given the past $(X_{t_1}, \ldots, X_{t_{n-2}})$ and the present $(X_{t_{n-1}})$, the future (X_{t_n}) depends only on the present. For a Brownian motion, (3.15) is easily verified as follows:

$$\begin{aligned}
\mathcal{P}(X_{t_n} &\leq x_n | X_{t_\nu} = x_\nu, \nu = 1, \ldots, n-1) \\
&= \int_{-\infty}^{x_n} p(\xi,t_n; x_1,t_1; \ldots ; x_{n-1},t_{n-1})\, d\xi \\
&= \int_{-\infty}^{x_n} \frac{p(x_1,t_1; \ldots ; \xi,t_n)}{p(x_1,t_1; \ldots ; x_{n-1},t_{n-1})}\, d\xi \\
&= \int_{-\infty}^{x_n} \frac{1}{\sqrt{2\pi(t_n - t_{n-1})}} \exp\left[-\frac{1}{2}\frac{(\xi - x_{n-1})^2}{t_n - t_{n-1}}\right] d\xi \\
&= \mathcal{P}(X_{t_n} \leq x_n | X_{t_{n-1}} = x_{n-1})
\end{aligned}$$

Indeed, a little reflection would show that any process with independent increments $\{X_t, t \in T\}$ is also a Markov process.

A Brownian motion $\{X_t, t \geq 0\}$ also has the property (see Exercises 8 and 9)

$$E(X_t | X_\tau, 0 \leq \tau \leq s) = X_s \qquad \text{a.s.}$$

for any $t \geq s$. A process having this property is called a martingale. We shall define a martingale a little more generally as follows.

Definition. Let $\{X_t, t \in T\}$ be a stochastic process, and let $\{\mathcal{A}_t, t \in T\}$ be an increasing family of σ algebras such that for each t, X_t is $\mathcal{A}_t$ measurable. $\{X_t, \mathcal{A}_t, t \in T\}$ is said to be a **martingale** if $t > s$ implies

$$E^{\mathcal{A}_s}X_t = X_s \qquad \text{a.s.} \tag{3.16}$$

The process is said to be a **submartingale (supermartingale)** if the equality (3.16) is replaced by $\geq$ (respectively, ≤ 0).

For any Brownian motion $\{X_t, t \geq 0\}$, if we take $\mathcal{A}_{xt}$ to be the smallest σ algebra with respect to which $\{X_s, s \leq t\}$ are all measurable, then $\{X_t, \mathcal{A}_{xt}, t \geq 0\}$ is a martingale. Suppose that $\{X_t, t \geq 0\}$ is a Brownian motion, and $\{\mathcal{A}_t, t \geq 0\}$ is an increasing family of σ algebras such that for each t, X_t is $\mathcal{A}_t$ measurable and for each t, $\{X_s - X_t, s \geq t\}$ is independent of $\mathcal{A}_t$. We shall emphasize this relationship between X_t and $\mathcal{A}_t$ by saying that $\{X_t, \mathcal{A}_t, t \geq 0\}$ is a Brownian motion. If $\{X_t, \mathcal{A}_t, t \geq 0\}$ is a Brownian motion, then it is a martingale. To prove this we merely note that for $t > s$,

$$\begin{aligned} E^{\mathcal{A}_s}X_t &= E^{\mathcal{A}_s}[X_s + (X_t - X_s)] \\ &= X_s + E(X_t - X_s) = X_s \qquad \text{a.s.} \end{aligned}$$

We should note that instead of taking the parameter set to be $[0, \infty)$, we can define a Brownian motion on $(-\infty, \infty)$ by replacing (3.13) with

$$EX_t = 0 \qquad EX_tX_s = \tfrac{1}{2}(|t| + |s| - |t - s|) \tag{3.17}$$

What results is a pair of independent Brownian motions $\{X_t, t \geq 0\}$ and $\{X_{-t}, t \geq 0\}$ pieced together at $t = 0$ (see Exercise 7).

If $\{X_t, t \geq 0\}$ is a Brownian motion, then $E(X_t - X_s)^2 = |t - s|$. Therefore, by virtue of the Chebyshev inequality, we have

$$\mathcal{P}(|X_{t+h} - X_t| \geq \epsilon) \leq \frac{|h|}{\epsilon^2} \xrightarrow[h \to 0]{} 0$$

for every $\epsilon > 0$. Thus, a Brownian motion is continuous in probability. It follows from Proposition 2.3 that every Brownian motion has a separable and measurable modification. A separable Brownian motion has some important sample function properties. The first of those that we shall consider is an inequality that it shares with all separable second-order martingales.

Proposition 3.2. Let $\{X_t, a \leq t \leq b\}$ be a separable martingale such that $EX_t^2 < \infty$ for every t in $[a,b]$. Then, for every positive ϵ,

$$\mathcal{P}(\sup_{a \leq t \leq b} |X_t| \geq \epsilon) \leq \frac{EX_b^2}{\epsilon^2} \tag{3.18}$$

Proof: Since $\{X_t, t \in [a,b]\}$ is separable, there is a countable set S such that

$$\mathcal{P}(\sup_{a\le t\le b} |X_t| \ge \epsilon) = \mathcal{P}(\sup_{t\in S} |X_t| \ge \epsilon)$$

Now, since S is countable, we can write $S = \bigcup_n S_n$ where $\{S_n\}$ is an increasing sequence of finite collections of points in $[a,b]$. Because $\mathcal{P}$ is monotone-sequential continuous,

$$\mathcal{P}(\sup_{t\in S} |X_t| \ge \epsilon) = \lim_{n\to\infty} \mathcal{P}(\max_{t\in S_n} |X_t| \ge \epsilon)$$

For a fixed n, let $t_1, t_2, \ldots, t_N$ be the points in S_n with $t_1 < t_2 < \cdots < t_N$. Let $\nu(\omega)$ be the first i (if any) such that $|X_{t_i}(\omega)| \ge \epsilon$, and write

$$\begin{aligned} EX_b^2 &= \sum_{k=1}^{N} \mathcal{P}(\nu = k)E(X_b^2|\nu = k) \\ &\qquad + \mathcal{P}(\max_{t\in S_n} |X_t| < \epsilon)E(X_b^2| \max_{t\in S_n} |X_t| < \epsilon) \\ &\ge \sum_{k=1}^{N} \mathcal{P}(\nu = k)E(X_b^2|\nu = k) \end{aligned}$$

Writing $X_b^2 = (X_b - X_{t_k} + X_{t_k})^2$, we find

$$\begin{aligned} E(X_b^2|\nu = k) &= E[(X_b - X_{t_k})^2|\nu = k] + E(X_{t_k}^2|\nu = k) \\ &\qquad + 2[X_{t_k}(X_b - X_{t_k})|\nu = k] \end{aligned}$$

From the definition of ν, we have $E(X_{t_k}^2|\nu = k) \ge \epsilon^2$. Because the event $\nu = k$ depends only on $X_{t_1}, X_{t_2}, \ldots, X_{t_k}$, we have

$$\begin{aligned} E[X_{t_k}(X_b - X_{t_k})|\nu = k] &= E\{X_{t_k}E[(X_b - X_{t_k})|X_{t_1}, \ldots, X_{t_k}]|\nu = k\} \\ &= 0 \end{aligned}$$

due to the martingale property. It follows that $E(X_b^2|\nu = k) \ge \epsilon^2$ and

$$\begin{aligned} EX_b^2 &\ge \epsilon^2 \sum_{k=1}^{N} \mathcal{P}(\nu = k) \\ &= \epsilon^2 \mathcal{P}(\max_{t\in S_n} |X_t| \ge \epsilon) \\ &\xrightarrow[n\to\infty]{} \epsilon^2 \mathcal{P}(\sup_{a\le t\le b} |X_t| \ge \epsilon) \end{aligned}$$

proving (3.18). ▮

Remarks:

(*a*) (3.18) holds for separable complex-valued martingales with $E|X_b|^2$ replacing EX_b^2. The proof is almost identical.

(*b*) If $\{X_t, a \leq t \leq b\}$ is a separable process with independent increments such that $\mathcal{P}(X_t - X_s > 0) = \mathcal{P}(X_t - X_s < 0)$, $a \leq t$, $s \leq b$. Then we can prove

$$\mathcal{P}(\sup_{a \leq t \leq b} |X_t| \geq \epsilon) \leq 2\mathcal{P}(|X_b| \geq \epsilon)$$

which is a formula in the same spirit as (3.18) and can be proved in a similar way.

Proposition 3.3. With probability 1, every sample function of a separable Brownian motion is uniformly continuous on every finite interval.

The proof of this proposition will be deferred until the next section where it can be easily proved using some sufficient conditions for sample continuity. The sample-continuity property of a separable Brownian motion can be interpreted in a different way as follows: Let C denote the space of all continuous real-valued functions on $[0, \infty)$. Let $X_t(\omega)$, $\omega \in C$, be the value of ω at t. Let $\mathcal{A}$ denote the smallest σ algebra of subsets of C such that every X_t is measurable. Then it can be shown that there exists a unique probability measure $\mathcal{P}$ on $(C,\mathcal{A})$ such that $\{X_t, 0 \leq t < \infty\}$ is a Brownian motion. So constructed, $\{X_t, t \geq 0\}$ is necessarily separable. Sample continuity in this context says no more than $\mathcal{P}(C) = 1$. The measure $\mathcal{P}$ on $(C,\mathcal{A})$ will be called the **Wiener measure.** Although the sample functions of a separable Brownian motion are almost surely continuous, they are highly irregular. For example, with probability 1, the sample functions of a separable Brownian motion are nowhere differentiable and are of unbounded variation on every interval. Roughly speaking, $(X_{t+\delta} - X_t)$ is of order $0(\sqrt{\delta})$, which is incompatible with either differentiability or bounded variation. This behavior of the Brownian motion is made precise by the following proposition.

Proposition 3.4. Let $T = [a,b]$ be a closed subinterval of $[0, \infty)$. Let $T_n = [a = t_0^{(n)} < t_1^{(n)} < \cdots < t_{N(n)}^{(n)} = b]$, $n = 1, 2, \ldots$, be a sequence of partitions of T such that

$$\Delta_n = \max_{1 \leq \nu \leq N(n)} (t_\nu^{(n)} - t_{\nu-1}^{(n)}) \xrightarrow[n \to \infty]{} 0$$

Let $\{X_t, t \geq 0\}$ be a Brownian motion, then

$$\sum_{\nu=1}^{N(n)} (X_{t_\nu^{(n)}} - X_{t_{\nu-1}^{(n)}})^2 \xrightarrow[n \to \infty]{\text{q.m.}} b - a$$

If $\sum_{n=1}^{\infty} \Delta_n < \infty$ then the convergence is also almost sure.

Proof: To prove quadratic-mean convergence, we write

$$S_n = \sum_{\nu=1}^{N(n)} (X_{t_\nu^{(n)}} - X_{t_{\nu-1}}^{(n)})^2 - (b-a)$$

$$= \sum_{\nu=1}^{N(n)} [(X_{t_\nu^{(n)}} - X_{t_{\nu-1}}^{(n)})^2 - (t_\nu^{(n)} - t_{\nu-1}^{(n)})]$$

Since S_n is a sum of independent random variables, each with zero mean, we have

$$ES_n = 0$$

$$ES_n^2 = \sum_{\nu=1}^{N(n)} E[(X_{t_\nu^{(n)}} - X_{t_{\nu-1}}^{(n)})^2 - (t_\nu^{(n)} - t_{\nu-1}^{(n)})]^2$$

$$= 2 \sum_{\nu=1}^{N(n)} (t_\nu^{(n)} - t_{\nu-1}^{(n)})^2 \leq 2\Delta_n(b-a)$$

Since $\Delta_n \xrightarrow[n\to\infty]{} 0$, this proves $S_n \xrightarrow[n\to\infty]{\text{q.m.}} 0$.

For the second part of the proposition, using the Chebyshev inequality (1.5.5), we find

$$\mathcal{P}(|S_n| > \epsilon) \leq \frac{ES_n^2}{\epsilon^2} \leq 2(b-a)\frac{\Delta_n}{\epsilon^2}$$

Since by assumption, $\sum_n \Delta_n < \infty$, we have for every $\epsilon > 0$,

$$\sum_n \mathcal{P}(|S_n| > \epsilon) < \infty$$

By the Borel-Cantelli lemma, $|S_n| > \epsilon$ for at most a finite number of times with probability 1, proving the theorem. ∎

Remark: The condition $\sum_n \Delta_n < \infty$ for a.s. convergence can be replaced by the condition that T_n be nested, that is, $T_{n+1} \supset T_n$ for each n [Doob, 1953, pp. 395–396].

As we said earlier, an intuitive interpretation of Proposition 3.4 is that $dX_t \sim \sqrt{dt}$. This has some important consequences. For example, we would expect the following:

$$df(X_t) = f(X_{t+dt}) - f(X_t)$$
$$= f'(X_t)\, dX_t + \tfrac{1}{2}f''(X_t)\, dt$$

This relationship will be made precise subsequently by means of the stochastic integral.

4. CONTINUITY

For a stochastic process $\{X_t, t \in T\}$ defined on an interval T, we distinguish the following types of continuity. The process is said to be

1. **Continuous in probability** at t if for any $\epsilon > 0$,

$$\mathcal{P}(|X_{t+h} - X_t| > \epsilon) \xrightarrow[h \to 0]{} 0 \tag{4.1}$$

2. **Continuous in νth mean** at t (in quadratic mean if $\nu = 2$), if

$$E|X_{t+h} - X_t|^\nu \xrightarrow[h \to 0]{} 0 \tag{4.2}$$

3. **Almost-surely continuous** at t, if

$$\mathcal{P}(\{\omega: \lim_{h \to 0} |X_{t+h}(\omega) - X_t(\omega)| = 0\}) = 1 \tag{4.3}$$

4. Almost-surely **sample continuous,** if $\{X_t, t \in T\}$ is separable and

$$\mathcal{P}\left(\bigcup_{t \in T} \{\omega: \lim_{h \to 0} |X_{t+h}(\omega) - X_t(\omega)| \neq 0\}\right) = 0 \tag{4.4}$$

If a process $\{X_t, t \in T\}$ is continuous in probability (in νth mean) at every t of T, then it is said to be continuous in probability (in νth mean). We have already encountered continuity in probability in connection with questions of separability. Continuity in quadratic mean ($\nu = 2$) will play an important role in our discussion of second-order properties in Chap. 3. In this section we are primarily interested in a.s. continuity and a.s. sample continuity.

First of all, we note that almost-sure continuity at every t of T is not equivalent to a.s. sample continuity. This is simply because $A_t = \{\omega: \lim_{h \to 0} |X_{t+h}(\omega) - X_t(\omega)| \neq 0\}$, being a null event for every t, does not imply that the uncountable union $\bigcup_{t \in T} A_t$ is a null event. The defining condition (4.4) for sample continuity is not one that can be easily verified in terms of finite-dimensional distributions. Our primary objective here is to replace it by simpler sufficient conditions. To begin with, if a process is a.s. sample continuous, then it is necessarily a.s. continuous at every t in T, which in turn implies that the process is continuous in probability. Thus, every sample-continuous process is both separable (by definition) and continuous in probability. From Proposition 2.2 we know that for every such process, every countable set S dense in T is a separating set. Hence, with probability 1

$$\sup_{\substack{t,s \in S \\ |t-s| < h}} |X_t - X_s| = \sup_{\substack{t,s \in T \\ |t-s| < h}} |X_t - X_s| \tag{4.5}$$

Assume T to be a closed interval. With no loss of generality we can take T to be [0,1]. Then S can be taken to be

$$S = \left\{\frac{k}{2^n}, k = 0, 1, \ldots, 2^n - 1; n = 0, 1, 2, \ldots\right\} \tag{4.6}$$

Proposition 4.1. Let S be given by (4.6) and define

$$Z_\nu(\omega) = \sup_{0 \le k \le 2^\nu} |X_{(k+1)/2^\nu}(\omega) - X_{k/2^\nu}(\omega)| \tag{4.7}$$

Then,

$$\sup_{\substack{t,s \in S \\ |t-s| < 2^{-n}}} |X_t - X_s| \le 2 \sum_{\nu=n+1}^{\infty} Z_\nu \tag{4.8}$$

Proof: If $|t - s| < 2^{-n}$, then we can find k such that $0 < k < 2^n$ and $|t - k/2^n| < 2^{-n}$, $|s - k/2^n| < 2^{-n}$. If $t \in S$ and $|t - k/2^n| < 2^{-n}$, t must be of the form

$$t = \frac{k}{2^n} \pm \sum_{\nu=1}^{m} t_\nu 2^{-(n+\nu)} \qquad (t_\nu = 0,1)$$

Thus,

$$|X_t - X_{k/2^n}| \le \sum_{\nu=n+1}^{n+m} Z_\nu \le \sum_{\nu=n+1}^{\infty} Z_\nu$$

Since, $|X_t - X_s| \le |X_t - X_{k/2^n}| + |X_s - X_{k/2^n}|$, we have

$$\sup_{\substack{t,s \in S \\ |t-s| < 2^{-n}}} |X_t - X_s| \le 2 \sum_{\nu=n+1}^{\infty} Z_\nu$$

which proves the proposition. ∎

Proposition 4.1 shows that if

$$\lim_{n \to \infty} \sum_{\nu=n}^{\infty} Z_\nu = 0 \tag{4.9}$$

with probability 1, then

$$\lim_{n \to \infty} \sup_{\substack{t,s \in S \\ |t-s| < 2^{-n}}} |X_t - X_s| = 0 \tag{4.10}$$

with probability 1, which proves that every sample function is uniformly continuous on T with probability 1, since in (4.10) S can be replaced by T.

An extremely useful condition which implies (4.9) is the Kolmogorov condition given in the following proposition.

Proposition 4.2. Let $\{X_t,\ t \in T\}$ be a separable process and let T be a finite interval. If there exist strictly positive constants α, β, C such that

$$E|X_{t+h} - X_t|^\alpha \le Ch^{1+\beta} \tag{4.11}$$

then

$$\sup_{\substack{t,s \in T \\ |t-s|<h}} |X_t - X_s| \xrightarrow[h\to 0]{\text{a.s.}} 0 \tag{4.12}$$

so that almost every sample function is uniformly continuous on T.

Remark: The great advantage of the Kolmogorov condition is simply that it only involves two-dimensional distributions and can usually be verified by direct computation.

Proof: From the Markov inequality (1.5.5), we have

$$\mathcal{P}(|X_{t+h} - X_t| \ge \epsilon) \le \frac{E|X_{t+h} - X_t|^\alpha}{\epsilon^\alpha} \qquad \epsilon > 0$$

Therefore,

$$\mathcal{P}(|X_{t+h} - X_t| \ge h^\gamma) \le Ch^{1+\beta-\alpha\gamma}$$

Let $0 \le \gamma < \beta/\alpha$, and set $\delta = \beta - \alpha\gamma > 0$. Then

$$\mathcal{P}(|X_{t+h} - X_t| \ge h^\gamma) \le Ch^{1+\delta}$$

Hence,

$$\begin{aligned}
\mathcal{P}\left(Z_\nu \ge \left(\frac{1}{2^\nu}\right)^\gamma\right) &= \mathcal{P}\left(\sup_{0\le k\le 2^\nu-1} |X_{k+1/2^\nu} - X_{k/2^\nu}| \ge \left(\frac{1}{2^\nu}\right)^\gamma\right) \\
&\le \sum_{k=0}^{2^\nu-1} \mathcal{P}\left(|X_{k+1/2^\nu} - X_{k/2^\nu}| \ge \left(\frac{1}{2^\nu}\right)^\gamma\right) \\
&\le C2^\nu \left(\frac{1}{2^\nu}\right)^{1+\delta} \\
&= C2^{-\delta\nu}
\end{aligned}$$

Since $\sum_{\nu=0}^{\infty} 2^{-\delta\nu} < \infty$, we have

$$\sum_{\nu=0}^{\infty} \mathcal{P}\left(Z_\nu \ge \left(\frac{1}{2^\nu}\right)^\gamma\right) < \infty$$

By the Borel-Cantelli lemma, $Z_\nu \geq 1/2^{\nu\gamma}$ for at most a finite number of ν's with probability 1. That is, there exists $N(\omega)$ almost-surely finite such that

$$Z_\nu(\omega) < \frac{1}{2^{\nu\gamma}} \qquad \text{for all } \nu \geq N(\omega)$$

and $\lim_{n\to\infty} \sum_{\nu=n+1}^{\infty} Z_\nu(\omega) = 0$ with probability 1. From Proposition 4.1,

$$\sup_{\substack{t,s\in S\\ |t-s|<2^{-n}}} |X_t - X_s| \xrightarrow[n\to\infty]{\text{a.s.}} 0$$

Since S is a separating set in T, this proves Proposition 4.2. ▮

The Kolmogorov condition (4.11) is sufficient to prove the sample continuity of Brownian motion, viz., Proposition 3.3. We recall that for a Brownian motion

$$\mathcal{P}(X_t - X_s < x) = \int_{-\infty}^{x} \frac{1}{\sqrt{2\pi|t-s|}} \exp\left(-\frac{u^2}{2|t-s|}\right) du$$

Therefore,

$$\begin{aligned} E|X_t - X_s|^4 &= \int_{-\infty}^{\infty} \frac{u^4}{\sqrt{2\pi|t-s|}} \exp\left(-\frac{1}{2}\frac{u^2}{|t-s|}\right) du \\ &= 3(t-s)^2 \end{aligned}$$

and

$$E|X_{t+h} - X_t|^4 = 3h^2$$

which verifies the Kolmogorov condition with $C = 3$, $\alpha = 4$, $\beta = 1$. Therefore, with probability 1, every sample function of a separable Brownian motion is uniformly continuous on every finite interval.

Let T be a finite and closed interval which we take to be [0,1] with no loss of generality. Let C be the space of all continuous functions on [0,1]. If $\omega \in C$, we define

$$X_t(\omega) = \text{value of } \omega \text{ at } t \qquad \begin{matrix} \omega \in C \\ t \in [0,1] \end{matrix}$$

The space C is a Banach space (complete normed linear space) with a norm given by

$$\|\omega\| = \max_{0\leq t\leq 1} |X_t(\omega)|$$

Let $\mathcal{B}$ be the minimal Boolean algebra with respect to which every X_t is measurable. $\mathcal{B}$ is not a σ algebra. Now, every consistent family of finite-dimensional distributions for $\{X_t,\ t \in [0,1]\}$ defines an elementary probability measure $\mathcal{P}'$ on $(C,\mathcal{B})$. However, $\mathcal{P}'$ is, in general, only finitely additive, but not σ additive. If $\mathcal{P}'$ is σ additive (equivalently, if $\mathcal{P}'$ is sequentially continuous), then it can be extended to a probability measure $\mathcal{P}$ on $(C,\mathcal{A})$ where $\mathcal{A}$ is the minimal σ algebra with respect to which every X_t is measurable. What results then is a stochastic process $\{X_t, t \in [0,1]\}$ defined on $(C,\mathcal{A},\mathcal{P})$. Since $\mathcal{P}(C) = 1$, every sample function of X_t is obviously continuous. We can interpret the Kolmogorov condition to mean that every consistent family of finite-dimensional distributions which satisfies (4.11) defines an elementary probability measure $\mathcal{P}'$ on $(C,\mathcal{B})$ which is σ *additive* [Prokhorov, 1956]. It is possible to show this directly, but we won't do it here. In the case of Brownian motion, the resulting $\mathcal{P}$ is known as the **Wiener measure.** We should note that every process $\{X_t, t \in T\}$ defined on $(C,\mathcal{A},\mathcal{P})$ in this way is necessarily separable and measurable.

If a process is not sample continuous, then it is clearly of interest to know the nature of its discontinuities. We shall say that a function $f(t), 0 \le t \le 1$ has only **discontinuities of the first kind** if (1) f is bounded, (2) at every $t \in (0,1)$ limit from the left $\lim_{s \uparrow t} f(s) = f^-(t)$ and limit from the right $\lim_{s \downarrow t} f(s) = f^+(t)$ exist. Clearly, $[0,1]$ can easily be replaced by $[a,b]$ in everything that we say. We give without proof a condition, similar to the Kolmogorov condition, which guarantees that a process $\{X_t, t \in [0,1]\}$ has sample functions which have only discontinuities of the first kind with probability 1 [Cramér, 1966].

Proposition 4.3. Let $\{X_t, 0 \le t \le 1\}$ be a separable process. If there exist strictly positive constants α, β, C such that

$$\sup_{t \le s \le t+h} E(|X_{t+h} - X_s|^\alpha |X_s - X_t|^\alpha) \le Ch^{1+\beta} \tag{4.13}$$

then with probability 1, every sample function of $\{X_t, 0 \le t \le 1\}$ has only discontinuities of the first kind.

Remark: Condition (4.13) is weaker than (4.12), as it should be. To show this, we need only to apply the Schwarz inequality and find

$$E(|X_{t+h} - X_s|^\alpha |X_s - X_t|^\alpha) \le \sqrt{E|X_{t+h} - X_s|^{2\alpha} E|X_s - X_t|^{2\alpha}}$$

Hence, if condition (4.12) is satisfied with α, β and C, then (4.13) is satisfied with $\alpha/2$, β and C.

As an example of the application of (4.13), consider the random telegraph process $\{X_t, -\infty < t < \infty\}$ defined as follows:

(*a*) $\mathcal{P}(X_t = 1) = \mathcal{P}(X_t = -1) = \frac{1}{2}$

(b) $\{X_t, \ -\infty < t < \infty\}$ is a process with independent increments, that is,

$$(X_{t_n} - X_{t_{n-1}}), (X_{t_{n-1}} - X_{t_{n-2}}), \ldots, (X_{t_2} - X_{t_1}), X_{t_1}$$

are independent whenever $t_n > t_{n-1} > \cdots > t_1$

(c) $$\begin{aligned} \mathcal{P}(X_t - X_s = 0) &= \tfrac{1}{2}(1 + e^{-|t-s|}) \\ \mathcal{P}(X_t - X_s = 2) &= \mathcal{P}(X_t - X_s = -2) = \tfrac{1}{4}(1 - e^{-|t-s|}) \end{aligned} \tag{4.14}$$

It is clear that with probability 1, $\{X_t, -\infty < t < \infty\}$ takes only values ± 1. It has independent increments,

$$\begin{aligned} E(|X_{t+h} - X_s|\,|X_s - X_t|) &= E|X_{t+h} - X_s|E|X_s - X_t| \\ &= (1 - e^{-|t+h-s|})(1 - e^{-|s-t|}) \\ &\le h^2/4 \qquad \text{for all } s \text{ in } [t, t+h] \end{aligned}$$

Hence, (4.13) is satisfied and with probability 1, every sample function has only discontinuities of the first kind.

As a final topic, we note that the proof of Proposition 4.2 contains an estimate of the modulus of continuity which we now make explicit. As in Proposition 4.2, we assume $\{X_t, 0 \le t \le 1\}$ to be separable and for some positive constants C, α, β

$$E|X_{t+h} - X_t|^\alpha \le Ch^{1+\beta}$$

We then found that there existed an a.s. finite random variable $N(\omega)$ so that

$$Z_\nu(\omega) = \sup_{0 \le k \le 2^\nu} |X_{(k+1)/2^\nu}(\omega) - X_{k/2^\nu}(\omega)| < 1/2^{\nu\gamma} \qquad \text{for all } \nu \ge N(\omega) \tag{4.15}$$

where γ is any constant in $[0, \beta/\alpha)$. If we take $S = \{k/2^n, k = 0, \ldots, 2^n - 1, n = 0, 1, \ldots\}$, then from Proposition 4.1

$$\sup_{\substack{t,s \in S \\ |t-s| < 2^{-n}}} |X_t(\omega) - X_s(\omega)| \le 2 \sum_{\nu=n+1}^{\infty} Z_\nu(\omega)$$

so that for $n \ge N(\omega)$

$$\sup_{\substack{t,s \in S \\ |t-s| < 2^{-n}}} |X_t(\omega) - X_s(\omega)| \le 2 \sum_{\nu=n+1}^{\infty} \frac{1}{2^{\nu\gamma}} = 2\left(\frac{1}{2^\gamma}\right)^{n+1}$$

for any γ in $[0, \beta/\alpha)$. For any $h < 2^{-N(\omega)}$,

$$\sup_{\substack{t,s \in S \\ |t-s| \le h}} |X_t(\omega) - X_s(\omega)| \le 2h^\gamma \qquad 0 \le \gamma < \frac{\beta}{\alpha} \tag{4.16}$$

Recalling that S is a separating set in [0,1], we have proven that

$$\lim_{h \downarrow 0} h^{-\beta/\alpha+\epsilon} \sup_{\substack{t,s \in [0,1] \\ |t-s| \le h}} |X_t - X_s| = 0 \tag{4.17}$$

with probability 1 for any $\epsilon > 0$. For a Brownian motion process, the largest β/α that we can take is $\frac{1}{2}$. Therefore, for a separable Brownian motion process,

$$\lim_{h \downarrow 0} \frac{\sup_{\substack{t,s \in [0,1] \\ |t-s| \le h}} |X_t - X_s|}{h^{\frac{1}{2}-\epsilon}} = 0 \text{ with probability 1} \tag{4.18}$$

for any $\epsilon > 0$.

5. MARKOV PROCESSES

In this section we shall give a preliminary discussion of Markov processes. In later chapters a more detailed treatment of sample-continuous Markov processes will be given in connection with the diffusion equation and stochastic differential equations. We recall the definition of a Markov process given in Sec. 3.

Definition. A process $\{X_t, t \in T\}$ is said to be a **Markov process** if for any increasing collection $t_1, t_2, \ldots, t_n$ in T

$$\begin{aligned} \mathcal{P}(X_{t_n} < x_n | X_{t_\nu} = x_\nu, \nu = 1, \ldots, n-1) \\ = \mathcal{P}(X_{t_n} < x_n | X_{t_{n-1}} = x_{n-1}) \end{aligned} \tag{5.1}$$

If the finite-dimensional distributions of a Markov process $\{X_t, t \in T\}$ can be expressed in terms of density functions, then whenever $t_1 < t_2 < \cdots < t_n$ we can write

$$\begin{aligned} p(x_1,t_1; x_2,t_2; \ldots ; x_n,t_n) \\ = p(x_n,t_n|x_1,t_1; \ldots ; x_{n-1},t_{n-1})p(x_1,t_1; \ldots ; x_{n-1},t_{n-1}) \\ = p(x_n,t_n|x_{n-1},t_{n-1})p(x_1,t_1; \ldots ; x_{n-1},t_{n-1}) \end{aligned} \tag{5.2}$$

Therefore, by using (5.2) repeatedly, we find for $t_1 < t_2 < \cdots < t_n$,

$$p(x_1,t_1; x_2,t_2; \ldots ; x_n,t_n) = p(x_1,t_1) \prod_{\nu=2}^{n} p(x_\nu,t_\nu|x_{\nu-1},t_{\nu-1}) \tag{5.3}$$

which means that all finite-dimensional distributions are completely determined by the two-dimensional distributions. This fact is true with all Markov processes and does not depend on the existence of density functions.

To obtain an expression relating the n-dimensional distribution function to the two-dimensional distribution, we first adopt the notation

$$P(x_1,t_1;\ x_2,t_2;\ \ldots\ ;\ x_n,t_n) = \mathcal{P}(X_{t_\nu} < x_\nu;\ \nu = 1, \ldots, n) \tag{5.4}$$
$$P(x_\nu,t_\nu|x_{\nu-1},t_{\nu-1}) = \mathcal{P}(X_{t_\nu} < x_\nu|X_{t_{\nu-1}} = x_{\nu-1}) \tag{5.5}$$

The conditional distribution $P(x,t|\xi,s)$, $t > s$, is often called the **transition function.** Suppose $t_1 < t_2 < t_3$, then we can write $P(x_1,t_1;\ x_2,t_2;\ x_3,t_3)$ as

$$P(x_1,t_1;\ x_2,t_2;\ x_3,t_3) = \int_{-\infty}^{x_1}\int_{-\infty}^{x_2} \mathcal{P}(X_{t_3} < x_3;\ X_{t_2} \in d\xi_2;\ X_{t_1} \in d\xi_1)$$

where $d\xi_i$ stands for $[\xi_i,\ \xi_i + d\xi_i)$. Using the Markov property, we get

$$\begin{aligned}P(x_1,t_1;\ x_2,t_2;\ x_3,t_3) &= \int_{-\infty}^{x_1}\int_{-\infty}^{x_2} P(x_3,t_3|\,\xi_2,t_2)\ dP(\xi_1,t_1;\ \xi_2,t_2)\\ &= \int_{-\infty}^{x_1}\int_{-\infty}^{x_2} P(x_3,t_3|\,\xi_2,t_2)\ dP(\xi_2,t_2|\,\xi_1,t_1)\ dP(\xi_1,t_1)\end{aligned}$$

More generally, we have for $t_1 < t_2 < \cdots < t_n$

$$\begin{aligned}&P(x_1,t_1;\ x_2,t_2;\ \ldots\ ;\ x_n,t_n)\\ &\quad= \int_{-\infty}^{x_1} \cdots \int_{-\infty}^{x_{n-1}} P(x_n,t_n|\,\xi_{n-1},t_{n-1})\ dP(\xi_{n-1},t_{n-1}|\,\xi_{n-2},t_{n-2})\\ &\qquad\cdots dP(\xi_2,t_2|\,\xi_1,t_1)\ dP(\xi_1,t_1)\end{aligned} \tag{5.6}$$

which expresses the n-dimensional distribution of a Markov process in terms of the one-dimensional distribution $P(x,t)$ and the conditional distribution $P(x,t|\,\xi,s)$, $t > s$, both of which are obtainable from $P(\xi,s,x,t)$.

However, we should note that we cannot construct the finite-dimensional distributions of a Markov process out of just any two-dimensional distribution. There are two consistency conditions that must be satisfied. The first one is obvious,

$$\begin{aligned}P(x,t) &= \mathcal{P}(X_t < x,\ -\infty < X_s < \infty)\\ &= \int_{-\infty}^{\infty} P(x,t|\,\xi,s)\ dP(\xi,s)\end{aligned} \tag{5.7}$$

This condition relates $P(x,t)$ to $P(x,t|\,\xi,s)$. The second condition is obtained by noting that if $t_0 < s < t$, then

$$\begin{aligned}P(x,t|x_0,t_0) &= \mathcal{P}(X_t < x|X_{t_0} = x_0)\\ &= \mathcal{P}(X_t < x,\ -\infty < X_s < \infty\,|X_{t_0} = x_0)\\ &= \int_{-\infty}^{\infty} \mathcal{P}(X_t < x|X_s = \xi,\ X_{t_0} = x_0)\mathcal{P}(X_s \in d\xi|X_{t_0} = x_0)\\ &= \int_{-\infty}^{\infty} P(x,t|\,\xi,s)\ dP(\xi,s|t_0,x_0)\end{aligned}$$

This yields a condition that must be satisfied by the transition function $P(x,t|\,\xi,s)$, $t > s$, namely,

$$P(x,t|x_0,t_0) = \int_{-\infty}^{\infty} P(x,t|\,\xi,s)\ dP(\xi,s|t_0,x_0) \qquad t_0 < s < t \tag{5.8}$$

Equation (5.8) is known as the **Chapman-Kolmogorov equation.** Given distribution functions $P(x,t)$ and $P(x,t|\xi,s)$ satisfying (5.7) and (5.8), the finite-dimensional distribution functions constructed by using (5.6) will indeed satisfy the Markov property. Therefore, (5.7) and (5.8) are both necessary and sufficient for $P(x,t)$ and $P(x,t|\xi,s)$ to be the one-dimensional distribution and the transition function of a Markov process.

As an example, suppose that we want to define a Markov process $\{X_t, -\infty < t < \infty\}$ with the following properties:

1. X_t takes only values $+1$ or -1
2. $\mathcal{P}(X_t = 1) = \frac{1}{2} = \mathcal{P}(X_t = -1)$
3. For $t \geq s$,

$$\mathcal{P}(X_t = X_s) = q(t-s)$$
$$\mathcal{P}(X_t = -X_s) = 1 - q(t-s)$$

where $q(t)$ is continuous and $q(0) = 1$

It is easy to verify that (5.7) is automatically satisfied. However, (5.8) imposes a constraint on q, namely,

$$q(t-t_0) = q(t-s)q(s-t_0) + [1-q(t-s)][1-q(s-t_0)] \tag{5.9}$$

A second equation can also be obtained, but it turns out to be redundant. Equation (5.9) looks a little simpler if we make a few changes of variables and write

$$q(t+s) = q(t)q(s) + [1-q(t)][1-q(s)]$$

By setting $q(t) = \frac{1}{2}[1+f(t)]$, we find

$$f(t+s) = f(t)f(s) \tag{5.10}$$

The only continuous solution of (5.10) not identically equal to 0 is

$$f(t) = e^{-\lambda t}$$

and λ must be nonnegative, because $f(t) \leq 1$. This means that $q(t)$ must be of the form

$$q(t) = \tfrac{1}{2}(1+e^{-\lambda t}) \tag{5.11}$$

The resulting process is precisely the **random telegraph process** that we introduced in Sec. 4 (4.14), except for the trivial scale factor λ.

As a second example of using (5.8), we shall derive a set of conditions which must be satisfied by the covariance function of a Gaussian-Markov process. We can always assume that $\{X_t, t \in T\}$ has zero mean, because X_t is Markov if and only if $X_t - \mu(t)$ is Markov. If $\{X_t, t \in T\}$ is Gaussian, zero mean, and if it has covariance function $R(t,s)$, then

$$E(X_t|X_s = \xi) = \frac{R(t,s)}{R(s,s)}\,\xi \tag{5.12}$$

To prove (5.12), we merely note that X_s and $X_t - [R(t,s)]/[R(s,s)]X_s$ are Gaussian and uncorrelated, hence are independent. Thus,

$$\begin{aligned} E(X_t|X_s = \xi) &= \frac{R(t,s)}{R(s,s)}\,\xi + E\left[X_t - \frac{R(t,s)}{R(s,s)}\,X_s\right] \\ &= \frac{R(t,s)}{R(s,s)}\,\xi \end{aligned}$$

Using (5.12) and (5.8) together, we get

$$\begin{aligned} E(X_t|X_{t_0} = x_0) &= \int_{-\infty}^{\infty} x\,dP(x,t|x_0,t_0) \\ &= \int_{-\infty}^{\infty} \frac{R(t,s)}{R(s,s)}\,\xi\,dP(\xi,s|x_0,t_0) \\ &= \frac{R(t,s)}{R(s,s)}\,\frac{R(s,t_0)}{R(t_0,t_0)}\,x_0 \\ &= \frac{R(t,t_0)}{R(t_0,t_0)}\,x_0 \qquad t > s > t_0 \end{aligned}$$

Therefore, for $\{X_t,\ t \in T\}$ to be Gaussian and Markov we must have

$$R(t,t_0) = \frac{R(t,s)R(s,t_0)}{R(s,s)} \qquad t > s > t_0 \tag{5.13}$$

Suppose that T is an interval and $R(t,t)$ is strictly positive for t in the interior of T. Then we can show that any solution of (5.13) must have the form

$$R(t,s) = f(\max\,(t,s))g(\min\,(t,s)) \qquad t,\ s \in T \tag{5.14}$$

It turns out that (5.14) is also a sufficient condition for a Gaussian process to be Markov. The simplest way to show this is as follows. First, we note that a Brownian motion is Gaussian, Markov, and the $R(t,s) = \min\,(t,s)$ (see Sec. 3). Set

$$\tau(t) = \frac{g(t)}{f(t)} \tag{5.15}$$

Because $R(t,s) \le \sqrt{R(t,t)R(s,s)}$, $\tau(t)$ must be monotone nondecreasing in t. Now, if X_t is a Gaussian process with zero mean and covariance function given by (5.14), then it can be represented as

$$X_t = f(t)Y_{\tau(t)} \tag{5.16}$$

where Y_t is a Brownian motion. Since $f(t)$ is a deterministic function, and $\tau(t)$ is nondecreasing, the fact that Y_t is Markov implies that X_t is Markov.

As was pointed out in Sec. 1.7, rather than to deal with conditional distributions, it is far simpler to deal with conditional expectations. The setting is however more abstract. We shall now reexamine the Markov property in terms of conditional expectations. First, we adopt some notations. Let $\{X_t,\ t \in T\}$ be a process defined on a fixed probability space $(\Omega,\mathcal{A},\mathcal{P})$. Let $\mathcal{A}_t$ denote the smallest σ algebra with respect to which every X_s, $s \leq t$, is measurable. Let $\mathcal{A}_t^+$ denote the smallest σ algebra with respect to which X_s is measurable for every $s \geq t$. Suppose that $\{X_t, t \in T\}$ is a Markov process, then the following relations hold: All equalities are understood to hold up to sets of zero probability.

1. The future and the past are conditionally independent given the present. That is, if Z is $\mathcal{A}_t^+$ measurable and Y is $\mathcal{A}_t$ measurable, then

$$E(ZY|X_t) = E(Z|X_t)E(Y|X_t) \tag{5.17}$$

2. The future, given the past and the present, is equal to the future given only the present. That is, if Z is $\mathcal{A}_t^+$ measurable then

$$E^{\mathcal{A}_t}Z = E(Z|X_t) \tag{5.18}$$

Either 1 or 2 can be taken as the defining condition for Markov process.

Roughly speaking, the counterparts of (5.17) and (5.18) in terms of density functions are, respectively, given as follows:

$$p(x,t;\ x_0,t_0|\xi,s) = p(x,t|\xi,s)p(x_0,t_0|\xi,s) \qquad t > s > t_0$$
$$p(x,t|x_0,t_0;\ \xi,s) = p(x,t|\xi,s) \qquad t > s > t_0$$

In this form, the equivalence between conditions 1 and 2 is intuitively clear. To get (5.17) from (5.18) is easy. We merely note that if Z is $\mathcal{A}_t^+$ measurable and Y is $\mathcal{A}_t$ measurable, then

$$\begin{aligned} E(ZY|X_t) &= E(E^{\mathcal{A}_t}ZY|X_t) = E(YE^{\mathcal{A}_t}Z|X_t) \\ &= E[YE(Z|X_t)|X_t] = E(Z|X_t)E(Y|X_t) \end{aligned}$$

To get (5.18) from (5.17), take an arbitrary set $B \in \mathcal{A}_t$ and compute

$$\begin{aligned} E[I_B E(Z|X_t)] &= E\{E[I_B E(Z|X_t)|X_t]\} \\ &= E[E(Z|X_t)E(I_B|X_t)] = E[E(I_B Z|X_t)] = E(I_B Z) \end{aligned}$$

Since $EI_B E(Z|X_t) = EI_B Z$ for every set B in $\mathcal{A}_t$ and $E(Z|X_t)$ is $\mathcal{A}_t$ measurable, (5.18) follows by the definition of conditional expectation.

There is still another consequence of the Markov property that is useful to express. Suppose that Z is $\mathcal{A}_t^+$ measurable and $t_0 < t$. Then we have

$$E(Z|X_{t_0}) = E[E(Z|X_t)|X_{t_0}] \tag{5.19}$$

To prove (5.19), we merely have to use (5.18) and write

$$E(Z|X_{t_0}) = E(E^{\mathcal{Q}_t}Z|X_{t_0}) = E[E(Z|X_t)|X_{t_0}]$$

Equation (5.19) is really the Chapman-Kolmogorov equation (5.8) in disguise. If we set $B = \{\omega\colon X_t < x\}$, then (5.8) can be rewritten as

$$E\{I_B|X_{t_0} = x_0\} = E\{E(I_B|X_s)|X_{t_0} = x_0\} \qquad t_0 < s < t \tag{5.20}$$

which follows immediately from (5.19). A comparison between (5.8) and (5.19) reveals the notational simplicity gained by using conditional expectation.

6. STATIONARITY AND ERGODICITY

Definition. A process $\{X_t, \ -\infty < t < \infty\}$ is said to be a **stationary process** if for any $(t_1, \ldots, t_n)$ the joint distribution of $\{X_{t_1+t_0}, X_{t_2+t_0}, \ldots, X_{t_n+t_0}\}$ does not depend on t_0.

Thus, if $\{X_t, \ t \in T\}$ is stationary, the n-dimensional distribution function $P(x_1,t_1; \ x_2,t_2; \ \ldots \ ; \ x_n,t_n)$ is equal to $P(x_1,0; \ x_2,t_2 - t_1; \ \ldots \ ; \ x_n,t_n - t_1)$, and hence, it depends only on the time differences $t_2 - t_1, \ldots, t_n - t_1$.

Definition. A process $\{X_t, \ -\infty < t < \infty\}$ is said to be a **wide-sense stationary process** if

(a) $EX_t^2 < \infty$

(b) $EX_t = \mu$ is a constant

(c) $E(X_t - \mu)(X_s - \mu)$ depends only on $t - s$

Because a stationary process may have $EX_t^2 = \infty$, a stationary process is also wide-sense stationary if and *only if* $EX_t^2 < \infty$. Wide-sense stationarity will be discussed in Chap. 3 in connection with second-order processes. In this section, we restrict our consideration to stationary processes, which are sometimes referred to as strict-sense stationary processes to emphasize the difference from wide-sense stationarity. The random telegraph process is an example of a stationary process. The Brownian motion is not stationary. A Gaussian process with zero mean and a covariance function given by

$$R(t,s) = e^{-|t-s|}$$

is both stationary and Markov. It is known as the Ornstein-Uhlenbeck process.

Let $\{X_t, \ -\infty < t < \infty\}$ be a stationary process defined on a completed probability space $(\Omega,\mathcal{A},\mathcal{P})$. We suppose that $\mathcal{A}$ is the completion of the smallest σ algebra with respect to which every X_t is measurable.

We also suppose that $\{X_t, -\infty < t < \infty\}$ is a separable and measurable process. Let S denote the space of all $\mathcal{A}$-measurable random variables. We agree that two random variables which are equal almost surely count as the same element of S. Then S is a linear space closed under almost-sure convergence. Now, we define a family $\{T_t, -\infty < t < \infty\}$ of linear mappings of S onto S as follows:

$$T_t X_s = X_{t+s} \qquad -\infty < t, s < \infty \tag{6.1a}$$

If $Z(\omega) = f(X_{t_1}(\omega), \ldots, X_{t_n}(\omega))$, then

$$T_t Z = f(T_t X_{t_1}, \ldots, T_t X_{t_n}) \tag{6.1b}$$

If $\{Z_n\}$ is a sequence in S converging almost surely to Z, then

$$T_t Z = \lim_{n\to\infty} \text{a.s. } T_t Z_n \tag{6.1c}$$

Since every Z in S can be approximated by (a.s. limit) a sequence of Borel functions of a finite number of X_t's, (6.1*a*) to (6.1*c*) adequately define $\{T_t, -\infty < t < \infty\}$.

Since $\{X_t, -\infty < t < \infty\}$ is stationary, $\{T_t X_{t_\nu}, \nu = 1, \ldots, n\}$ and $\{X_{t_\nu}, \nu = 1, \ldots, n\}$ have the same distribution. Therefore, $T_t Z$ and Z have identical distributions, whenever Z is of the form $f(X_{t_1}, \ldots, X_{t_n})$. By virtue of the fact that convergence a.s. implies convergence in distribution, $T_t Z$ and Z are always identically distributed. It is also easy to see that $\{T_t, -\infty < t < \infty\}$ is a translation group, that is, $T_{t+s} = T_t T_s$, $T_0 = I$, $T_{-t} = T_t^{-1}$. Summarizing these properties of T_t, we have the following:

$$T_t \text{ is a linear mapping of } S \text{ onto } S \tag{6.2a}$$
$$T_t \text{ preserves a.s. convergence} \tag{6.2b}$$
$$T_{t+s} = T_t T_s,\ T_0 = I,\ T_{-t} = T_t^{-1} \tag{6.2c}$$
$$T_t Z \text{ and } Z \text{ are identically distributed}, \quad -\infty < t < \infty, Z \in S \tag{6.2d}$$

Now, suppose a random variable Z in S is such that for each t, $T_t Z = Z$ almost surely. That is, for each t, $T_t Z$ and Z differ only on a set Λ_t such that $\mathcal{P}(\Lambda_t) = 0$. Then, Z is said to be an **invariant random variable** of the process $\{X_t, -\infty < t < \infty\}$.

Definition. A stationary process $\{X_t, -\infty < t < \infty\}$ is said to be **ergodic** if every invariant random variable of the process is almost surely equal to a constant.

We note that a random variable almost surely equal to a constant is always invariant. The great interest in ergodic processes, from the point of view of applications, is largely due to the following theorem.

Proposition 6.1. Let $\{X_t, -\infty < t < \infty\}$ be a separable and measurable ergodic process. Let f be any Borel function such that $E|f(X_0)| < \infty$. Then

$$\lim_{T\to\infty} \frac{1}{2T} \int_{-T}^{T} f(X_t)\, dt = Ef(X_0) \qquad \text{almost surely} \tag{6.3}$$

Conversely, if (6.3) holds for every such f, $\{X_t, -\infty < t < \infty\}$ is ergodic.

Remark: We note that because of stationarity, $Ef(X_t) = Ef(X_0)$ for every t. We interpret Proposition 6.1 as saying that for an ergodic process, time average equals ensemble average.

The proof of Proposition 6.1 will be omitted. We consider, instead, an example illustrating the foregoing discussion. Suppose

$$X_t(\omega) = A(\omega) \cos [2\pi t + \theta(\omega)] \qquad -\infty < t < \infty \tag{6.4}$$

where A and θ are independent random variables, and θ is uniformly distributed on $[0,2\pi)$. It is easy to show (e.g., by computing the characteristic functions) that $\{X_t, -\infty < t < \infty\}$ is stationary. Now, for this simple example, every Z in S is some Borel function of the pair (A,θ), and

$$T_t f(A,\theta) = f(A, \operatorname{mod}_{2\pi} (\theta + 2\pi t)) \tag{6.5}$$

Every Z in S which depends only on A and not on θ is an invariant random variable. Thus $\{X_t, -\infty < t < \infty\}$ is ergodic if and only if A is almost surely a constant. This result can also be inferred from Proposition 6.1. For an arbitrary Borel function f such that $E|f(X_0)| < \infty$, we have

$$\begin{aligned}
\lim_{T\to\infty} \frac{1}{2T} \int_{-T}^{T} f(X_t(\omega))\, dt &= \lim_{N\to\infty} \frac{1}{2N} \int_{-N}^{N} f(A(\omega) \cos [2\pi t + \theta(\omega)])\, dt \\
&= \lim_{N\to\infty} \frac{1}{2N} \int_{-N}^{N} f(A(\omega) \cos 2\pi t)\, dt \\
&= \int_0^1 f(A(\omega) \cos 2\pi t)\, dt
\end{aligned} \tag{6.6}$$

where we made repeated use of the fact that $f(A \cos (2\pi t + \theta))$ is periodic in t with period 1. On the other hand

$$\begin{aligned}
Ef(X_0) &= Ef(A \cos \theta) \\
&= \int_{-\infty}^{\infty} \left[\int_0^{2\pi} \frac{1}{2\pi} f(a \cos \varphi)\, d\varphi \right] dP_A(a) \\
&= E \int_0^1 f(A \cos 2\pi t)\, dt
\end{aligned} \tag{6.7}$$

If we denote $\int_0^1 f(A \cos 2\pi t)\, dt$ by $\hat{f}(A)$, then the time average is $\hat{f}(A(\omega))$. These two are equal if and only if $A(\omega)$ is almost surely equal to a constant.

In general, it is not easy to give a simple condition which would ensure ergodicity. For Gaussian processes, however, we have the following sufficient condition for ergodicity:[1] Assume that $\{X_t, -\infty < t < \infty\}$ is Gaussian and stationary, with mean μ and covariance function

$$R(\tau) = E(X_{t+\tau} - \mu)(X_t - \mu)$$

Then $\{X_t, -\infty < t < \infty\}$ is ergodic if

$$\int_{-\infty}^{\infty} |R(\tau)|\, d\tau < \infty \tag{6.8}$$

The question of possible equality between time average and ensemble average can be asked in a different and more direct way. Suppose for a fixed stationary process $\{X_t, -\infty < t < \infty\}$ and a fixed function f, we seek a sufficient condition for

$$\frac{1}{2T} \int_{-T}^{T} f(X_t)\, dt \xrightarrow[T\to\infty]{\text{q.m.}} Ef(X_0) \tag{6.9a}$$

that is,

$$\lim_{T\to\infty} E\left[\frac{1}{2T}\int_{-T}^{T} f(X_t)\, dt - Ef(X_0)\right]^2 = 0 \tag{6.9b}$$

We note that $Ef(X_t) = Ef(X_0)$ for every t. Therefore,

$$E\left[\frac{1}{2T}\int_{-T}^{T} f(X_t)\, dt - Ef(X_0)\right]^2$$

$$= \frac{1}{(2T)^2} \iint_{-T}^{T} E\{[f(X_t) - Ef(X_t)][f(X_s) - Ef(X_s)]\}\, dt\, ds$$

$$= \frac{1}{(2T)^2} \iint_{-T}^{T} R_f(t-s)\, dt\, ds \tag{6.10}$$

where we have denoted the covariance function of $f(X_t)$ by R_f. By a change in the variables of integration $\tau = t - s$, $\sigma = t + s$, the last integral in (6.10) can be rewritten as

$$\frac{1}{(2T)^2}\frac{1}{2}\int_{-2T}^{2T}\left[\int_{\max(2\tau - T, \tau - T)}^{\min(2\tau + T, \tau + T)} d\sigma\right] R_f(\tau)\, d\tau$$

$$= \frac{1}{4T}\int_{-2T}^{2T} R_f(\tau)\left(1 - \frac{|\tau|}{2T}\right) d\tau \le \frac{1}{4T}\int_{-\infty}^{\infty} |R_f(\tau)|\, d\tau$$

[1] A necessary and sufficient condition is for the spectral-distribution function, defined by (3.5.21) and (3.5.22), to be continuous [Grenander, 1950, pp. 257–260].

Therefore, a sufficient condition for (6.9*b*) is given by

$$\int_{-\infty}^{\infty} |R_f(\tau)|\, d\tau < \infty \tag{6.11}$$

Unlike ergodicity, condition (6.11) has to be verified for each f separately.

EXERCISES

1. Suppose that $\{X_t,\ 0 \le t \le 1\}$ is a family of independent random variables, i.e., every finite subcollection $X_{t_1}, \ldots, X_{t_n}$ is mutually independent. Show that $\{X_t,\ 0 \le t \le 1\}$ cannot be continuous in probability unless there is a continuous function $f(\cdot)$ such that for each t, $X_t(\omega) = f(t)$ for almost all ω.

2. Suppose that X is a Gaussian random variable with $EX = 0$ and $EX^2 = 1$. Let $X_t = t + X,\ -\infty < t < \infty$.

   (*a*) Let T_n be any countable set in $(-\infty, \infty)$, say $T_n = \{t_1, t_2, \ldots\}$. Show that $\mathcal{P}(X_t = 0 \text{ for at least one } t \text{ in } T_n) = 0$
   (*b*) Let T be an interval, say [0,1]. Show that
   $\mathcal{P}(X_t = 0 \text{ for at least one } t \text{ in } T) = \mathcal{P}\left(\bigcup_{t \in T} \{\omega\colon X_t(\omega) = 0\}\right) > 0$

   Note that even though $\{\omega\colon X_t(\omega) = 0 \text{ for at least one } t \text{ in } T\}$ is an uncountable union, it is an event in this particular case.

3. For an extremely explicit example of stochastic process consider the following. Let $\Omega = [0,1]$, let $\mathcal{A}$ be the σ algebra of Borel sets in [0,1], and let $\mathcal{P}$ be the Lebesgue measure so that $\mathcal{P}$ (interval) = length. Define

   $X_t(\omega) = t\omega \qquad 0 \le t \le 1$

   Find the one-dimensional distribution function P_t for this process. Repeat for the two-dimensional distribution function $P_{t,s}$.

4. Find the mean function $\mu(t) = EX_t$ and the covariance function $R(t,s) = E[X_t - \mu(t)][X_s - \mu(s)]$ for the process defined in Exercise 3.

5. Verify that the process defined in Exercise 3 is both separable and measurable.

6. Let Z and θ be independent random variables such that Z has a density function p_Z given by

   $$p_Z(z) = \begin{cases} 0 & z < 0 \\ ze^{-\frac{1}{2}z^2} & z \ge 0 \end{cases}$$

   and θ is uniformly distributed in the interval $[0,2\pi)$. Define

   $X_t = Z \cos(2\pi t + \theta)$

   Show that $\{X_t,\ -\infty < t < \infty\}$ is a Gaussian process.

7. Let $\{X_t,\ -\infty < t < \infty\}$ be a Gaussian process with $EX_t = 0$ and

   $EX_sX_t = \frac{1}{2}(|t| + |s| - |t - s|)$

Now define $Y_t = X_{-t}$ for $t \le 0$. Show that $\{Y_t, 0 \le t < \infty\}$ and $\{X_t, 0 \le t < \infty\}$ are two independent processes.

8. Let $\{X_t, t \in T\}$ be a stochastic process such that $E|X_t| < \infty$ for every $t \in T$, and let $\mathcal{A}_{xt}$ denote the smallest σ algebra with respect to which X_s is measurable for every $s \le t$. Suppose that

$$E(X_{t_n}|X_{t_1}, X_{t_2}, \ldots, X_{t_{n-1}}) = X_{t_{n-1}} \qquad \text{a.s.}$$

whenever $t_1 < t_2 < \cdots < t_n$. Prove that whenever $t \ge s$

$$E^{\mathcal{A}_{xs}} X_t = X_s \qquad \text{a.s.}$$

Note: $E^{\mathcal{A}_{xs}} X_t$ can also be written in a more suggestive way as $E(X_t|X_\tau, \tau \le s)$.

9. Let $\{X_t, t \ge 0\}$ be a Brownian motion. Use the result of Exercise 8 to prove that for $t \ge s$,

$$E(X_t|X_\tau, \tau \le s) = X_s \qquad \text{a.s.}$$

10. Let $\{X_t, t \in [0,1]\}$ be a separable Gaussian process with zero mean. If

$$E(X_{t+h} - X_t)^2 = h^\gamma \qquad \gamma > 0$$

show that $\{X_t \in [0,1]\}$ must be sample continuous no matter how small γ may be.

11. Suppose that $\{X_t, -\infty < t < \infty\}$ is a Gaussian process with zero mean and $EX_tX_s = e^{-|t-s|}$. Express X_t in the form

$$X_t = f(t)W_{g(t)/f(t)}$$

where $\{W_s, 0 \le s < \infty\}$ is a standard Brownian motion.

12. Let $\{X_t, -\infty < t < \infty\}$ be a stationary Markov process which assumes only a finite number of values $x_1, x_2, \ldots, x_n$. Let $\mathbf{p}(\tau)$ be an $n \times n$ matrix with elements

$$p_{ij}(\tau) = \mathcal{P}(X_{t+\tau} = x_i|X_t = x_j)$$

(*a*) Suppose that $\lim_{\tau \downarrow 0} (1/\tau)[\mathbf{p}(\tau) - \mathbf{I}] = \mathbf{A}$ exists and is finite. Show that $\mathbf{p}(\tau)$ must have the form

$$\mathbf{p}(\tau) = e^{\tau \mathbf{A}}$$

(*b*) Let $\mathbf{q}$ be an n vector with

$$q_i = \text{prob}\,(X_t = x_i)$$

and let $\mathbf{I}$ denote the n vector $\begin{bmatrix} 1 \\ 1 \\ \cdot \\ \cdot \\ \cdot \\ 1 \end{bmatrix}$. Show that

$$\mathbf{p}(\tau)\mathbf{q} = \mathbf{q}$$

and

$$\mathbf{p}'(\tau)\mathbf{l} = \mathbf{l} \qquad (\mathbf{p}' = \text{transpose of } \mathbf{p})$$

Suppose that $n = 2$ and $\mathbf{q} = \begin{bmatrix} \frac{1}{2} \\ \frac{1}{2} \end{bmatrix}$. Show that $\mathbf{p}(\tau)$ must be of the form

$$\mathbf{p}(\tau) = \begin{bmatrix} \dfrac{1 + e^{-\lambda\tau}}{2} & \dfrac{1 - e^{-\lambda\tau}}{2} \\ \dfrac{1 - e^{-\lambda\tau}}{2} & \dfrac{1 + e^{-\lambda\tau}}{2} \end{bmatrix} \qquad \lambda \geq 0$$

13. Let T be an interval and $R(t,s)$, $t, s \in T$, be a continuous covariance function such that $R(t,t) > 0$ for every t in the interior of T. Suppose that R satisfies

$$R(t,t_0) = \frac{R(t,s)R(s,t_0)}{R(s,s)} \qquad t_0 < s < t$$

(*a*) Let $\rho(t,s) = R(t,s)/\sqrt{R(t,t)R(s,s)}$ and show that ρ satisfies

$$\rho(t,t_0) = \rho(t,s)\rho(s,t_0) \qquad \begin{array}{l} t_0 < s < t \\ t_0,\ s,\ t \text{ in int } (T) \end{array}$$

(*b*) Show that $\rho(t,s) > 0$ for all t and s in the interior of T.

(*c*) Let a be a fixed point in int (T) and define

$$\alpha(t) = \begin{cases} \rho(t,a) & t \leq a \\ \dfrac{1}{\rho(t,a)} & t \geq a \end{cases}$$

Show that

$$\rho(t,s) = \frac{\alpha(\min\ (t,s))}{\alpha(\max\ (t,s))} \qquad t, s \in \text{int } (T)$$

Note: This *proves* (5.14).

14. Let $\{X_t, -\infty < t < \infty\}$ be a q.m. continuous and stationary Gaussian-Markov process. Find its covariance function.

15. Suppose that $\{X_t, -\infty < t < \infty\}$ is given by the form

$$X_t(\omega) = A(\omega) \cos [2\pi t + \theta(\omega)]$$

where A and θ are independent. A is nonnegative and θ is uniformly distributed on $[0,2\pi)$.

(*a*) Show that $\{X_t, -\infty < t < \infty\}$ is stationary.
(*b*) Show that $EX_t = 0$ for all t provided that $EA < \infty$. Does

$M_T = \dfrac{1}{2T} \displaystyle\int_{-T}^{T} X_t\, dt$ converge in probability to EX_t as $T \to \infty$?

(*c*) Show that $\{X_t, -\infty < t < \infty\}$ is a Gaussian process if and only if A has a Rayleigh distribution, that is, A has a density p_A given by

$$p_A(r) = \frac{r}{\sigma^2} \exp\left(-\frac{1}{2}\frac{r^2}{\sigma^2}\right) \qquad r \geq 0$$

Is $\{X_t, -\infty < t < \infty\}$ Markov then?

16. Suppose that $\{X_t, t \geq 0\}$ and $\{Y_t, t \geq 0\}$ are two independent standard Brownian motions. Let $Z_t = \sqrt{X_t^2 + Y_t^2}$.
(*a*) Suppose that X_t, Y_t are observed at $t = 1, 2, 3$ and the data are summarized as follows:

t	1	2	3
x	.3	1	−2
y	−.1	−.2	1

Find $E\{Z_4^2 | \text{observed data}\}$ and show that

$$\sqrt{5} \leq E\{Z_4 | \text{observed data}\} \leq \sqrt{7}$$

(*b*) Is $\{Z_t, t \geq 0\}$ a Markov process?

3
Second-Order Processes

1. INTRODUCTION

In this chapter it will actually be easier to deal with complex-valued random variables and stochastic processes. A complex-valued random variable X is a complex-valued function on Ω such that its real and imaginary parts are real random variables.

Definition. A **second-order random variable** X is one which satisfies $E|X|^2 < \infty$. A **second-order stochastic process** $\{X_t,\ t \in T\}$ is a one-parameter family of second-order random variables.

If $\{X_t,\ t \in T\}$ is a second-order process, then we can define the **mean, correlation function,** and **covariance function,** respectively, as follows:

$$EX_t = \mu(t) \tag{1.1}$$
$$EX_t\bar{X}_s = \mathcal{R}(t,s) \tag{1.2}$$
$$E[X_t - \mu(t)]\overline{[X_s - \mu(s)]} = R(t,s) \tag{1.3}$$

where the overbar denotes complex conjugate. As far as second-order properties are concerned, we are primarily interested in "linear operations" on processes. Since the output of a linear operation on $\{X_t,\ t \in T\}$ is equal to the sum of the outputs of the same operation on $X_t - \mu(t)$ and $\mu(t)$ separately, we can always assume zero mean with little loss in generality. In this chapter we shall always assume that the mean is zero. If the mean is zero, the correlation function and the covariance function are the same thing. We use the term covariance function exclusively. Roughly speaking, second-order properties of a process are those properties that can be deduced knowing only its covariance function.

A covariance function satisfies a number of important properties, some of which are listed below.

1. *Nonnegative definite*
A complex-valued function $f(t,s)$ defined on a square $T \times T$ is said to be **nonnegative definite,** if for any finite collection $t_1, t_2, \ldots, t_n$ in T and any complex constants $\alpha_1, \alpha_2, \ldots, \alpha_n$

$$\sum_{j=1}^{n} \sum_{k=1}^{n} \alpha_j \bar{\alpha}_k f(t_j,t_k) \geq 0 \tag{1.4}$$

Every covariance function is nonnegative definite, because

$$\sum_{j=1}^{n} \sum_{k=1}^{n} \alpha_j \bar{\alpha}_k R(t_j,t_k) = E \left| \sum_{j=1}^{n} \alpha_j X_{t_j} \right|^2 \geq 0$$

Conversely, if $R(t,s)$ is a nonnegative definite function on $T \times T$, then we can always find a second-order process $\{X_t,\ t \in T\}$ whose covariance function is $R(t,s)$. In fact, we can always find a pair of real processes $\{Y_t,\ Z_t,\ t \in T\}$ such that $Y_t + iZ_t$ has covariance function $R(t,s)$, and Y_t, Z_t are jointly Gaussian, i.e., any real linear combination $\alpha Y_t + \beta Z_t$ is again a Gaussian process. Summarizing, *a function $R(t,s)$ on $T \times T$ is a covariance function if and only if it is nonnegative definite.*

2. *Hermitian symmetry*
A covariance function $R(t,s)$ always satisfies

$$R(t,s) = \bar{R}(s,t) \tag{1.5}$$

because $EX_t\bar{X}_s = \overline{E\bar{X}_tX_s}$.

3. *Schwarz inequality*
If Y and Z are a pair of second-order random variables, then

$$|EY\bar{Z}| \leq \sqrt{E|Y|^2E|Z|^2} \tag{1.6}$$

which is a special case of the Schwarz inequality. To prove (1.6) we need only note that

$$E|Y|^2E|Z|^2 - E|Y\bar{Z}|^2 = E|Z|^2E|Y - \frac{EY\bar{Z}}{E|Z|^2}Z|^2 \geq 0 \tag{1.7}$$

Applying (1.6) to a second-order process, we get immediately

$$|R(t,s)| \leq \sqrt{R(t,t)R(s,s)} \tag{1.8}$$

whenever $R(t,s)$ is a covariance function.

4. *Closure properties*

(*a*) *Multiplication.* If $R_1(t,s)$ and $R_2(t,s)$ are two covariance functions, then we can always find two independent Gaussian processes Y_t and Z_t with covariance functions $R_1(t,s)$ and $R_2(t,s)$, respectively. The covariance function of $X_t = Y_tZ_t$ is given by $R_1(t,s)R_2(t,s)$. Therefore, the product of two covariance functions is again a covariance function.

(*b*) *Addition.* The sum $R_1(t,s) + R_2(t,s)$ of two covariance functions is again a covariance function. The argument for this assertion is almost identical to that of (*a*).

(*c*) *Positive sums.* A positive constant C is always a covariance function, simply because we can always take $X_t = X$ to be a Gaussian random variable with zero mean and $EX^2 = C$. It follows that if $R_1(t,s)$, $R_2(t,s)$, . . . , $R_n(t,s)$ are covariance functions on $T \times T$ and C_1, C_2, . . . , C_n are positive constants, then

$$R(t,s) = \sum_{\nu=1}^{n} C_\nu R_\nu(t,s) \tag{1.9}$$

is again a covariance function on $T \times T$.

(*d*) *Pointwise limit.* If $\{R_\nu(t,s),\ \nu = 1, \ldots\}$ is a sequence of covariance functions on $T \times T$ converging pointwise to $R(t,s)$ at every point of $T \times T$, then $R(t,s)$ is a covariance function. We only need to verify the definition of a nonnegative definite function,

$$\sum_{j=1}^{n}\sum_{k=1}^{n} \alpha_j\bar{\alpha}_k R(t_j,t_k) = \lim_{\nu\to\infty} \sum_{j=1}^{n}\sum_{k=1}^{n} \alpha_j\bar{\alpha}_k R_\nu(t_j,t_k) \geq 0$$

(*e*) *Bilinear forms.* For any function $\sigma(t)$, $\sigma(t)\bar{\sigma}(s)$ is a covariance function, simply because we can always set $X_t = \sigma(t)\,X$ where X is normal $N(0,1)$. It follows that any finite bilinear sum

$$\sum_{\nu=1}^{n} \bar{\sigma}_\nu(t)\sigma_\nu(s) \tag{1.10}$$

is a covariance function, and so is any convergent infinite series

$\sum_{\nu=1}^{\infty} \sigma_\nu(t)\bar{\sigma}_\nu(s)$. More generally, any pointwise limit of a sequence of bilinear sums of the form (1.10) is a covariance function. This includes not only infinite sum such as $\sum_{\nu=1}^{\infty} \sigma_\nu(t)\bar{\sigma}_\nu(s)$, but also integrals of the form $\int_a^b \sigma(t,\lambda)\bar{\sigma}(s,\lambda)\, d\lambda$. It will be seen later that most covariance functions can be represented in the form of bilinear sums and/or integrals, and these representations play an extremely useful role in the application of stochastic processes.

2. SECOND-ORDER CONTINUITY

Definition. A second-order process $\{X_t, t \in T\}$ is said to be **continuous in quadratic mean** (q.m. continuous) at t if

$$E|X_{t+h} - X_t|^2 \xrightarrow[h\to 0]{} 0$$

If a process is q.m. continuous at every t of T, we shall say that it is a q.m. continuous process.

Q.m. continuity of a second-order process must, of course, be closely related to questions of continuity of the covariance function of the process. The following proposition summarizes succinctly the relationship between q.m. continuity and continuity of the covariance function.

Proposition 2.1. Let $\{X_t, t \in T\}$ be a second-order process on an interval T, and let $R(t,s) = EX_t\bar{X}_s$ denote its covariance function.
(*a*) $\{X_t, t \in T\}$ is q.m. continuous at t if and only if $R(\cdot,\cdot)$ is continuous at the diagonal point (t,t).
(*b*) If $\{X_t, t \in T\}$ is q.m. continuous at every $t \in T$, then $R(\cdot,\cdot)$ is continuous at every point of the square $T \times T$.
(*c*) If a nonnegative definite function on $T \times T$ is continuous at every diagonal point, then it is continuous everywhere on $T \times T$.

Proof:
(*a*) If R is continuous at (t,t), then

$$\begin{aligned} E|X_{t+h} - X_t|^2 &= R(t+h, t+h) - R(t, t+h) \\ &\qquad - R(t+h, t) + R(t,t) \\ &= [R(t+h, t+h) - R(t,t)] - [R(t, t+h) - R(t,t)] \\ &\qquad - [R(t+h, t) - R(t,t)] \xrightarrow[h\to 0]{} 0 \end{aligned}$$

Conversely, if $\{X_t, t \in T\}$ is q.m. continuous at t, then

$$\begin{aligned} R(t+h, t+h') - R(t,t) &= EX_{t+h}\bar{X}_{t+h'} - EX_t\bar{X}_t \\ &= E(X_{t+h} - X_t)\bar{X}_{t+h'} + EX_t\overline{(X_{t+h'} - X_t)} \end{aligned}$$

From the Schwarz inequality (1.6), we have

$$|R(t+h,\, t+h') - R(t,t)| \le [E|X_{t+h} - X_t|^2 E|\bar{X}_{t+h'}|^2]^{\frac{1}{2}} + [E|X_t|^2 E|X_{t+h'} - X_t|^2]^{\frac{1}{2}} \xrightarrow[h,h'\to 0]{} 0$$

(*b*) If $\{X_t,\, t \in T\}$ is q.m. continuous at every $t \in T$, then

$$\begin{aligned} |R(t+h,\, s+h') - R(t,s)| &= |EX_{t+h}\bar{X}_{s+h'} - EX_t\bar{X}_s| \\ &= |E(X_{t+h} - X_t)\bar{X}_{s+h'} + EX_t(\overline{X_{s+h'} - X_s})| \\ &\le \sqrt{E|X_{t+h} - X_t|^2 E|X_{s+h'}|^2} + \sqrt{E|X_t|^2 E|X_{s+h'} - X_s|^2} \xrightarrow[h,h'\to 0]{} 0 \end{aligned}$$

(*c*) Since every nonnegative definite function on $T \times T$ is the covariance function of some second-order process on T, part (*c*) follows immediately from (*a*) and (*b*). ∎

3. LINEAR OPERATIONS AND SECOND-ORDER CALCULUS

Let $\{X_t,\, t \in T\}$ be a second-order process. A random variable Y is said to be derived from a **linear operation** on $\{X_t,\, t \in T\}$ if either

$$(a)\quad Y(\omega) = \sum_{\nu=1}^{N} \alpha_\nu X_{t_\nu}(\omega) \tag{3.1}$$

or

(*b*) Y is the q.m. limit of a sequence of such finite linear combinations

We denote the collection of all such random variables derived from a given process $\{X_t,\, t \in T\}$ by $\mathcal{H}_X$. Two elements Y and Y' are not distinguished whenever $E|Y - Y'|^2 = 0$.[1] Equality between second-order random variables will always be understood to be up to zero q.m. differences. The space $\mathcal{H}_X$ becomes an inner product space if we define the inner product $\langle Y,Z\rangle$ by

$$\langle Y,Z\rangle = EY\bar{Z} \tag{3.2}$$

The inner product automatically defines a norm

$$\|Y\| = \sqrt{\langle Y,Y\rangle} \tag{3.3}$$

and a metric

$$d(Y,Z) = \|Y - Z\| \tag{3.4}$$

A sequence $\{Y_n\}$ in $\mathcal{H}_X$ is said to be a **Cauchy sequence** if

$$\|Y_m - Y_n\| \xrightarrow[m,n\to\infty]{} 0$$

[1] Strictly speaking, $\mathcal{H}_X$ is a space of equivalence class, as is generally the case of L_2 spaces.

Since $\|Y_m - Y_n\|^2 = E|Y_m - Y_n|^2$, a Cauchy sequence in $\mathcal{H}_X$ is a mutually q.m. convergent sequence, which must converge in q.m. That is, for every Cauchy sequence $\{Y_n\}$ in $\mathcal{H}_X$, there exists a Y such that $\|Y_n - Y\| \xrightarrow[n\to\infty]{} 0$. It is easy to show that Y is in $\mathcal{H}_X$. This means that $\mathcal{H}_X$ is an inner product space which, as a metric space, is *complete*. By definition, $\mathcal{H}_X$ is a **Hilbert space.**

Second-order properties of a process $\{X_t, t \in T\}$ are those properties that are derivable from its covariance function. Every property of $\mathcal{H}_X$ is a second-order property, and conversely. It is easy to see, therefore, that there is inherently a close relationship between linear operations (defined by membership in $\mathcal{H}_X$) and second-order properties. Often, questions on second-order properties are most lucidly answered in the framework of the Hilbert space theory. A good example of this is the series representation of X_t to be treated in the next section.

Quadratic-mean derivative of a process $\{X_t, t \in T\}$ at a point t is defined by

$$\dot{X}_t = \lim_{h\to 0} \text{in q.m.} \frac{1}{h} (X_{t+h} - X_t) \tag{3.5}$$

Because of the equivalence between mutual q.m. convergence and q.m. convergence, $\dot{X}_t$ exists if and only if

$$\lim_{h,h'\to 0} E \left| \frac{1}{h} (X_{t+h} - X_t) - \frac{1}{h'} (X_{t+h'} - X_t) \right|^2 = 0 \tag{3.6}$$

which holds if the mixed partial derivative of the covariance function, that is, $\partial^2 R(t_1,t_2)/\partial t_1\, \partial t_2$, exists at the point (t,t). We note that

$$E\dot{X}_t\bar{\dot{X}}_s = \frac{\partial^2 R(t,s)}{\partial t\, \partial s} \tag{3.7}$$

Since if $\dot{X}_t$ exists, it is equal to

$$\dot{X}_t = \lim_{n\to\infty} \text{in q.m.}\ n\, (X_{t+1/n} - X_t) \tag{3.8}$$

$\dot{X}_t \in \mathcal{H}_X$ by definition.

Quadratic-mean integrals arise even more frequently than q.m. derivatives. Let $\{X_t, t \in T\}$ be a second-order process, and let $f(t)$ be a complex-valued function defined on the interval T. We define the q.m. integral $\int_T f(t) X_t\, dt$ as an element in $\mathcal{H}_X$ as follows. Let $\{T_n\}$ be a sequence of partitions of T such that as $n \to \infty$ T_n becomes dense in T. To be specific, let T be a finite interval $[a,b]$ and

$$T_n = \{a = t_0^{(n)} < t_1^{(n)} < \cdots < t_n^{(n)} = b\} \tag{3.9}$$

$$\max_{1 \le \nu \le n} (t_\nu^{(n)} - t_{\nu-1}^{(n)}) \xrightarrow[n \to \infty]{} 0 \tag{3.10}$$

We define the integral $\int_a^b f(t) X_t \, dt$ as

$$\int_a^b f(t) X_t \, dt = \lim_{n \to \infty} \text{in q.m.} \sum_{\nu=0}^{n-1} f(t'_{\nu n}) X_{t'_{\nu n}} (t_{\nu-1}^{(n)} - t_\nu^{(n)}) \tag{3.11}$$

where $t'_{\nu n}$ are any sequence of points satisfying $t_\nu^{(n)} \le t'_{\nu n} < t_{\nu+1}^{(n)}$. The integral $\int_a^b f(t) X_t \, dt$ is well defined by this procedure, provided that the q.m. limit exists and is independent of the choice of $\{T_n\}$ and, for each $\{T_n\}$, is independent of the choice of $\{t'_{\nu n}\}$. In that case, we say that *the integral* $\int_a^b f(t) X_t \, dt$ *exists.*

Proposition 3.1. The q.m. integral $\int_a^b f(t) X_t \, dt$ exists if and only if $\int_a^b \int_c^b f(t) \bar{f}(s) R(t,s) \, dt \, ds$ exists as a Riemann integral.

Remark:

(*a*) We note that if X_t is q.m. continuous, that is, if $R(t,s)$ is continuous, then $f(t)$ being piecewise continuous on $[a,b]$ is sufficient to ensure the existence of $\int_a^b f(t) X_t \, dt$.

(*b*) The q.m. integral can easily be generalized to include infinite intervals by taking q.m. limit as one or both of the endpoints goes to infinity.

(*c*) A common way in which q.m. integrals arise is when a process $\{Y_t, t \in T'\}$ is generated from a second process $\{X_t, t \in T\}$ according to the formula

$$Y_t = \int_T h(t,s) X_s \, ds \qquad t \in T' \tag{3.12}$$

Equation (3.12) often admits the interpretation of a linear system with input X_t, output Y_t, and impulse response $h(t,s)$.

(*d*) From the point of view of applications one might well prefer $\int_T f(t) X_t(\omega) \, dt$ to be defined as the Lebesgue integral of a sample function $\{X_t(\omega), t \in T\}$. The existence of such a Lebesgue integral is ensured if the process is a measurable process and if

$$\int_T |f(t)| E|X_t| \, dt < \infty$$

(cf. discussion at the end of Sec. 2.2).

4. ORTHOGONAL EXPANSIONS

A family $\mathfrak{F}$ of elements of $\mathcal{H}_X$ is said to be an **orthonormal** (O-N) family if any two distinct elements Y and Z of $\mathfrak{F}$ satisfy

$$\begin{aligned} \|Y\| &= 1 = \|Z\| \\ \langle Y,Z\rangle &= EY\bar{Z} = 0 \end{aligned} \tag{4.1}$$

The second of these conditions is called orthogonality. An O-N family $\mathfrak{F}$ is said to be **complete** in $\mathcal{H}_X$ if there exists no element of $\mathcal{H}_X$, except the zero element, which is orthogonal to every element of $\mathfrak{F}$.

Suppose that $\{X_t, t \in T\}$ is a q.m. continuous process, where T is an interval, finite or infinite. Let T' be the set of all rational points in T. For every $t \in T$, there exists a sequence $\{t_n\}$ in T' such that $t_n \xrightarrow[n\to\infty]{} t$. Since $\{X_t, t \in T\}$ is q.m. continuous, we have for every t,

$$X_t = \lim_{n\to\infty} \text{ in q.m. } X_{t_n}$$

It follows that every element in $\mathcal{H}_X$ is a linear combination of $\{X_t, t \in T'\}$ or the q.m. limit of a sequence of such linear combinations. In short, the countable family $\{X_t, t \in T'\}$ is dense in $\mathcal{H}_X$. It follows that every O-N family in $\mathcal{H}_X$ is at most countable [see, e.g., Taylor, 1961, Sec. 3.2].

For a q.m. continuous process $\{X_t, t \in T\}$, let $\{Z_n, n = 1, 2, \ldots\}$ be an O-N family in $\mathcal{H}_X$. If $Y \in \mathcal{H}_X$, then

$$E\left| Y - \sum_{n=1}^{N} \langle Y,Z_n\rangle Z_n \right|^2 = E|Y|^2 - \sum_{n=1}^{N} |\langle Y,Z_n\rangle|^2 \geq 0$$

Therefore,

$$\infty > E|Y|^2 \geq \sum_{n=1}^{\infty} |\langle Y,Z_n\rangle|^2$$

so that $\sum_{n=1}^{\infty} \langle Y,Z_n\rangle Z_n$ is well defined and

$$Y - \sum_{n=1}^{\infty} \langle Y,Z_n\rangle Z_n$$

is orthogonal to every Z_n. It follows that if $\{Z_n\}$ is complete in $\mathcal{H}_X$, then every Y in $\mathcal{H}_X$ has the representation

$$Y = \sum_{n=1}^{\infty} \langle Y,Z_n\rangle Z_n \tag{4.2}$$

Suppose that $\{Z_n, n = 1, 2, \ldots\}$ is a given complete O-N family in $\mathcal{H}_X$ and we set

$$\sigma_n(t) = \langle X_t,Z_n\rangle \tag{4.3}$$

Then, from (4.2) we have

$$X_t = \sum_{n=1}^{\infty} \sigma_n(t) Z_n \qquad t \in T \tag{4.4}$$

The functions $\sigma_n(t)$, $t \in T$, are continuous, because $\{X_t, t \in T\}$ is q.m. continuous. Further, the set of functions $\{\sigma_n, n = 1, 2, \ldots\}$ is also **linearly independent,** i.e., for every N

$$\sum_{n=1}^{N} \alpha_n \sigma_n(t) = 0 \qquad \text{for all } t \in T$$

implies $\alpha_n = 0$, $n = 1, \ldots, N$. The linear independence of $\{\sigma_n, n = 1, 2, \ldots\}$ is due to the fact that $\sum_{n=1}^{N} \alpha_n \sigma_n(t) = 0$ for all $t \in T$ implies that $\sum_{n=1}^{N} \alpha_n Z_n$ is orthogonal to X_t for every $t \in T$, hence, also orthogonal to every Z_n which implies $\alpha_n = 0$ for every n. It follows from (4.4) that

$$EX_t\bar{X}_s = R(t,s) = \sum_{n=1}^{\infty} \sigma_n(t)\bar{\sigma}_n(s) \qquad t, s \in T \tag{4.5}$$

Conversely, suppose that $\{\sigma_n(t), t \in T, n = 1, 2, \ldots\}$ is a linearly independent family of continuous functions such that

$$R(t,s) = \sum_{n=1}^{\infty} \sigma_n(t)\bar{\sigma}_n(s) \qquad t, s \in T$$

Then, it follows from a very general representation theorem of Karhunen [Karhunen, 1947] that there exists a complete O-N family $\{Z_n, n = 1, 2, \ldots\}$ in $\mathcal{H}_X$ such that

$$X_t(\omega) = \sum_{n=1}^{\infty} \sigma_n(t) Z_n(\omega) \qquad t \in T$$

Thus, (4.4) and (4.5) imply each other. Representations of the form (4.4) are useful because they permit the continuum of random variables $\{X_t, t \in T\}$ to be represented by a countable number of orthonormal random variables $\{Z_n\}$. However, their use is, in general, limited by the fact that it is usually difficult to express the random variables Z_n explicitly in terms of $\{X_t, t \in T\}$. An exceptional case is when $\{\sigma_n\}$ are orthogonal, that is, $\int_T \sigma_m(t)\bar{\sigma}_n(t)\, dt = 0$ whenever $m \neq n$. This motivates the expression widely known as the **Karhunen-Loève expansion.**

Consider a q.m. continuous process $\{X_t, a \le t \le b\}$, where the parameter set is explicitly assumed to be a closed and finite interval.

Suppose that there exists an expansion of the form

$$X_t(\omega) = \sum_{n=1}^{\infty} \sigma_n(t) Z_n(\omega) \qquad a \le t \le b \tag{4.6}$$

where $\{Z_n\}$ and $\{\sigma_n\}$ satisfy

$$EZ_m \bar{Z}_n = \delta_{mn} \tag{4.7}$$

$$\int_a^b \sigma_m(t)\bar{\sigma}_n(t)\, dt = \lambda_n \,\delta_{mn} \tag{4.8}$$

Now, from (4.5), the covariance function R must satisfy

$$R(t,s) = \sum_{n=1}^{\infty} \sigma_n(t)\bar{\sigma}_n(s) \tag{4.9}$$

for each (t,s) in $[a,b] \times [a,b]$. Now from the Schwarz inequality and the fact that R is continuous on $[a,b] \times [a,b]$ we have

$$\sup_{a \le t,s \le b} \left| \sum_{n=1}^{N} \sigma_n(t)\bar{\sigma}_n(s) \right| \le \sup_{a \le t \le b} \sum_{n=1}^{N} |\sigma_n(t)|^2 \le \sup_{a \le t \le b} R(t,t) < \infty \tag{4.10}$$

Therefore, the convergence in (4.9) is bounded. It follows that for every m

$$\int_a^b R(t,s)\sigma_m(s)\, ds = \lambda_m \sigma_m(t) \qquad a \le t \le b \tag{4.11}$$

What we have shown is that if an expansion (4.6), satisfying (4.7) and (4.8), exists, then $\{\sigma_n\}$ must be solutions to the integral equation (4.11). We shall see that under our assumptions, such an expansion always exists.

The above considerations suggest that we investigate integral equations of the form of (4.11). Fortunately, such equations are well known. We shall now summarize some of the important facts concerning it. First, we shall introduce a few definitions. Consider the integral equation

$$\int_a^b R(t,s)\varphi(s)\, ds = \lambda\varphi(t) \qquad a \le t \le b \tag{4.12}$$

where we have explicitly denoted the interval T by $[a,b]$. We assume that $[a,b]$ is finite and $R(t,s)$ is continuous on $[a,b] \times [a,b]$. A nonzero number λ, for which there exists a φ satisfying both (4.12) and the condition $\int_a^b |\varphi(t)|^2\, dt < \infty$, is called an **eigenvalue** of the integral equation. The corresponding φ is called an **eigenfunction.**

1. Any eigenvalue of (4.12) must be real and positive. The fact that λ is real follows from the Hermitian symmetry $R(t,s) = \bar{R}(s,t)$. The fact that λ is positive follows from the nonnegative definiteness of R.

2. There is at least one eigenvalue for (4.12), if R is not identically zero. The largest eigenvalue λ_0 is given by

$$\lambda_0 = \max_{\|\varphi\|=1} \iint_a^b R(t,s)\varphi(s)\bar{\varphi}(t)\, ds\, dt$$
$$\|\varphi\| = \left[\int_a^b |\varphi(t)|^2\, dt\right]^{\frac{1}{2}} \tag{4.13}$$

This fact is not easily proved. It depends on both the nonnegative definiteness of R and its continuity [see, e.g., Taylor, 1961, pp. 334–336].

3. Let $\varphi_0(t)$ denote the normalized eigenfunction corresponding to λ_0, then $\varphi_0(t)$ is continuous on $[a,b]$. This is because we can write

$$\varphi_0(t) = \frac{1}{\lambda_0}\int_a^b R(t,s)\varphi_0(s)\, ds \qquad a \le t \le b \tag{4.14}$$

and the continuity of $\varphi_0(t)$ follows from the continuity of $R(t,s)$ in t.

4. Let $R_1(t,s) = R(t,s) - \lambda_0\varphi_0(t)\bar{\varphi}_0(s)$. Then $R_1(t,s)$ is both continuous and nonnegative definite. The continuity of R_1 is obvious from 3. To show nonnegative definiteness, let

$$Y_t = X_t - \varphi_0(t)\int_a^b X_\tau\varphi_0(\tau)\, d\tau \tag{4.15}$$

Then, we have

$$\begin{aligned} EY_t\bar{Y}_s = EX_t\bar{X}_s &- E\left[X_t\bar{\varphi}_0(s)\int_a^b \bar{X}_\sigma\varphi_0(\sigma)\, d\sigma\right] \\ &- E\left[\bar{X}_s\varphi_0(t)\int_a^b X_\tau\bar{\varphi}_0(\tau)\, d\tau\right] \\ &+ E\left[\varphi_0(t)\bar{\varphi}_0(s)\iint_a^b X_\tau X_\sigma\bar{\varphi}_0(\tau)\varphi_0(\sigma)\, d\tau\, d\sigma\right] \\ = R(t,s) &- \lambda_0\varphi_0(t)\bar{\varphi}_0(s) \\ = R_1(t,s) & \end{aligned} \tag{4.16}$$

Hence, $R_1(t,s)$, being a covariance function, must be nonnegative definite.

5. We observe that $\int_a^b R_1(t,s)\varphi_0(s)\, ds = 0$. Therefore, if we repeat step 2 and obtain λ_1, φ_1, then

$$\int_a^b R_1(t,s)\varphi_1(s)\, ds = \lambda_1\varphi_1(t) \qquad a \le t \le b$$

It follows that $\int_a^b \varphi_1(t)\bar{\varphi}_0(t)\, dt = \frac{1}{\lambda}\int_a^b \varphi_1(s)\left[\overline{\int_a^b R_1(s,t)\varphi_0(t)\, dt}\right] ds = 0$

so that φ_1 is orthogonal to φ_0. In addition,

$$\begin{aligned} \int_a^b R(t,s)\varphi_1(s)\, ds &= \int_a^b R_1(t,s)\varphi_1(s)\, ds - \lambda_0\varphi_0(t)\int_a^b \varphi_1(s)\bar{\varphi}_0(s)\, ds \\ &= \lambda_1\varphi_1(t) \qquad a \le t \le b \end{aligned} \tag{4.17}$$

In other words, λ_1 and φ_1 are eigenvalue and eigenfunction of (4.12).

6. It is clear that the above procedure can be iterated, and we get a nonincreasing sequence of eigenvalues $\lambda_0, \lambda_1, \ldots$ and a corresponding sequence of eigenfunctions $\varphi_0, \varphi_1, \ldots$ which are orthonormal, that is,

$$\int_a^b \varphi_n(t)\bar{\varphi}_m(t)\, dt = \delta_{mn} \tag{4.18}$$

7. The sequence $\lambda_0, \lambda_1, \ldots,$ may terminate after a finite number of terms, in which case we have

$$R(t,s) = \sum_{n=0}^{N} \lambda_n \varphi_n(t)\bar{\varphi}_n(s) \tag{4.19}$$

If the sequence $\lambda_0, \lambda_1, \ldots$ does not terminate, then $\lambda_n \xrightarrow[n\to\infty]{} 0$ (see Exercise 2).

8. If the number of eigenvalues is infinite, then

$$\lim_{N\to\infty} \sup_{a\le t,s\le b} \left| R(t,s) - \sum_{n=0}^{N} \lambda_n \varphi_n(t)\bar{\varphi}_n(s) \right| = 0 \tag{4.20}$$

In other words,

$$\lim_{N\to\infty} \sum_{n=0}^{N} \lambda_n \varphi_n(t)\bar{\varphi}_n(s) = R(t,s) \qquad \text{uniformly on } [a,b]^2 \tag{4.21}$$

This important result is known as the **Mercer's theorem** [Riesz and Sz.-Nagy, 1952, p. 245]. The fact that the convergence is uniform is a strong result and immediately implies convergence in mean square, that is,

$$\lim_{N\to\infty} \iint_a^b \left| R(t,s) - \sum_{n=0}^{N} \lambda_n \varphi_n(t)\bar{\varphi}_n(s) \right|^2 dt\, ds = 0 \tag{4.22}$$

9. In general, the O-N family $\{\varphi_n\}$ is not complete in the space $L_2(a,b)$. The most that can be said is that given $f \in L_2(a,b)$ (that is,

$$\int_a^b |f(t)|^2\, dt < \infty)$$

we can write

$$f(t) = f_0(t) + \underset{N\to\infty}{\text{l.i.m.}} \sum_{n=0}^{N} \varphi_n(t) \int_a^b f(s)\bar{\varphi}_n(s)\, ds \tag{4.23}$$

where l.i.m. means limit in mean square and f_0 satisfies

$$\int_a^b R(t,s) f_0(s)\, ds = 0 \qquad \text{for all } t \in [a,b] \tag{4.24}$$

It follows that $\{\varphi_n\}$ is complete in $L_2(a,b)$, if and only if (4.24) implies $\int_a^b |f_0(t)|^2\, dt = 0$.

We are now in a position to state precisely the theorem concerning the biorthogonal series for a second-order process. This expansion is often referred to as the Karhunen-Loève expansion [Loève, 1963, pp. 478–479].

Proposition 4.1. Let $\{X(\omega,t),\ t \in [a,b]\}$ be a q.m. continuous second-order process with covariance function $R(t,s)$.

(*a*) If $\{\varphi_n\}$ are the orthonormal eigenfunctions of

$$\int_a^b R(t,s)\varphi(s)\,ds = \lambda\varphi(t) \qquad a \le t \le b \tag{4.25}$$

and $\{\lambda_n\}$ the eigenvalues, then

$$X(\omega,t) = \lim_{N\to\infty} \text{in q.m.} \sum_{n=0}^{N} \sqrt{\lambda_n}\,\varphi_n(t)b_n(\omega) \qquad \text{uniformly for } a \le t \le b \tag{4.26}$$

where $\{b_n\}$ satisfy

$$b_n(\omega) \overset{\text{q.m.}}{=} \sqrt{\lambda_n} \int_a^b \bar{\varphi}_n(t)X(\omega,t)\,dt \tag{4.27}$$

and

$$Eb_m\bar{b}_n = \delta_{mn} \tag{4.28}$$

(*b*) Conversely, if $X(\omega,t)$ has an expansion of the form (4.26) with $\int_a^b \varphi_m(t)\bar{\varphi}_n(t)\,dt = \delta_{mn} = Eb_m\bar{b}_n$, then $\{\varphi_n\}$ and $\{\lambda_n\}$ must be eigenfunctions and eigenvalues of (4.25).

Proof:

(*a*) By direct computation we have

$$E\left| X_t - \sum_{n=0}^{N} \sqrt{\lambda_n}\,\varphi_n(t)b_n \right|^2 = R(t,t) - \sum_{n=0}^{N} \lambda_n|\varphi_n(t)|^2$$

which goes to zero as $N \to \infty$ uniformly in t by virtue of the Mercer's theorem.

(*b*) Suppose X_t has the stated expansion. Then, we have

$$R(t,s) = EX_t\bar{X}_s = \sum_{n=0}^{\infty} \lambda_n\varphi_n(t)\bar{\varphi}_n(s)$$

Hence,

$$\int_a^b R(t,s)\varphi_m(s)\,ds = \int_a^b \sum_{n=0}^{\infty} \lambda_n\varphi_n(t)\bar{\varphi}_n(s)\varphi_m(s)\,ds$$
$$= \lambda_n\varphi_m(t) \qquad a \le t \le b$$

The proof is complete. ▮

Consider now an example of how an integral equation of the form of (4.25) might be solved. Let $R(t,s) = \min(t,s)$ and consider

$$\int_0^T \min(t,s)\varphi(s)\,ds = \lambda\varphi(t) \qquad 0 \leq t \leq T \tag{4.29}$$

or

$$\int_0^t s\varphi(s)\,ds + t\int_t^T \varphi(s)\,ds = \lambda\varphi(t)$$

Differentiating once with respect to t yields

$$\int_t^T \varphi(s)\,ds = \lambda\dot{\varphi}(t) \qquad 0 < t < T \tag{4.30}$$

Differentiating again, we find

$$-\varphi(t) = \lambda\ddot{\varphi}(t) \qquad 0 < t < T \tag{4.31}$$

We also have the obvious boundary conditions $\varphi(0) = 0$ and $\dot{\varphi}(T) = 0$. Equation (4.31) with initial condition $\varphi(0) = 0$ yields

$$\varphi(t) = A \sin \frac{1}{\sqrt{\lambda}}\,t$$

Applying the condition $\dot{\varphi}(T) = 0$ obtained from (4.30), we find

$$\cos \frac{1}{\sqrt{\lambda}}\,T = 0$$

In other words the eigenvalues are given by

$$\lambda_n = \frac{T^2}{(n + \frac{1}{2})^2\pi^2} \qquad n = 0, 1, 2, \ldots \tag{4.32}$$

The normalized eigenfunctions are given by

$$\varphi_n(t) = \sqrt{\frac{2}{T}} \sin\left[(n + \tfrac{1}{2})\pi\left(\frac{t}{T}\right)\right] \tag{4.33}$$

It is rather interesting to note that because of Mercer's theorem, we have

$$\min(t,s) = \frac{2}{T}\sum_{n=0}^{\infty} \frac{T^2}{\pi^2(n + \frac{1}{2})^2} \sin\left[(n + \tfrac{1}{2})\pi\left(\frac{t}{T}\right)\right] \sin\left[(n + \tfrac{1}{2})\pi\left(\frac{s}{T}\right)\right] \qquad \text{uniformly on } [0,T]^2 \tag{4.34}$$

which is by no means an obvious result.

Analytical solution to the integral equation (4.25) is, in general, difficult. For some special cases, the integral equation can be reduced to a differential equation and then solved in a way similar to the example above. In particular, if the covariance function $R(t,s)$ has the form

$$R(t,s) = \int_{-\infty}^{\infty} e^{i2\pi\lambda(t-s)}\phi(\lambda)\, d\lambda \tag{4.35}$$

where $\phi(\lambda)$ is a ratio of polynomials in λ, then the integral equation can be reduced to a differential equation with constant coefficient. The details are given in Davenport and Root [1958, pp. 241–242].

Before leaving the subject of orthogonal expansion, consider a second example. Suppose $EX_t\bar{X}_s = R(t,s) = \cos 2\pi(t-s)$. Then we can show that the Karhunen-Loève expansion has only two terms. In fact, any expansion in terms of an orthonormal basis in $\mathcal{H}_X$ has only two terms. Furthermore, in this instance, the integral equation (4.25) can be solved rather easily. To show all this, we first observe that

$$EX_0\bar{X}_{\frac{1}{4}} = 0$$
$$E|X_0|^2 = E|X_{\frac{1}{4}}|^2 = 1$$

Therefore $\{X_0, X_{\frac{1}{4}}\}$ is an O-N family in $\mathcal{H}_X$. Next, we observe that

$$E|X_t - X_0 \cos 2\pi t - X_{\frac{1}{4}} \sin 2\pi t|^2 = 0$$

Therefore, $\{X_0, X_{\frac{1}{4}}\}$ is also complete. This means that $\mathcal{H}_X$ is two dimensional. Hence, the Karhunen-Loève expansion has only two terms. The eigenvalues and eigenfunctions depend on the interval T. Suppose the interval is $[0,k]$, where k = integer. Then, the eigenfunction can be taken to be $\{(1/\sqrt{k})e^{i2\pi t}, (1/\sqrt{k})e^{-i2\pi t}\}$ or $\{(2/\sqrt{k}) \cos 2\pi t, (2/\sqrt{k}) \sin 2\pi t\}$. If the interval $[0,T]$ is not integral, then the eigenfunctions can be taken to be suitable linear combinations of $\sin 2\pi t$ and $\cos 2\pi t$.

5. WIDE-SENSE STATIONARY PROCESSES

A second-order process $\{X_t, -\infty \le t \le \infty\}$ is said to be wide-sense stationary if its covariance function is a function of only the time difference, that is,

$$EX_t\bar{X}_s = R(t-s) \tag{5.1}$$

We recall once again that the mean EX_t is assumed to be zero throughout this chapter. Wide-sense stationarity means that

$$EX_{t+t_0}\bar{X}_{s+t_0} = EX_t\bar{X}_s \tag{5.2}$$

for any t_0. As before, let $\mathcal{H}_X$ denote the Hilbert space containing all linear combinations of $\{X_t, -\infty < t < \infty\}$. Suppose we define a linear operator U_t for t, mapping $\mathcal{H}_X$ onto $\mathcal{H}_X$, as follows:

$$U_tX_s = X_{t+s} \tag{5.3a}$$
$$\|Z_n - Z\| \xrightarrow[n\to\infty]{} 0 \quad \text{implies} \quad \|U_tZ_n - U_tZ\| \xrightarrow[n\to\infty]{} 0 \tag{5.3b}$$

It is clear from the definition of $\mathcal{H}_X$ (3.1) that U_tZ is well defined for every Z in $\mathcal{H}_X$ by (5.3). Equation (5.2) can now be rewritten as

$$\langle U_{t_0}X_t, U_{t_0}X_s\rangle = \langle X_t, X_s\rangle \qquad \forall t_0, t, s$$

which is easily extended to

$$\langle U_{t_0}Z, U_{t_0}Y\rangle = \langle Z, Y\rangle \qquad \begin{matrix} -\infty < t_0 < \infty \\ Y, Z \in \mathcal{H}_X \end{matrix} \tag{5.4}$$

Equation (5.4) means that for each t, U_t is a unitary operator. It is also easy to see that $U_tU_s = U_{t+s}$, $U_0 = I$, $U_t^{-1} = U_{-t}$, so that $\{U_t, -\infty < t < \infty\}$ is a **translation group.** Summarizing these results, we can state the following.

Every wide-sense stationary process $\{X_t, -\infty < t < \infty\}$ can be represented as

$$X_t = U_tX_0 \tag{5.5}$$

where $\{U_t, -\infty < t < \infty\}$ is a translation group of unitary operators mapping $\mathcal{H}_X$ onto $\mathcal{H}_X$. This result should be compared with the corresponding result for the strictly stationary process given in Sec. 2.6.

The invariance of covariance function with respect to time shifts suggests that harmonic analysis (i.e., representation in terms of complex sinusoids $e^{i2\pi\nu t}$) should play a useful role in the theory of wide-sense stationary processes. To those who are familiar with application of Fourier integrals to the analysis of time-invariant linear systems, this is certainly no surprise. We begin our discussion with a brief review of Fourier integrals and their application to the analysis of linear time-invariant systems. Let L_p, C_0, and $\mathcal{S}$ denote, respectively, the following function spaces of Lebesgue measurable complex functions defined on $(-\infty, \infty)$:

L_p is the space of all functions satisfying[1]

$$\int_{-\infty}^{\infty} |f(t)|^p\, dt < \infty \qquad 1 \le p < \infty \tag{5.6a}$$

C_0 is the space of all bounded-continuous functions such that

$$f(t) \xrightarrow[|t|\to\infty]{} 0 \tag{5.6b}$$

$\mathcal{S}$ is the space of all infinitely differentiable functions such that for any integers m and n,

$$|t|^m \left|\frac{d^nf(t)}{dt^n}\right| \xrightarrow[|t|\to\infty]{} 0 \tag{5.6c}$$

[1] Strictly speaking, L_p is a space of equivalence classes. Two functions f_1 and f_2 belong to the same equivalence class if and only if $\int_{-\infty}^{\infty} |f_1(t) - f_2(t)|^p\, dt = 0$.

The spaces L_p and C_0 are complete normed linear spaces (Banach spaces) with respective norms $\left[\int_{-\infty}^{\infty} |f(t)|^p \, dt\right]^{1/p}$ and $\sup_t |f(t)|$. The space $\mathcal{S}$ is **dense** in both L_p and C_0. That is, the completion of $\mathcal{S}$ with respect to the norm of L_p is L_p, and its completion with respect to the norm of C_0 is C_0. Therefore, for every t in L_p, we can find a sequence $\{f_n\}$ in $\mathcal{S}$ such that

$$\int_{-\infty}^{\infty} |f(t) - f_n(t)|^p \, dt \xrightarrow[n\to\infty]{} 0$$

and for every f in C_0, we can find $\{f_n\}$ in $\mathcal{S}$ such that

$$\sup_t |f_n(t) - f(t)| \xrightarrow[n\to\infty]{} 0$$

For $f \in L_1$, the **Fourier integral** (or Fourier transform)

$$\hat{f}(\nu) = \int_{-\infty}^{\infty} e^{-i2\pi\nu t} f(t) \, dt \tag{5.7}$$

is an absolutely convergent integral for each $\nu \in (-\infty, \infty)$ and defines a function $\hat{f}$ in C_0. If $f \in L_1$ and is of bounded variation in a neighborhood of t, then the **inversion formula** is given by

$$\frac{f(t-0) + f(t+0)}{2} = \lim_{N\to\infty} \int_{-N}^{N} e^{i2\pi\nu t} \hat{f}(\nu) \, d\nu \tag{5.8}$$

where $f(t-0)$ and $f(t+0)$ denote, respectively, the limit from the left and the limit from the right of f at t. The right-hand side of (5.8) is often written simply as $\int_{-\infty}^{\infty} e^{i2\pi\nu t} \hat{f}(\nu) \, d\nu$, although the integral is not absolutely convergent unless $\hat{f}$ is also in L_1.

Roughly speaking, if $f \in L_1$, L_2, or $\mathcal{S}$, we have a pair of equations relating a function f and its Fourier transform $\hat{f}$:

$$\begin{aligned} \hat{f}(\nu) &= \int_{-\infty}^{\infty} e^{-i2\pi\nu t} f(t) \, dt \\ f(t) &= \int_{-\infty}^{\infty} e^{i2\pi\nu t} \hat{f}(\nu) \, d\nu \end{aligned} \tag{5.9}$$

These two equations are nearly identical, the only difference being the terms $e^{\pm i2\pi\nu t}$ in the integrals. By convention, $\hat{f}$ is called the Fourier transform of f, and f is called the **inverse Fourier transform** of $\hat{f}$. Depending on f, one or both of the integrals in (5.9) may have to be defined as the limit in some sense of a sequence of finite integrals, and the equality may only hold in a restricted sense. For example, if $f \in L_1$ and is of bounded variation, then the first integral is absolutely convergent, but the second equation, strictly speaking, should be replaced by (5.8). If $f \in \mathcal{S}$, then $\hat{f}$ is also in $\mathcal{S}$. In this case both integrals are absolutely convergent, and equality holds for every ν and t in $(-\infty, \infty)$. If $f \in L_2$, then the first equation really says that there exists $\hat{f} \in L_2$ such that

$$\int_{-\infty}^{\infty} \left| \hat{f}(\nu) - \int_{-T_1}^{T_2} f(t) e^{-i2\pi\nu t}\, dt \right| d\nu \xrightarrow[T_1, T_2 \to \infty]{} 0$$

The second equation is also interpreted in a similar way. When we use (5.9), it is understood that these equations are subject to such qualifications.

The convolution product between two functions f and h, when it exists, is defined by[1]

$$(f * h)(t) = \int_{-\infty}^{\infty} f(t-s) h(s)\, ds \tag{5.10}$$

We note that if f and h are in $\mathcal{S}$, then $f * h$ is also in $\mathcal{S}$. The convolution product is symmetric in f and h. This is easily seen by a change in the variable of integration on the right-hand side of (5.10). If we denote Fourier transformation by $\hat{\ }$, then a most important property of convolution is given by the relationship

$$\widehat{f * h} = \hat{f}\hat{h} \tag{5.11}$$

Equations (5.10) and (5.11) are well-known representations of linear time-invariant filtering operations. Roughly speaking, a linear filter is a linear mapping of some input function space V_i into some output function space V_0. Suppose we define time shift T_α by

$$(T_\alpha f)(t) = f(t + \alpha) \tag{5.12}$$

A linear filter A is said to be time invariant if for every $\alpha \in (-\infty, \infty)$,

$$AT_\alpha f = T_\alpha A f$$

If the input space V_i is L_1, and a filter A is defined by $Af = h * f$, then A is linear and time invariant. The function h is known as the impulse response of the filter A. The Fourier transform $\hat{h}$ of the impulse response is known as the transfer function. More generally, suppose a filter A is defined by

$$(Af)(t) = \int_{-\infty}^{\infty} e^{-i2\pi\nu t} \hat{h}(\nu) f(\nu)\, d\nu \tag{5.13}$$

Then, A is again time invariant and linear. The function $\hat{h}$ is again called a transfer function. In general, $\hat{h}(\nu)$ may not be the Fourier integral of any impulse response. To put it in another way, the inverse Fourier transform of $\hat{h}$ may not exist except as a generalized function. We have been deliberately vague in specifying the input space V_i and the output space V_0, because they depend very much on the filter. For example,

[1] If f, $h \in L_1$, then $f * h \in L_1$. If f, $h \in L_2$, then $f * h$ is bounded. If $f \in L_1$ and h is bounded, then $f * h$ is again bounded.

$\hat{h}(\nu) = 2\pi i\nu$ represents the transfer function of a differentiator. Clearly, the input to a differentiator must be differentiable. In any event, our main interest here is not the filtering of known t functions, but stochastic processes.

Let $\{X_t, \ -\infty < t < \infty\}$ be a q.m. continuous and wide-sense stationary process with covariance function $R(\tau)$. Assume $R \in L_1$, then clearly R also belongs to $L_1 \cap C_0$ because of the q.m. continuity of X_t. Let S be the Fourier transform of R defined by

$$S(\nu) = \int_{-\infty}^{\infty} e^{-i2\pi\nu t} R(\tau)\, d\tau \qquad -\infty < \nu < \infty \tag{5.14}$$

The function S is nonnegative because of the nonnegative definiteness of R (to be proved), and if $R \in L_1 \cap C_0$ as is assumed, then $S \in L_1 \cap C_0$, also. The inversion integral gives us

$$R(\tau) = \int_{-\infty}^{\infty} e^{i2\pi\nu t} S(\nu)\, d\nu \tag{5.15}$$

The function S is called the **spectral-density function** for the process $\{X_t, \ -\infty < t < \infty\}$, and has the interpretation of being the density of average power distribution per unit frequency. To make this notion precise, we need to introduce the concept of linear time-invariant filtering of wide-sense stationary processes.

Imagine a time-invariant linear filter characterized by an impulse response function h, and suppose that the input is a sample function $X_t(\omega)$, $-\infty < t < \infty$, of a wide-sense stationary process. It is natural to view the output $Y_t(\omega)$, $-\infty < t < \infty$, as a sample function of another second-order process and write

$$Y_t(\omega) = \int_{-\infty}^{\infty} h(t - \tau) X_\tau(\omega)\, d\tau \tag{5.16}$$

The difficulty is that if the convolution integral is to be viewed as an absolutely convergent Lebesgue integral involving a sample function of the X process, then we need to impose conditions on the X process which are not in the province of second-order properties. Mathematically, it is much neater to regard the integral in (5.16) as a q.m. integral as defined in Sec. 3. From Proposition 3.1 we have that for the integral in (5.16) to exist as a q.m. integral, it is necessary and sufficient that

$$\iint_{-\infty}^{\infty} h(t - \tau)\bar{h}(t - \sigma) R(\tau - \sigma)\, d\tau\, d\sigma$$

exists as a Riemann integral. If $R \in L_1 \cap C_0$ as was assumed, and h

is square integrable, then

$$\iint_{-\infty}^{\infty} h(t-\tau)h(t-\sigma)R(\tau-\sigma)\,d\tau\,d\sigma$$
$$= \int_{-\infty}^{\infty} |\hat{h}(\nu)|^2 S(\nu)\,d\nu < \infty \tag{5.17}$$

and the existence of (5.16) as a q.m. integral is ensured. We should not lose sight of the fact that to treat (5.16) as a q.m. integral is a mathematical convenience. In applications, since we never observe more than one sample function at a time, filtering operations should be interpreted as being performed on a sample function. Thus, in practice, we really want (5.16) to be interpreted both as a q.m. integral, which is the limit of a q.m. convergent sequence of random variables, and as a Lebesgue integral for almost all sample functions. In this chapter, we focus our attention only on the former interpretation (cf. Sec. 2.2).

With $\{Y_t, -\infty < t < \infty\}$ defined by (5.16) with h square integrable, then

$$\begin{aligned} EY_t\bar{Y}_s &= \iint_{-\infty}^{\infty} h(t-\tau)\bar{h}(s-\tau)R(\tau-\sigma)\,d\tau\,d\sigma \\ &= \int_{-\infty}^{\infty} e^{i2\pi\nu(t-s)}|\hat{h}(\nu)|^2 S(\nu)\,d\nu \end{aligned} \tag{5.18}$$

which means that (1) $\{Y_t, -\infty < t < \infty\}$ is again wide-sense stationary, and (2) the spectral-density function of $\{Y_t, -\infty < t < \infty\}$ is given by

$$S_Y(\nu) = |\hat{h}(\nu)|^2 S(\nu) \tag{5.19}$$

If we take $\hat{h}(\nu)$ to be the indicator function for $[\nu_1,\nu_2]$, then from the usual interpretation of filtering, Y process is just those components of X process lying in the frequency range $\nu_1 \le \nu \le \nu_2$. Since

$$\begin{aligned} E|Y_t|^2 &= \int_{-\infty}^{\infty} |h(\nu)|^2 S(\nu)\,d\nu \\ &= \int_{\nu_1}^{\nu_2} S(\nu)\,d\nu \end{aligned} \tag{5.20}$$

it follows that $\int_{\nu_1}^{\nu_2} S(\nu)\,d\nu$ is just the average power of the X process in $[\nu_1,\nu_2]$. This justifies our earlier assertion that S is nonnegative and that it has the interpretation of average power per unit frequency.

Equation (5.15) is a representation of the covariance function as a positive linear combination of sinusoids. This idea can be generalized to situations where $R \notin L_1$ and the spectral-density function may not exist. Intuitively, the situation is quite simple. When $R \in L_1 \cap C_0$, the average power is smoothly distributed over the continuum of all frequencies $(-\infty,\infty)$, with no single frequency having a finite amount of power.

In the more general situation, there may be spectral lines, i.e., distinct frequencies with finite amount of power. Even more complicated situations involving continuous, but not absolutely continuous, distributions may arise. The general statement concerning the harmonic representation of a stationary covariance function is given by the **Bochner's theorem** stated below.

Proposition 5.1. A function $R(\tau)$, $-\infty < \tau < \infty$, is the covariance function of a q.m. continuous and wide-sense stationary process if and only if it is of the form

$$R(\tau) = \int_{-\infty}^{\infty} e^{i2\pi\nu\tau} F(d\nu) \tag{5.21}$$

where F is a finite Borel measure on the real line $(-\infty, \infty)$ and is called the **spectral measure.**

Remarks:

(*a*) We note that $[1/R(0)]F$ is a probability measure defined on the σ algebra of Borel sets of the real line.

(*b*) The integral in (5.21) is defined exactly as the expectation was defined in (Sec. 1.5).

(*c*) Alternatively, if we define a **spectral-distribution function** ϕ by

$$\phi(\nu) = F((-\infty, \nu)) \tag{5.22}$$

then ϕ is a bounded nondecreasing function, and the integral in (5.21) can be replaced by a Stieltjes integral

$$\int_{-\infty}^{\infty} e^{i2\pi\nu\tau}\, d\phi(\nu)$$

(*d*) Finally, we note that if ϕ is absolutely continuous (that is, F is an absolutely continuous measure with respect to Lebesgue measure), then there exists a nonnegative function S such that

$$\phi(\lambda) = \int_{-\infty}^{\lambda} S(\nu)\, d\nu \tag{5.23}$$

Naturally, $S(\nu)$ is called the **spectral-density function,** agreeing with and generalizing our earlier definition.

Proof: First, if F is a finite Borel measure, then

$$\int_{-\infty}^{\infty} e^{i2\pi\nu(t-s)} F(d\nu) \tag{5.24}$$

is a nonnegative definite function, because it is the limit of a sequence of nonnegative definite functions of the form $\sum_{n=0}^{N} \sigma_n(t)\bar{\sigma}_n(s)$. It is continuous at $t = s$, because

$$\left| \int_{-\infty}^{\infty} (e^{i2\pi\nu\tau} - 1)F(d\nu) \right| \xrightarrow[\tau\to 0]{} 0$$

Hence, it is continuous everywhere on $(-\infty,\infty) \times (-\infty,\infty)$. Therefore, (5.22) defines the covariance function of some q.m. continuous wide-sense stationary process. Thus, the first half of Proposition 5.1 is proved.

Next, suppose $R(\tau)$ is the covariance function of some q.m. continuous and wide-sense stationary process. That is, $R(t - s)$ is a continuous nonnegative definite function on $(-\infty,\infty) \times (-\infty,\infty)$. Now, define a sequence of continuous covariance functions $R_n(\tau)$ by

$$R_n(\tau) = \begin{cases} \left(1 - \dfrac{|\tau|}{2n}\right) R(\tau) & |\tau| \le 2n \\ 0 & |\tau| > 2n \end{cases} \tag{5.25}$$

The fact that $R_n(\tau)$ is a nonnegative definite function follows from the fact that max $(0, 1 - |\tau|/2n)$ is a nonnegative definite function. Now, clearly $R_n \in L_1 \cap C_0$. Therefore, there corresponds a sequence of spectral-density functions $\{S_n\}$ defined by

$$\begin{aligned} S_n(\nu) &= \int_{-\infty}^{\infty} R_n(\tau)e^{-i2\pi\nu\tau}\,d\tau \\ &= \int_{-2n}^{2n} \left(1 - \frac{|\tau|}{2n}\right) R(\tau)e^{-i2\pi\nu\tau}\,d\tau \end{aligned} \tag{5.26}$$

Now, let f and $\hat{f}$ be functions in $\mathcal{S}$ which are Fourier transform pairs,

$$\hat{f}(x) = \int_{-\infty}^{\infty} e^{-i2\pi xy} f(y)\,dy$$

$$f(y) = \int_{-\infty}^{\infty} e^{i2\pi xy} \hat{f}(x)\,dx$$

Define a linear functional ρ on $\mathcal{S}$ by

$$\rho(f) = \int_{-\infty}^{\infty} R(\tau)\hat{f}(\tau)\,d\tau \tag{5.27}$$

Using the dominated-convergence theorem, we have

$$\begin{aligned} |\rho(f)| &= \lim_{n\to\infty} \left| \int_{-\infty}^{\infty} R_n(\tau)\hat{f}(\tau)\,d\tau \right| \\ &= \lim_{n\to\infty} \left| \int_{-\infty}^{\infty} S_n(\nu)f(\nu)\,d\nu \right| \\ &\le \sup_\nu |f(\nu)| \lim_{n\to\infty} \int_{-\infty}^{\infty} S_n(\nu)\,d\nu \\ &= \sup_\nu |f(\nu)| R(0) \end{aligned}$$

This means that ρ is a linear functional defined on a set $\mathcal{S}$ dense in C_0 and bounded with respect to the supremum norm. Thus, ρ can be extended to a bounded-linear functional on C_0, and the Riesz representation theorem [see, e.g., Rudin, 1966, p. 131] states that there exists a bounded σ additive set function F defined on the Borel sets such that

$$\rho(f) = \int_{-\infty}^{\infty} f(\nu)F(d\nu) \tag{5.28}$$

Because R is nonnegative definite, F is nonnegative and is thus a finite Borel measure. Comparing (5.27) and (5.28), we conclude that

$$\int_{-\infty}^{\infty} R(\tau)\hat{f}(\tau)\,d\tau = \int_{-\infty}^{\infty} f(\nu)F(d\nu) \qquad f \in \mathcal{S} \tag{5.29}$$

Now, for a fixed t define

$$f_n(\nu) = \exp\left[-\frac{1}{2}\frac{(2\pi\nu)^2}{n}\right]e^{i2\pi\nu t} \tag{5.30a}$$

then

$$\hat{f}_n(\tau) = \sqrt{\frac{n}{2\pi}}\,e^{-\frac{1}{2}n(t-\tau)^2} \tag{5.30b}$$

It is easy to verify that f_n, $\hat{f}_n$ are in $\mathcal{S}$. If R is continuous, then

$$\int_{-\infty}^{\infty} R(\tau)\hat{f}_n(\tau)\,d\tau \xrightarrow[n\to\infty]{} R(t)$$

and

$$\int_{-\infty}^{\infty} f_n(\nu)F(d\nu) \xrightarrow[n\to\infty]{} \int_{-\infty}^{\infty} e^{i2\pi\nu t}F(d\nu)$$

which proves the second half of Proposition 5.1. ▮

Proposition 5.1 gives an expression of a covariance function in terms of the corresponding spectral measure (alternatively, the spectral-distribution function), but gives no inversion formula expressing the spectral measure F in terms of the covariance function R. Actually, the proof already contains the heart of both the direct and the inversion formulas in (5.29). We need only to choose an approximating sequence of f's or $\hat{f}$'s. For example, suppose $b > a$ are two continuity points of F, that is, a and b are such that

$$F([a,b]) = F((a,b)) = F([a,b)) = F((a,b])$$

Denote the common value of these four quantities by F_{ab}, then F_{ab} can be found from (5.29) by approximating the indicator function of the interval a to b by a sequence in $\mathcal{S}$. Specifically, let

$$\begin{aligned} f_n(\nu) &= \sqrt{\frac{n}{2\pi}}\int_a^b e^{-\frac{1}{2}n(\nu-\nu')^2}\,d\nu' \\ \hat{f}_n(\tau) &= \exp\left[-\frac{1}{2}\frac{(2\pi\tau)^2}{n}\right]\frac{e^{-i2\pi\tau b} - e^{-i2\pi\tau a}}{-2\pi i\tau} \end{aligned} \tag{5.31}$$

Then, from (5.29) we have

$$F_{ab} = \lim_{n\to\infty} \int_{-\infty}^{\infty} f_n(\nu)F(d\nu)$$
$$= \lim_{n\to\infty} \int_{-\infty}^{\infty} \hat{f}_n(\tau)R(\tau)\, d\tau$$

Thus, we have in final form the inversion formula

$$F_{ab} = \lim_{n\to\infty} \int_{-\infty}^{\infty} R(\tau) \frac{e^{-i2\pi\tau b} - e^{-i2\pi\tau a}}{-2\pi i\tau} \exp\left[-\frac{1}{2}\frac{(2\pi\tau)^2}{n}\right] d\tau \qquad (5.32)$$

Equation (5.32) holds only at continuity points of F. However, because $R(0) = F((-\infty,\infty))$ is finite, the number of discontinuities of F which have jumps greater than $1/n$ is, at most, $nR(0)$. Hence, the discontinuity points (mass points) of F are, at most, countable. Thus, the continuity points of F are always dense in $(-\infty,\infty)$. It follows that F is completely determined by (5.32).

We can now repeat the arguments relating to linear time-invariant filtering operations on a wide-sense stationary process. Instead of (5.18), we have for the more general case

$$EY_t\bar{Y}_s = \int_{-\infty}^{\infty} e^{i2\pi\nu(t-s)}|\hat{h}(\nu)|^2F(d\nu)$$

where Y_t is the output of a linear time-invariant filter and F is the spectral measure of the input. We postpone a detailed discusson of this point until later. First, we need to develop a spectral representation, not just for the covariance function as is given by Proposition 5.1, but for the process itself.

6. SPECTRAL REPRESENTATION

Let $\{X_t, -\infty < t < \infty\}$ be a q.m. continuous and wide-sense stationary process. If X_t had a Fourier integral, then the inversion formula would immediately give a representation of X_t as a linear combination of sinusoids. However, X_t does not have a Fourier integral. That is, $\int_{-\infty}^{\infty} e^{-i2\pi\nu t}X_t\, dt$ fails to exist as a q.m. integral or any other kind of integral. To obtain a spectral representation, we introduce a kind of integrated Fourier integral $\{\hat{X}_\lambda, -\infty < \lambda < \infty\}$ as follows:

$$\lim_{a\to-\infty} \text{in q.m. } \hat{X}_a = 0 \qquad (6.1a)$$

If a and b are continuity points of the spectral measure F, we set

$$\hat{X}_b - \hat{X}_a = \int_{-\infty}^{\infty} \left(\int_a^b e^{-i2\pi\nu t}\, d\nu\right) X_t\, dt \qquad (6.1b)$$

At discontinuity points of F, $\hat{X}_\lambda$ is defined so that $\{\hat{X}_\lambda, -\infty < \lambda < \infty\}$ is q.m. continuous from the left (6.1c)

Proposition 6.1. Defined as above, the process $\{\hat{X}_\lambda, -\infty < \lambda < \infty\}$ satisfies

$$E(\hat{X}_b - \hat{X}_a)\overline{(\hat{X}_d - \hat{X}_c)} = F([a,b) \cap [c,d)) \tag{6.2}$$

so that $\{\hat{X}_\lambda, -\infty < \lambda < \infty\}$ is a **process with orthogonal increments.**

Remark: Denoting $d\hat{X}_\lambda = \hat{X}_{\lambda+d\lambda} - \hat{X}_\lambda$ and $F(d\lambda) = F([\lambda, \lambda + d\lambda))$, we can write (6.2) in a differential form as

$$E\, d\hat{X}_\lambda\, \overline{d\hat{X}_\mu} = \delta_{\lambda\mu}F(d\lambda) \qquad \delta_{\lambda\mu} = \begin{cases} 1, & \lambda = \mu \\ 0, & \lambda \neq \mu \end{cases} \tag{6.3}$$

The process $\{\hat{X}_\lambda, -\infty < \lambda < \infty\}$ will be called the **spectral process** of $\{X_t, -\infty < t < \infty\}$. We shall show that

$$X_t = \int_{-\infty}^{\infty} e^{i2\pi\lambda t}\, d\hat{X}_\lambda$$

which is the spectral-representation formula that we are after. However, first we have to define integrals such as $\int_{-\infty}^{\infty} f(\lambda)\, d\hat{X}_\lambda$.

Definition. Let $\{\hat{X}_\lambda, -\infty < \lambda < \infty\}$ be a process with orthogonal increments which is q.m. continuous from the left. Let

$$E\, d\hat{X}_\lambda\, \overline{d\hat{X}_\mu} = \delta_{\lambda\mu}F(d\lambda).$$

(a) If $f = I_{ab}$, the indicator function of $[a,b)$, we set

$$\int_{-\infty}^{\infty} f(\lambda)\, d\hat{X}_\lambda = \hat{X}_b - \hat{X}_a \tag{6.4}$$

(b) If $f = \sum_{\nu=1}^{n} a_\nu f_\nu$, we require

$$\int_{-\infty}^{\infty} f(\lambda)\, d\hat{X}_\lambda = \sum_{\nu=1}^{n} a_\nu \int_{-\infty}^{\infty} f_\nu(\lambda)\, d\hat{X}_\lambda \tag{6.5}$$

(c) If $\int_{-\infty}^{\infty} |f_n(\lambda) - f(\lambda)|^2 F(d\lambda) \xrightarrow[n\to\infty]{} 0$, we require

$$\int_{-\infty}^{\infty} f(\lambda)\, d\hat{X}_\lambda = \lim_{n\to\infty} \text{in q.m.} \int_{-\infty}^{\infty} f_n(\lambda)\, d\hat{X}_\lambda \tag{6.6}$$

It is clear that $\int_{-\infty}^{\infty} f(\lambda)\, d\hat{X}_\lambda$ is well defined for every f for which there exists a sequence of step functions $\{f_n\}$ (linear combinations of indicator functions) such that

$$\int_{-\infty}^{\infty} |f_n(\lambda) - f(\lambda)|^2 F(d\lambda) \xrightarrow[n\to\infty]{} 0 \tag{6.7}$$

Although we won't prove it, the class of all such f is precisely $L_2(F)$, that is, the class of functions satisfying $\int_{-\infty}^{\infty} |f(\lambda)|^2 F(d\lambda) < \infty$. For a continuous function f in $L_2(F)$, a suitable approximating sequence of step functions can be constructed by sampling f at continuity points of F. For an arbitrary f in $L_2(F)$, the construction of an approximating sequence is somewhat more complicated [Doob, 1953, pp. 439–441].

Proposition 6.2. Let $\{\hat{X}_\lambda, -\infty < \lambda < \infty\}$ be a process with orthogonal increments assumed to be q.m. continuous from the left. Let F be a finite Borel measure so that $E\hat{X}_\lambda\overline{\hat{X}}_\mu = \delta_{\lambda\mu}F(d\lambda)$. Then,

$$E\left[\int_{-\infty}^{\infty} f(\lambda)\, d\hat{X}_\lambda\right]\left[\overline{\int_{-\infty}^{\infty} g(\lambda)\, d\hat{X}_\lambda}\right] = \int_{-\infty}^{\infty} f(\lambda)\,\bar{g}(\lambda)F(d\lambda) \qquad f, g \in L_2(F) \tag{6.8}$$

Proof: Equation (6.8) is certainly true if f and g are step functions. Let $\{f_n\}$ and $\{g_n\}$ be approximating sequences of step functions for f and g, respectively, and write $A(f) = \int_{-\infty}^{\infty} f(\lambda)\, d\hat{X}_\lambda$. Then,

$$\begin{aligned} EA(f)\overline{A(g)} &= E[A(f - f_n) + A(f_n)][\overline{A(g - g_n) + A(g_n)}] \\ &= \int_{-\infty}^{\infty} f_n(\lambda)\overline{g_n(\lambda)}F(d\lambda) + EA(f - f_n)\overline{A(g - g_n)} \\ &\qquad + EA(f_n)\overline{A(g - g_n)} + EA(f - f_n)\overline{A(g_n)} \end{aligned}$$

Therefore,

$$\begin{aligned} &\left| EA(f)\overline{A(g)} - \int_{-\infty}^{\infty} f_n(\lambda)\bar{g}_n(\lambda)F(d\lambda)\right| \\ &\leq \sqrt{E|A(f - f_n)|^2}\sqrt{E|A(g - g_n)|^2} + \sqrt{E|A(f_n)|^2}\sqrt{E|A(g - g_n)|^2} \\ &\qquad + \sqrt{E|A(f - f_n)|^2}\sqrt{E|A(g_n)|^2} \end{aligned}$$

and (6.8) follows by letting $n \to \infty$. ∎

Proposition 6.3. Let $\{X_t, -\infty < t < \infty\}$ be a q.m. continuous and wide-sense stationary process, and let $\{\hat{X}_\lambda, -\infty < \lambda < \infty\}$ be its spectral process as defined by (6.1). Then

$$\int_{-\infty}^{\infty} f(\lambda)\, d\hat{X}_\lambda = \int_{-\infty}^{\infty} \hat{f}(t)X_t\, dt \qquad f \in \mathcal{S} \tag{6.9}$$

where $\hat{f}$ denotes the Fourier transform of f and $\mathcal{S}$, as usual, denotes the space of infinitely differentiable functions of rapid descent.

Proof: The definition of $\hat{X}_\lambda$ as given by (6.1) yields

$$\int_{-\infty}^{\infty} I_{ab}(\lambda)\, d\hat{X}_\lambda = \int_{-\infty}^{\infty} \hat{I}_{ab}(t) X_t\, dt$$

where I_{ab} denotes the indicator function of $[a, b)$. Let $\{f_n\}$ be a sequence of step functions obtained by sampling f at continuity points of the spectral measure F. As the sampling points become dense in $(-\infty, \infty)$, f_n converges to f uniformly and in L_1 metric. Therefore, $\{\hat{f}_n\}$ converges to $\hat{f}$ uniformly. Now, for each n,

$$\int_{-\infty}^{\infty} f_n(\lambda)\, d\hat{X}_\lambda = \int_{-\infty}^{\infty} \hat{f}_n(t) X_t\, dt$$

Therefore, if $f \in \mathcal{S}$,

$$\begin{aligned} E\left|\int_{-\infty}^{\infty} f(\lambda)\, d\hat{X}_\lambda - \int_{-\infty}^{\infty} \hat{f}(t) X_t\, dt\right|^2 &\le 2E\left|\int_{-\infty}^{\infty} [f(\lambda) - f_n(\lambda)]\, d\hat{X}_\lambda\right|^2 \\ + 2E\left|\int_{-\infty}^{\infty} [\hat{f}(t) - \hat{f}_n(t)] X_t\, dt\right|^2 &= 2\int_{-\infty}^{\infty} |f(\lambda) - f_n(\lambda)|^2 F(d\lambda) \\ &\quad + 2 \iint_{-\infty}^{\infty} R(t-s)[\hat{f}(t) - \hat{f}_n(t)]\overline{[\hat{f}(s) - \hat{f}_n(s)]}\, dt\, ds \end{aligned}$$

The first of these integrals goes to zero as $n \to \infty$ by virtue of the construction of f_n. The second integral also goes to zero, because

$$\int_{T}^{T} [\hat{f}(t) - \hat{f}_n(t)] e^{i2\pi\nu t}\, dt \xrightarrow[T\to\infty]{} f(\nu) - f_n(\nu)$$

and the convergence is bounded. Hence, from (5.21), we have

$$\begin{aligned} \lim_{T\to\infty} \iint_{-T}^{T} R(t-s)[\hat{f}(t) - \hat{f}_n(t)]\overline{[\hat{f}(s) - \hat{f}_n(s)]}\, dt\, ds \\ = \int_{-\infty}^{\infty} |f(\nu) - f_n(\nu)|^2 F(d\nu) \xrightarrow[n\to\infty]{} 0 \quad \blacksquare \end{aligned}$$

Corollary. Under the hypotheses of Proposition 6.3,

$$\int_{-\infty}^{\infty} f(\lambda)\bar{g}(\lambda)\, F(d\lambda) = \iint_{-\infty}^{\infty} \hat{f}(t)\bar{\hat{g}}(s) R(t-s)\, dt\, ds \qquad f, g \in \mathcal{S} \quad (6.10)$$

Remark: Note the similarity between (6.10) and (5.29). Indeed, (5.29) can be obtained from (6.10) by using

$$g_n(\lambda) = \exp\left(-\frac{1}{2}\frac{\lambda^2}{n}\right)$$

$$\hat{g}_n(s) = \sqrt{\frac{n}{2\pi}} \exp\left(-\frac{n}{2} s^2\right)$$

and letting $n \to \infty$.

Proposition 6.4 (Spectral Representation). A q.m. continuous process $\{X_t, -\infty < t < \infty\}$ is wide-sense stationary if and only if there exists a process with orthogonal increments $\{\hat{X}_\lambda, -\infty < \lambda < \infty\}$ such that $E|\hat{X}_\lambda|^2$ is bounded and

$$X_t = \int_{-\infty}^{\infty} e^{i2\pi\lambda t}\, d\hat{X}_\lambda \tag{6.11}$$

Proof: If $\{\hat{X}_\lambda, -\infty < \lambda < \infty\}$ is as described, we can always assume that it is continuous from the left. Then,

$$F([a,b)) = E|\hat{X}_b - \hat{X}_a|^2$$

defines a finite Borel measure, and $E\, d\hat{X}_\lambda\, \overline{d\hat{X}_\mu} = \delta_{\lambda\mu} F(d\lambda)$. Since

$$E\left(\int_{-\infty}^{\infty} e^{i2\pi\lambda t}\, dX_\lambda\right)\left(\overline{\int_{-\infty}^{\infty} e^{i2\pi\nu s}\, d\hat{X}_\nu}\right) = \int_{-\infty}^{\infty} e^{i2\pi\lambda(t-s)} F(d\lambda)$$

the process defined by $\int_{-\infty}^{\infty} e^{i2\pi\lambda t}\, d\hat{X}_\lambda$ must be q.m. continuous and wide-sense stationary by virtue of Bochner's theorem (Proposition 5.1).

Conversely, let $\{X_t, -\infty < t < \infty\}$ be q.m. continuous and wide-sense stationary and define $\hat{X}_\lambda$ by (6.1). Then, (6.11) follows from (6.9) by using the familiar approximations

$$f_n(\lambda) = e^{i2\pi\lambda t_0} \exp\left(\frac{1}{2}\frac{\lambda}{n}\right)$$

$$\hat{f}_n(t) = \sqrt{\frac{n}{2\pi}} \exp\left[-\frac{1}{2} n(t - t_0)^2\right]$$

and letting $n \to \infty$. ▮

Equation (6.11) finally provides us with a representation of wide-sense stationary processes as a linear combination of sinusoids. The spectral process $\{\hat{X}_\lambda, -\infty < \lambda < \infty\}$ also provides a complete characterization of the Hilbert space $\mathcal{H}_X$ generated by the process $\{X_t, -\infty < t < \infty\}$, and a complete characterization of every process that can be said to be derived by a time-invariant linear operation on $\{X_t, -\infty < t < \infty\}$.

Proposition 6.5. Let $\{X_t, -\infty < t < \infty\}$ be a q.m. continuous and wide-sense stationary process with spectral process $\{\hat{X}_\lambda, -\infty < \lambda < \infty\}$ and spectral measure F. Let $\mathcal{H}_X$ be the Hilbert space generated by $\{X_t, -\infty < t < \infty\}$. Then, a random variable Y belongs to $\mathcal{H}_X$ if and only if there exists $\eta \in L_2(F)$ such that

$$Y = \int_{-\infty}^{\infty} \eta(\lambda)\, d\hat{X}_\lambda \tag{6.12}$$

Proof: The proof that every Y of the form of (6.12) is in $\mathcal{H}_X$ is elementary. We shall only consider the other half. First, suppose Y is of the form $\sum_{\nu=1}^{n} a_\nu X_{t_\nu}$; then from (6.11), we have

$$Y = \int_{-\infty}^{\infty} \left(\sum_{\nu=1}^{n} a_\nu e^{i2\pi\lambda t_\nu} \right) d\hat{X}_\lambda$$

hence (6.12) holds. In general, $Y \in \mathcal{H}_X$ is the q.m. limit of a sequence $\{Y_n\}$, each of which is a finite sum. Hence

$$Y = \lim_{n\to\infty} \text{in q.m. } Y_n = \lim_{n\to\infty} \text{in q.m. } \int_{-\infty}^{\infty} \eta_n(\lambda)\, d\hat{X}_\lambda$$

Since $\{Y_n\}$ is q.m. convergent, it is also mutually q.m. convergent, and from (6.8) we have

$$E|Y_m - Y_n|^2 = \int_{-\infty}^{\infty} |\eta_m(\lambda) - \eta_n(\lambda)|^2 F(d\lambda) \xrightarrow[m,n\to\infty]{} 0$$

This means that $\{\eta_n\}$ is a Cauchy sequence in $L_2(F)$, and the completeness of $L_2(F)$ implies the existence of $\eta \in L_2(F)$ such that

$$\int_{-\infty}^{\infty} |\eta_n(\lambda) - \eta(\lambda)|^2 F(d\lambda) \xrightarrow[n\to\infty]{}$$

But this is equivalent to saying that $\int_{-\infty}^{\infty} \eta_n(\lambda)\, d\hat{X}_\lambda$ converges in q.m. to $\int_{-\infty}^{\infty} \eta(\lambda)\, d\hat{X}_\lambda$. The proof is complete. ∎

Proposition 6.5 provides an explicit representation for elements of $\mathcal{H}_X$ in terms of functions in $L_2(F)$. Indeed, it provides a one-to-one mapping of $L_2(F)$ onto $\mathcal{H}_X$ which preserves the inner products, that is,

$$EY_1\bar{Y}_2 = \int_{-\infty}^{\infty} \eta_1(\lambda)\bar{\eta}_2(\lambda) F(d\lambda) \tag{6.13}$$

That is, (6.12) is an isometry between $\mathcal{H}_X$ and $L_2(F)$. Equation (6.12) is extremely useful in representing an arbitrary linear operation on a wide-sense stationary process.

Proposition 6.6. Let $\{X_t, -\infty < t < \infty\}$ be as in Proposition 6.5. Let $\{U_t, -\infty < t < \infty\}$ be the corresponding translation group of unitary operators as defined by (5.3). Suppose $\{Y_t, -\infty < t < \infty\}$ is a process such that for each t, $Y_t \in \mathcal{H}_X$. Then, the following conditions are equivalent:

(*a*) There exists $\eta \in L_2(F)$ such that

$$Y_t = \int_{-\infty}^{\infty} e^{i2\pi\lambda t} \eta(\lambda)\, d\hat{X}_\lambda \qquad t \in (-\infty, \infty) \tag{6.14}$$

(b) For each $t \in (-\infty, \infty)$,

$$Y_t = U_t Y_0 \tag{6.15}$$

(c) For arbitrary t and s in $(-\infty, \infty)$,

$$EY_t \bar{X}_s = EY_0 \bar{X}_{s-t} \tag{6.16}$$

Proof: It is nearly obvious that (a) implies (c). By direct computation we have

$$EY_t \bar{X}_s = \int_{-\infty}^{\infty} e^{i2\pi\lambda(t-s)} \eta(\lambda) F(d\lambda) = EY_0 \bar{X}_{s-t}$$

To prove that (a) implies (b), we make use of the fact that

$$U_t \, d\hat{X}_\lambda = e^{i2\pi\lambda t} \, d\hat{X}_\lambda \tag{6.17}$$

More precisely, this means that

$$E|U_t(\hat{X}_{\lambda+\epsilon} - \hat{X}_\lambda) - e^{i2\pi\lambda t}(\hat{X}_{\lambda+\epsilon} - \hat{X}_\lambda)|^2 \xrightarrow[\epsilon \to 0]{} 0$$

Finally, if (6.14) holds, then we can write

$$U_t Y_0 = \int_{-\infty}^{\infty} \eta(\lambda) U_t \, d\hat{X}_\lambda = \int_{-\infty}^{\infty} e^{i2\pi\lambda t} \eta(\lambda) \, d\hat{X}_\lambda = Y_t$$

This step can be made precise by approximating the stochastic integral in (6.14) by sums.

Next, we prove that each of (b) and (c) implies (a). Suppose (b) is satisfied. Because of Proposition 6.4, we can write

$$Y_0 = \int_{-\infty}^{\infty} \eta(\lambda) \, d\hat{X}_\lambda$$

Thus, (6.14) follows upon using (6.17). To prove that (c) implies (a), we first note that from Proposition 6.4 for each t there exists $\eta(\cdot,t) \in L_2(F)$ such that

$$Y_t = \int_{-\infty}^{\infty} \eta(\lambda,t) \, d\hat{X}_\lambda$$

If (c) is satisfied, then from (6.16),

$$\begin{aligned} EY_t \bar{X}_s &= \int_{-\infty}^{\infty} \eta(\lambda,t) \, e^{-i2\pi\lambda s} F(d\lambda) \\ &= EY_0 \bar{X}_{s-t} = \int_{-\infty}^{\infty} \eta(\lambda,0) e^{i2\pi\lambda(t-s)} F(d\lambda) \end{aligned}$$

This means that for arbitrary t,s in $(-\infty, \infty)$, we have

$$\int_{-\infty}^{\infty} e^{-i2\pi\lambda s} [\eta(\lambda,t) - e^{i2\pi\lambda t} \eta(\lambda,0)] F(d\lambda) = 0$$

If we denote $\tilde{Y}_t = \int_{-\infty}^{\infty} e^{i2\pi\lambda t}\eta(\lambda,0)\, d\hat{X}_\lambda$, then the above equation is equivalent to

$$E(Y_t - \tilde{Y}_t)\bar{X}_s = 0 \qquad t,s \in (-\infty,\infty)$$

Since $(Y_t - \tilde{Y}_t)$, being in $\mathcal{H}_X$, is the q.m. limit of a sequence of finite sums of the form $\sum_\nu a_\nu X_{t_\nu}$, we have $E|Y_t - \tilde{Y}_t|^2 = 0$. The proof is now complete. ▮

Remark: The process $\{Y_t, -\infty < t < \infty\}$ is necessarily wide-sense stationary. More than that, $\{X_t, Y_t, -\infty < t < \infty\}$ may be said to be **jointly stationary** (wide-sense) in the sense that every linear combination $\alpha X_t + \beta Y_t$ defines a wide-sense stationary process. On the other hand, suppose that for every t in $(-\infty,\infty)$, $Z_t \in \mathcal{H}_X$ and $\{Z_t, -\infty < t < \infty\}$ is wide-sense stationary. The process $\{Z_t, -\infty < t < \infty\}$ does not necessarily satisfy the conditions (6.14) to (6.16) and is not necessarily jointly stationary with $\{X_t, -\infty < t < \infty\}$ (see Exercise 12).

If $h(t)$, $-\infty < t < \infty$, is a function such that its Fourier transform $\hat{h}$ is in $L_2(F)$, then

$$\begin{aligned} Y_t &= \int_{-\infty}^{\infty} h(t-\tau)X_\tau\, d\tau \\ &= \int_{-\infty}^{\infty} \hat{h}(\lambda)e^{i2\pi\lambda t}\, d\hat{X}_\lambda \end{aligned} \tag{6.18}$$

Therefore, every process which can be expressed as a convolution of an impulse response and a wide-sense stationary process has a spectral representation of the form (6.14), and η is just the transfer function. However, not every process of the form (6.14) can be represented as a convolution. Hence, (6.14) is a more general representation. We shall call every process of the form (6.14) a process derivable by a linear and time-invariant filtering on the X process. Naturally, η will be referred to as the transfer function of the filter. We note that if $\{Y_t, -\infty < t < \infty\}$ is given by (6.14), then its spectral process $\{\hat{Y}_\lambda, -\infty < \lambda < \infty\}$ is defined by

$$d\hat{Y}_\lambda = \eta(\lambda)\, d\hat{X}_\lambda \tag{6.19}$$

The Y process is automatically wide-sense stationary and q.m. continuous with

$$EY_t\bar{Y}_s = \int_{-\infty}^{\infty} e^{i2\pi\lambda(t-s)}|\eta(\lambda)|^2F(d\lambda) \tag{6.20}$$

This means that

$$F_Y(d\lambda) = |\eta(\lambda)|^2F(d\lambda) \tag{6.21}$$

Equation (6.21) generalizes (5.19). Equation (6.19) should be considered the random-process counterpart of the familiar formula relating the Fourier transforms of the input and output of a linear time-invariant filter.

It may be useful to consider some examples. First, if $\{X_t, -\infty < t < \infty\}$ is a wide-sense stationary process, then its q.m. derivative as defined by (3.5) can be expressed as

$$\dot{X}_t = \int_{-\infty}^{\infty} (i2\pi\lambda)e^{i2\pi\lambda t}\, d\hat{X}_\lambda \tag{6.22}$$

whenever it exists. It is rather obvious from (6.22) that the q.m. derivative exists if and only if $\int_{-\infty}^{\infty} |\lambda|^2 F(d\lambda) < \infty$. As a second example, consider the process $\{Y_t, -\infty < t < \infty\}$ defined by

$$Y_t = \int_{-\infty}^{\infty} (-i \operatorname{sgn} \lambda)e^{i2\pi\lambda t}\, d\hat{X}_\lambda \tag{6.23}$$

where $\operatorname{sgn} \lambda = 1$ or -1 according as $\lambda \geq 0$ or $\lambda < 0$. Thus defined, Y_t is known as the **Hilbert transform** of X_t. The Y process has exactly the same covariance function as the X process. Also, since

$$X_t + iY_t = 2\int_0^{\infty} e^{i2\pi\lambda t}\, d\hat{X}_\lambda$$

we can say that $X_t + iY_t$ has no negative frequencies. Hilbert transforms will be made use of again in the next section.

7. LOWPASS AND BANDPASS PROCESSES

Definition. A wide-sense stationary and q.m. continuous process $\{X_t, -\infty < t < \infty\}$ is said to be **bandlimited** to frequency W if its spectral measure F satisfies

$$F((-\infty, -W]) = 0 = F([W, \infty)) \tag{7.1}$$

In other words, the X process has no average power for frequencies $\leq -W$ and $\geq W$. We note that the points $-W$ and W are required to be continuity points of F. A process bandlimited to frequency W can always be written as

$$X_t = \int_{-W}^{W} e^{i2\pi\nu t}\, d\hat{X}_\nu \tag{7.2}$$

where it makes no difference if the limits are replaced by $W \pm 0$ and $-W \pm 0$. Equation (7.2) results from the fact that

$$\begin{aligned} E\left|X_t - \int_{-W+0}^{W-0} e^{i2\pi\nu t}\, d\hat{X}_\nu\right|^2 &= E\left|\int_{|\nu|\geq W} e^{i2\pi\nu t}\, d\hat{X}_\nu\right|^2 \\ &= F((-\infty, -W]) + F([W, \infty)) = 0 \end{aligned}$$

The concept of a bandlimited process is an important one in practice, because such a process is specified for all t by its sampled values spaced $1/2W$ apart in a formula which is known as the **sampling theorem.**

Proposition 7.1 (Sampling Theorem). Let $\{X_t, -\infty < t < \infty\}$ be a process bandlimited to frequency W. Then

$$X_t = \lim_{N\to\infty} \text{ in q.m.} \sum_{k=-N}^{N} X_{\alpha+k/2W} \frac{\sin 2\pi W(t-\alpha-k/2W)}{2\pi W(t-\alpha-k/2W)} \tag{7.3}$$

for all $t \in (-\infty, \infty)$, where α is an arbitrary constant.

Remark: Proposition 7.1 implies that $\mathcal{H}_X$ has a countable basis, namely $\{X_{\alpha+k/2W}, k = 0, \pm 1, \pm 2, \ldots\}$.

Proof: We begin with the representation (7.2) and for emphasis rewrite it as

$$X_t = \int_{-W+0}^{W-0} e^{i2\pi\nu t}\, d\hat{X} \tag{7.4}$$

For a fixed t, the function $f(\nu) = e^{i2\pi\nu t}$ is square integrable over $(-W,W)$. Since $\{\varphi^k(\nu) = e^{i2\pi\nu k/2W}, k = 0, \pm 1, \pm 2, \ldots\}$ forms an orthonormal basis in $L_2(-W,W)$, we have

$$e^{i2\pi\nu t} = \underset{N\to\infty}{\text{l.i.m.}} \sum_{k=-N}^{N} e^{i2\pi\nu k/2W} \frac{1}{2W}\int_{-W}^{W} e^{i2\pi\nu(t-k/2W)}\, d\nu$$

where l.i.m. denotes limit in L_2 mean. The Fourier series

$$\sum_{k=-N}^{N} e^{i2\pi\nu k/2W} \frac{1}{2W}\int_{-W}^{W} e^{i2\pi\nu(t-k/2W)}\, d\nu = \sum_{k=-N}^{N} e^{i2\pi\nu k/2W} \frac{\sin 2\pi W(t-k/2W)}{2\pi W(t-k/2W)}$$

also converges pointwise to $e^{i2\pi\nu t}$ for $\nu \in (-W,W)$, but not at the endpoints $\nu = \pm W$. At the endpoints, it converges to $\frac{1}{2}(e^{i2\pi Wt} + e^{-i2\pi Wt}) = \cos 2\pi Wt$. Therefore,

$$E\left| X_t - \sum_{k=-N}^{N} X_{k/2W} \frac{\sin 2\pi W(t-k/2W)}{2\pi W(t-k/2W)} \right|^2$$

$$= E\left| \int_{-W+0}^{W-0} \left[e^{i2\pi\nu t} - \sum_{k=-N}^{N} e^{i2\pi\nu k/2W} \frac{\sin 2\pi W(t-k/2W)}{2\pi W(t-k/2W)} \right] d\hat{X}_\lambda \right|^2$$

$$= \int_{-W+0}^{W-0} \left| e^{i2\pi\nu t} - \sum_{k=-N}^{N} e^{i2\pi\nu k/2W} \frac{\sin 2\pi W(t-k/2W)}{2\pi W(t-k/2W)} \right|^2 F(d\lambda) \xrightarrow[N\to\infty]{} 0$$

which proves the theorem for $\alpha = 0$. For $\alpha \neq 0$, we note that $Y_t = X_{t+\alpha}$ is also bandlimited to W and (7.3) is merely the expansion for $Y_{t-\alpha}$. ∎

Remark: The proof fails if F merely satisfies $F((-\infty, -W)) = 0$ and $F((W, \infty)) = 0$ but has a finite mass at one or both of the points $\pm W$. Then, we would have to write

$$X_t = \int_{-W-0}^{W+0} e^{i2\pi\nu t}\, d\hat{X}_\nu$$

and would find

$$E\left| X_t - \sum_{k=-N}^{N} X_{k/2W} \frac{\sin 2\pi W(t - k/2W)}{2\pi W(t - k/2W)} \right|^2 \xrightarrow[N\to\infty]{} \sin^2 2\pi Wt(F^+ + F^-) \tag{7.5}$$

where F^+ and F^- are, respectively, the mass of F at $+W$ and $-W$. As an example, suppose $X_t = \cos(2\pi Wt + \varphi)$ where φ is a random variable uniformly distributed in $[0,2\pi)$. Then, $\{X_t, -\infty < t < \infty\}$ is a wide-sense stationary and q.m. continuous process with

$$EX_tX_s = R(t - s) = \tfrac{1}{2} \cos 2\pi W(t - s)$$

Therefore, F is concentrated at $\pm W$, assigning a mass of $\frac{1}{4}$ to each point. The sampling theorem obviously fails, because in this case,

$$X_{\alpha+k/2W} = (-1)^k X_\alpha \qquad k = 0, \pm 1, \ldots$$

which means that for every t, the sum

$$\sum_k X_{\alpha+k/2W} \frac{\sin 2\pi W(t - \alpha - k/2W)}{2\pi W(t - \alpha - k/2W)}$$

is proportional to X_α. On the other hand $X_{\alpha+1/4W}$ is orthogonal to X_α. So, $X_{\alpha+1/4W}$ obviously cannot be represented by the sampling formula.

In communication problems, we frequently encounter the so-called bandpass processes. Here, we define a **bandpass process** with **center frequency** W_0 and **bandwidth** $2W (W_0 > W)$ as q.m. continuous and a wide-sense stationary process with

$$\begin{aligned} F((-\infty, -W_0 - W]) &= F([-W_0 + W, W_0 - W]) \\ &= F([+W_0 + W, \infty)) = 0 \end{aligned} \tag{7.6}$$

In other words, the average power is concentrated in the frequency ranges $(-W_0 - W, -W_0 + W)$ and $(W_0 - W, W_0 + W)$. Of course, a bandpass process is bandlimited to frequency $W_0 + W$. Therefore, the sampling theorem can be applied. However, a straightforward application of the sampling theorem requires a sampling rate of $2W_0 + 2W$, which does not reflect the fact that there is no average power in the frequency range $[-W_0 + W, W_0 - W]$. From the point of view of bandwidth, a bandpass has the same bandwidth as a process bandlimited to $2W$. Thus, we should expect a sampling rate of $4W$ rather than $2(W_0 + W)$. In practice W_0 is much greater than W, and there is a substantial difference between these two rates of sampling. It turns out that a bandpass process with bandwidth $2W$ can be represented in terms of the samples of itself and the Hilbert transform, each being sampled at a rate of $2W$ to give a combined rate of $4W$. Let $\{X_t, -\infty < t < \infty\}$ be a bandpass process with center frequency W_0 and bandwidth $2W$. Since $W_0 > W$, $\nu = 0$ is a continuity point of F. We define the Hilbert transform by

$$Y_t = \int_{-\infty}^{\infty} (-i \operatorname{sgn} \nu) e^{i2\pi\nu t}\, d\hat{X}_\nu \tag{7.7}$$

where $\operatorname{sgn} \nu = 1$ or -1 according as $\nu > 0$ or $\nu < 0$. The value of sgn 0 is immaterial, since (7-7) can be rewritten as

$$Y_t = i \int_{-W_0-W}^{-W_0+W} e^{i2\pi\nu t}\, d\hat{X}_\nu - i \int_{+W_0-W}^{W_0+W} e^{i2\pi\nu t}\, d\hat{X}_\nu \tag{7.8}$$

Again, we note that the limits in the integrals in (7.8) can be replaced by their limits from the left or the right without any material effect. The process $\{X_t, -\infty < t < \infty\}$ can be written as

$$X_t = \int_{-W_0-W}^{-W_0+W} e^{i2\pi\nu t}\, d\hat{X}_\nu + \int_{W_0-W}^{W_0+W} e^{i2\pi\nu t}\, d\hat{X}_\nu \tag{7.9}$$

It follows that we can write

$$\begin{aligned} X_t + iY_t &= 2 \int_{W_0-W}^{W_0+W} e^{i2\pi\nu t}\, d\hat{X}_\nu \\ &= 2e^{i2\pi W_0 t} \int_{-W}^{W} e^{i2\pi\nu t}\, d\hat{X}_{\nu+W_0} \\ &= 2e^{i2\pi W_0 t}\xi_t \end{aligned} \tag{7.10}$$

$$\begin{aligned} X_t - iY_t &= 2e^{-i2\pi W_0 t} \int_{-W}^{W} e^{i2\pi\nu t}\, d\hat{X}_{\nu-W_0} \\ &= 2e^{-i2\pi W_0 t}\eta_t \end{aligned} \tag{7.11}$$

It is clear that both $\{\xi_t, -\infty < t < \infty\}$ and $\{\eta_t, -\infty < t < \infty\}$ are bandlimited to W. Hence, by applying (7.3) to ξ_t and η_t, we get

$$\begin{aligned} X_t &= e^{-i2\pi W_0 t}\eta_t + e^{i2\pi W_0 t}\xi_t \\ &= \sum_{n=-\infty}^{\infty} (\eta_{\alpha+k/2W} e^{-i2\pi W_0 t} + \xi_{\alpha+k/2W} e^{i2\pi W_0 t}) \frac{\sin 2\pi W(t - k/2W - \alpha)}{2\pi W(t - k/2W - \alpha)} \end{aligned}$$

Reexpressing $\eta_{\alpha+k/2W}$ and $\xi_{\alpha+k/2W}$ in terms of $X_{\alpha+k/2W}$ and $Y_{\alpha+k/2W}$, we get the following result.

Proposition 7.2. Let $\{X_t, \ -\infty < t < \infty\}$ be a bandpass process with center frequency W_0 and bandwidth $2W$. Let Y_t be its Hilbert transform defined by (7.7). Then

$$X_t = \lim_{N\to\infty} \text{in q.m.} \sum_{k=-N}^{N} \frac{\sin 2\pi W(t - k/2W - \alpha)}{2\pi W(t - k/2W - \alpha)} [X_{k/2W+\alpha} \cos 2\pi W_0(t - k/2W - \alpha) - Y_{k/2W+\alpha} \sin 2\pi W_0(t - k/2W - \alpha)] \tag{7.12}$$

Equation (7.12) involves both X_t and Y_t being sampled at $2W$, giving a total sampling rate of $4W$.

It is interesting to note that

$$|X_t|^2 + |Y_t|^2 = |\xi_t|^2 + |\eta_t|^2 \tag{7.13}$$

Since ξ_t and η_t are both bandlimited to W, we can expect $|\xi_t|^2 + |\eta_t|^2$ to be relatively slowly varying when $W \ll W_0$. Therefore, $|X_t|^2 + |Y_t|^2$ is also slowly varying (relative to a sinusoid with frequency W_0). On the other hand, most of the average power of X_t is concentrated near $\pm W_0$, so X_t itself is rapidly varying. If X_t is real valued, then it can be written as

$$X_t = \sqrt{X_t^2 + Y_t^2} \cos [2\pi W_0 t + \theta(t)]$$

where both $\sqrt{X_t^2 + Y_t^2}$ and $\theta(t)$ are slowly varying. Thus, $\sqrt{X_t^2 + Y_t^2}$ has the interpretation of the envelope and $\theta(t)$ the phase of a sinusoid being slowly modulated in both its amplitude and phase. Thus, we see that the Hilbert transform plays a rather important role in bandpass processes. It permits the envelope of such a process to be expressed simply.

8. WHITE NOISE AND WHITE-NOISE INTEGRALS

A **white noise** is usually described as a wide-sense stationary process with a spectral-density function given by

$$S(\nu) = S_0 \qquad \text{for all } \nu \tag{8.1}$$

Since for this case $\int_{-\infty}^{\infty} S(\nu)\, d\nu = R(0) = \infty$, a white noise is not really a second-order process at all, and the spectral-density function is not well defined. However, if we proceed heuristically, (8.1) suggests that $R(\tau)$ must be given by

$$R(\tau) = EX_{t+\tau}\bar{X}_t = \delta(\tau)S_0$$

where $\delta(\tau)$ is the Dirac δ function. The reason for this is that

$$S(\nu) = \int_{-\infty}^{\infty} e^{i2\pi\nu\tau} R(\tau)\, d\tau$$

would then give (8.1). These considerations are purely formal and require elaboration and substantiation. At the outset, we should distinguish between the problem of handling white noise in a mathematically consistent way and the problem of interpreting white noise as an abstraction of physical phenomena. As far as the calculus of white noise is concerned, the problem is not difficult, at least for linear problems. Nonlinear problems involving Gaussian white noise are substantially more complex and will be dealt with in a later chapter. The principal tool that we shall use in establishing a self-consistent calculus for white noise is the second-order stochastic integral that we introduced in Sec. 6 in connection with spectral representation. There remains the problem of interpretation.

Since $R(0) = \infty$ implies an infinite average power, a white noise cannot be a physical process. If a white noise is not a physical process, and if it leads to mathematical complications, then why is it used at all? First, even though the calculus of white noise requires justification, once justified it leads to a tremendous analytical simplification in many problems. Secondly, many processes that one encounters in practice are well approximated by white noise, but this statement requires amplification. Because a white noise is not a second-order process (indeed, it is not a stochastic process at all!), no sequence of processes $\{X_n(t), t \in (-\infty, \infty)\}$ which is q.m. convergent for each t can converge to a white noise. The way out of this difficulty is to recall that just as a δ function is never used outside of an integral, the same is true with white noise.

Definition. A sequence of q.m. continuous processes $\{X_t^{(n)}, t \in (-\infty, \infty)\}$ is said to "converge to a white noise" if

(*a*) For each $f \in L_2$, $\{X_n(f)\}$ is a q.m. convergent sequence, where $X_n(f)$ is defined by

$$X_n(f) = \int_{-\infty}^{\infty} f(t) X_t^{(n)}\, dt \tag{8.2}$$

(*b*) There exists a positive constant S_0 such that

$$\lim_{n\to\infty} EX_n(f)\overline{X_n(g)} = S_0 \int_{-\infty}^{\infty} f(t)\bar{g}(t)\, dt \tag{8.3}$$

This definition helps to make clear the idea that a process $\{X_t, -\infty < t < \infty\}$ is approximately a white noise. What we really mean is that for all function f that we are concerned with, the quantity $EX(f)\overline{X(f)}$ is very nearly equal to $S_0 \int_{-\infty}^{\infty} |f(t)|^2\, dt$.

Suppose $\{X_t^{(n)}, t \in (-\infty, \infty)\}$ is a sequence of processes converging to a white noise. By definition, for every $f \in L_2$, there exists a second-order random variable $X(f)$ such that

$$EX(f)\overline{X(f)} = S_0 \int_{-\infty}^{\infty} |f(t)|^2\, dt \tag{8.4}$$

It is common practice to write $X(f)$ as

$$X(f) = \int_{-\infty}^{\infty} f(t) X_t\, dt \tag{8.5}$$

where X_t is a white noise. We shall do so on many occasions. It should not be forgotten, however, that the right-hand side of (8.5) is nothing more than a symbolic way of writing $X(f)$, and there exists no stochastic process X_t for which the right-hand side of (8.5) is an integral. Although (8.5) is merely formal, $X(f)$ does admit a representation as a second-order stochastic integral as is indicated by the following proposition.

Proposition 8.1. Let $\{X_t^{(n)}, -\infty < t < \infty\}$ be a sequence of q.m. continuous converging to a white noise. Let

$$X(f) = \lim_{n\to\infty} \text{ in q.m.} \int_{-\infty}^{\infty} f(t) X_t^{(n)}\, dt \qquad f \in L_2 \tag{8.6}$$

Then

$$X(f) = \int_{-\infty}^{\infty} f(t)\, dZ_t \tag{8.7}$$

where $\{Z_t \; -\infty < t < \infty\}$ is a process with orthogonal increments defined by

$$Z_t = \lim_{n\to\infty} \text{ in q.m.} \int_0^t X_s^{(n)}\, ds \tag{8.8}$$

Proof: Define $\{Z_t, -\infty < t < \infty\}$ by (8.8). If we denote the indicator function of $[a,b)$ by I_{ab}, then we can write

$$Z_b - Z_a = \lim_{n\to\infty} \text{ in q.m. } X_n(I_{ab}) \tag{8.9}$$

From (8.3) we have

$$E(Z_b - Z_a)(Z_d - Z_c) = S_0 \int_{-\infty}^{\infty} I_{ab}(t) I_{cd}(t)\, dt \tag{8.10}$$

Hence, $\{Z_t, -\infty < t < \infty\}$ has orthogonal increments and

$$E\, dZ_t\, \overline{dZ_s} = S_0\, \delta_{ts}\, dt \tag{8.11}$$

Define the second-order stochastic integral $\int_{-\infty}^{\infty} f(t)\, dZ_t$ as in Sec. 6 for $f \in L_2$. If f is a step function, it is obvious from (8.9) that

$$X(f) = \int_{-\infty}^{\infty} f(t)\, dZ_t$$

If $f \in L_2$ is not a step, there exists a sequence of step functions converging to f in L_2 distance. Therefore

$$X(f) = \int_{-\infty}^{\infty} f_n(t)\, dZ_t + X(f - f_n) \xrightarrow[n\to\infty]{} \int_{-\infty}^{\infty} f(t)\, dZ_t + X(0)$$
$$= \int_{-\infty}^{\infty} f(t)\, dZ_t$$

which proves the theorem. ∎

Manipulations of **white-noise integrals** $\int_{-\infty}^{\infty} X_t f(t)\, dt$ are justified by replacing it by the stochastic integral $\int_{-\infty}^{\infty} f(t)\, dZ_t$. In the future, when we speak of being given a white noise $\{X_t, -\infty < t < \infty\}$, we shall take it to mean that we are given a process with orthogonal increments $\{Z_t, -\infty < t < \infty\}$ which satisfies $E\, dZ_t\, \overline{dZ_s} = S_0\, \delta_{ts}\, dt$. From now on we shall also omit the constant S_0 which is irrelevant to most considerations. A most interesting property of a white noise is that its Fourier transform is again a white noise. Of course, Fourier transform has to be defined properly here.

Proposition 8.2. Let $\{X_t, -\infty < t < \infty\}$ be a white noise. Then there exists a second white noise $\{\hat{X}_\nu, -\infty < \nu < \infty\}$ such that

$$\int_{-\infty}^{\infty} \hat{h}(t) X_t\, dt = \int_{-\infty}^{\infty} h(\nu) \hat{X}_\nu\, d\nu \tag{8.12}$$

for all $h, \hat{h} \in L_2$.

Remark: We repeat once again that the integrals in (8.12) are merely symbolic representations of $\int_{-\infty}^{\infty} \hat{h}(t)\, dZ_t$ and $\int_{-\infty}^{\infty} h(\nu)\, d\hat{Z}_\nu$.

Proof: Define $\{\hat{Z}_\nu, -\infty < \nu < \infty\}$ as follows:

(a) $\hat{Z}_0 = 0$

(b) $\hat{Z}_b - \hat{Z}_a = \int_{-\infty}^{\infty} \hat{I}_{ab}(t)\, dZ_t$

Then,

$$E(\hat{Z}_b - \hat{Z}_a)\overline{(\hat{Z}_d - \hat{Z}_c)} = \int_{-\infty}^{\infty} \hat{I}_{ab}(t) \overline{\hat{I}}_{cd}(t)\, dt$$
$$= \int_{-\infty}^{\infty} I_{ab}(\nu) \bar{I}_{cd}(\nu)\, d\nu$$

Hence, $E\, d\hat{Z}_\nu\, \overline{d\hat{Z}_\mu} = \delta_{\nu\mu}\, d\nu$, and $\{\hat{Z}_\nu, -\infty < \nu < \infty\}$ has the same second-order properties as $\{Z_t, -\infty < t < \infty\}$. For an arbitrary $h \in L_2$, by approximating h by step functions in the familiar way, we find

$$\int_{-\infty}^{\infty} \hat{h}(t)\, dZ_t = \int_{-\infty}^{\infty} h(\nu)\, d\hat{Z}_\nu \tag{8.13}$$

proving the theorem. ∎

The spectral process of a white noise can now be used to define linear time-invariant filtering operations on white noise. Let $\eta \in L_2$, and let $\{X_t, -\infty < t < \infty\}$ be a white noise. Define

$$Y_t = \int_{-\infty}^{\infty} \eta(\nu) e^{i2\pi\nu t} \hat{X}_\nu \, d\nu \tag{8.14}$$

that is,

$$Y_t = \int_{-\infty}^{\infty} \eta(\nu) e^{i2\pi\nu t} \, d\hat{Z}_\nu$$

Then, Y_t is a q.m. continuous and wide-sense stationary process with a spectral-density function given by

$$S_Y(\nu) = |\eta(\nu)|^2 \tag{8.15}$$

Equation (8.15) justifies the interpretation that a white noise has a spectral-density function which is equal to a constant for all frequencies. This is because if we filter a white noise $\{X_t, -\infty < t < \infty\}$ with an ideal filter having a transfer function

$$\eta(\nu) = \begin{cases} 1 & a \le \nu \le b \\ 0 & \text{otherwise} \end{cases}$$

then the total average output power is given by

$$\begin{aligned} E|Y_t|^2 &= E\left| \int_a^b e^{i2\pi\nu t} \hat{X}_\nu \, d\nu \right|^2 \\ &= \int_a^b d\nu = b - a \end{aligned}$$

Equation (8.15) also has another interpretation. It suggests that every process with a spectral density $S(\nu)$ can be represented as white noise filtered by a time-invariant linear filter whose transfer function $\hat{h}$ satisfies $|\hat{h}(\nu)|^2 = S(\nu)$. This interpretation plays an important role in Wiener filtering problems. Closely related to linear time-invariant filtering is the idea of a differential equation driven by a white noise. Consider the following example. Suppose X_t is a white noise and Y_t satisfies

$$\dot{Y}_t = -Y_t + X_t \tag{8.16}$$

A good guess is that a solution of this equation is given by

$$Y_t = \int_{-\infty}^{\infty} \frac{1}{1 + i2\pi\nu} e^{i2\pi\nu t} \hat{X}_\nu \, d\nu \tag{8.17}$$

Before we can ascertain this, we need to give a precise interpretation to an equation such as (8.16).

Even if X_t is not a white noise, but a q.m. continuous process, an equation like (8.16) still cannot be interpreted as an ordinary differential equation involving sample functions of the two processes without assumptions such as separability, sample differentiability, etc. However, these assumptions would not be necessary if we interpret (8.16) as an integral equation. More generally, let $\{X_t, -\infty < t < \infty\}$ be a white noise with $\{Z_t, -\infty < t < \infty\}$ as the corresponding process with orthogonal increments. A second-order process $\{Y_t, t \in [a,b]\}$ is said to be the solution of the differential equation

$$\dot{Y}_t = \alpha(t)Y_t + \beta(t)X_t \tag{8.18}$$

if it satisfies

$$Y_t = Y_a + \int_a^t \alpha(s)Y_s\,ds + \int_a^t \beta(s)\,dZ_s \qquad a \le t \le b \tag{8.19}$$

Proposition 8.3. Let α and $\beta \in L_2(a,b)$. Then, (8.19) has one and only one solution with the same initial condition Y_a, provided that $E|Y_a|^2 < \infty$.

Remark: The proof will be omitted, because it is identical to the corresponding proof for stochastic differential equations of the Ito type. At this point we merely point out the following facts concerning the solution of (8.19).

(*a*) If we set $Y_t^{(0)} = Y_a$, $a \le t \le b$, and

$$Y_t^{(n+1)} = \int_a^t \alpha(s)Y_s^{(n)}\,ds + \int_a^t \beta(s)\,dZ_s$$

then $\{Y_t^{(n)}, a \le t \le b, n = 0, 1, \ldots\}$ converges in q.m. to the unique solution of (8.19).

(*b*) The solution of (8.19) is q.m. continuous on $[a,b]$.

If $\{X_t^{(n)}, -\infty < t < \infty, n = 1, 2, \ldots\}$ is a sequence of q.m. continuous processes, then for each n, we can consider the differential equation

$$\dot{Y}_t^{(n)} = \alpha(t)Y_t^{(n)} + \beta(t)X_t^{(n)} \tag{8.20}$$

where $\dot{Y}_t^{(n)}$ is the q.m. derivative of $Y_t^{(n)}$. Equation (8.20) is equivalent to the integral equation

$$Y_t^{(n)} = Y_a^{(n)} + \int_a^t \alpha(s)Y_s^{(n)}\,ds + \int_a^t \beta(s)X_s^{(n)}\,ds$$

where the integrals are q.m. integrals. Suppose $Y_a^{(n)} \xrightarrow[n\to\infty]{\text{q.m.}} Y_a$, and $\{X_t^{(n)}\}$ converges to a white noise X_t. Then we can show that $\{Y_t^{(n)}, a \le t \le b\}$ converges to a q.m. continuous process $\{Y_t, a \le t \le b\}$ for any b such that $\alpha, \beta \in L_2(a,b)$. Further, $\{Y_t, a \le t \le b\}$ satisfies the equation

$$Y_t = Y_a + \int_a^t \alpha(s)Y_s\,ds + \int_a^t \beta(s)X_s\,ds \qquad a \le t \le b$$

These considerations justify the use of (8.19), even when the driving force is "not quite white." Finally, we note that for our earlier example, we can easily show that

$$Y_t = \int_{-\infty}^{\infty} \frac{1}{1 + i2\pi\nu} e^{i2\pi\nu t} \, d\hat{Z}_\nu \tag{8.21}$$

is the unique solution to the equation

$$Y_t = Y_a - \int_a^t Y_s \, ds + (Z_t - Z_a) \qquad t \geq a \tag{8.22}$$

corresponding to the initial condition

$$Y_a = \int_{-\infty}^{\infty} \frac{1}{1 + i2\pi\nu} e^{i2\pi\nu a} \, d\hat{Z}_\nu \tag{8.23}$$

For an arbitrary initial condition Y_a, the general solution of (8.22) is of the form

$$Y_t = Y_a e^{-(t-a)} + \int_a^t e^{-(t-s)} \, dZ_s \tag{8.24}$$

Hence, if $E|Y_a|^2$ stays finite as $a \to -\infty$, then

$$Y_t \xrightarrow[a \to -\infty]{\text{q.m.}} \int_{-\infty}^t e^{-(t-s)} \, dZ_s$$

By Proposition 8.2,

$$\int_{-\infty}^t e^{-(t-s)} \, dZ_s = \int_{-\infty}^{\infty} \frac{1}{2\pi i\nu + 1} e^{i2\pi\nu t} \, d\hat{Z}_\nu$$

Therefore, under rather general conditions, every solution of (8.22) approaches (8.21) as the initial time $a \to -\infty$.

It is easy to generalize the preceding considerations to vector differential equations

$$\dot{\mathbf{Y}}_t = \mathbf{A}(t)\mathbf{Y}_t + \mathbf{B}(t)\mathbf{X}_t$$

where $\mathbf{A}$ and $\mathbf{B}$ are now matrices and $\mathbf{Y}_t$ and $\mathbf{X}_t$ are column vectors of possibly different dimensions whose components are, respectively, second-order processes and orthogonal white-noise processes. Proposition 8.3 is easily generalized to this case. The required conditions on A and B are now

$$\sum_{i,j} \int_a^b |A_{ij}(t)|^2 \, dt < \infty$$

$$\sum_{i,j} \int_a^b |B_{ij}(t)|^2 \, dt < \infty$$

9. LINEAR PREDICTION AND FILTERING

Linear prediction and filtering are special cases of a general estimation problem that can be formulated as follows: Let $\{X_t, t \in T\}$ be a second-order process and let Y be a second-order random variable. Let $\mathcal{H}_X$ denote Hilbert space generated by $\{X_t, t \in T\}$. A random variable $\hat{Y}$ is said to be a **linear least-squares estimator** of Y given $\{X_t, t \in T\}$ if (1) $\hat{Y} \in \mathcal{H}_X$, and (2)

$$E|\hat{Y} - Y|^2 = \min_{Z \in \mathcal{H}_X} E|Z - Y|^2 \tag{9.1}$$

Filtering and prediction problems are estimation problems involving two second-order processes $\{X_t\}$ and $\{Y_t\}$, where for each t we want to find the linear least-squares estimator of Y_t given $\{X_s, s \leq t\}$. A pure prediction problem is one where $Y_t = X_{t+\alpha}$ for some positive α. In practice, the parameter space for the two processes is usually the same, the most common cases being $[0, \infty)$ or $(-\infty, \infty)$. The fact that an estimator is required for each t makes it especially important that the implementation of the estimator be simple. Both the Wiener filter and Kalman-Bucy filter, in different circumstances, achieve this goal extremely well. This fact is responsible for the importance and the widespread use of these filters. Before discussing these problems in detail, we first prove the following important characterization of linear least-squares estimators.

Proposition 9.1. Let $\hat{Y} \in \mathcal{H}_X$. Then, $\hat{Y}$ is a linear least-squares estimator of Y given $\{X_t, t \in T\}$ if and only if

$$E(\hat{Y} - Y)\bar{X}_t = 0 \qquad \text{for every } t \in T \tag{9.2}$$

and equivalently,

$$E(\hat{Y} - Y)\bar{Z} = 0 \qquad \text{for every } Z \in \mathcal{H}_X$$

Proof: For any $Z \in \mathcal{H}_X$, $\hat{Y} - Z$ is again in $\mathcal{H}_X$. Hence, if (9.2) holds, then

$$\begin{aligned} E|Z - Y|^2 &= E|Z - \hat{Y} + \hat{Y} - Y|^2 \\ &= E|Z - \hat{Y}|^2 + E|\hat{Y} - Y|^2 \end{aligned} \tag{9.3}$$

Therefore, (9.1) is satisfied.

Conversely, suppose $\hat{Y}$ satisfies (9.1). Set

$$Z = Y - \frac{E(\hat{Y} - Y)\bar{X}_t}{E|X_t|^2} X_t$$

Then, $Z \in \mathcal{H}_X$, and from (9.1), we have

$$E|Z - Y|^2 - E|\hat{Y} - Y|^2 \geq 0$$

However, by direct computation, we find

$$E|Z - Y|^2 = E|Y - \hat{Y}|^2 - \frac{|E(\hat{Y} - Y)X_t|^2}{E|X_t|^2} \geq E|Y - \hat{Y}|^2$$

Therefore, (9.2) follows.

Remark: Equation (9.3) proves the uniqueness of linear least-squares estimators. If Z and $\hat{Y}$ are both linear least-squares estimators of Y given $\{X_t, t \in T\}$, then $E|Z - Y|^2 = E|\hat{Y} - Y|^2$, so (9.3) implies $E|Z - Y|^2 = 0$. Hence, we shall refer to *the* linear least-squares estimator of Y given $\{X_t, t \in T\}$. Furthermore, from Proposition 9.1 we see that the linear least-squares estimator of Y given $\{X_t, t \in T\}$ is the projection of Y into the smallest Hilbert space $\mathcal{H}_X$ spanned by $\{X_t, t \in T\}$. We shall denote this projection by a more explicit notation

$$\hat{E}(Y|\mathcal{H}_X)$$

The Wiener theory of filtering and prediction is distinguished by two facts. (1) It deals with wide-sense stationary processes with well-defined spectral-density functions, and (2) the given information on which the estimator is to be constructed is the infinite past. These conditions will be made more precise later. Under these conditions, it is often possible to express the estimator as the output of a linear time-invariant filtering operation on the observed process. This is an extremely convenient form for the solution from the point of view of implementation.

Let $\{X_t, -\infty < t < \infty\}$ and $\{Y_t, -\infty < t < \infty\}$ be wide-sense stationary processes. We assume that there exist bounded functions S_x, S_y, and S_{xy} such that

$$EX_{t+\tau}\bar{X}_t = \int_{-\infty}^{\infty} S_x(\nu)e^{i2\pi\nu\tau}\,d\nu \tag{9.4a}$$

$$EX_{t+\tau}\bar{Y}_t = \int_{-\infty}^{\infty} S_{xy}(\nu)e^{i2\pi\nu\tau}\,d\nu \tag{9.4b}$$

$$EY_t\bar{Y}_{t+\tau} = \int_{-\infty}^{\infty} S_y(\nu)e^{i2\pi\nu\tau}\,d\nu \tag{9.4c}$$

The spectral-density functions S_x and S_y are necessarily real and positive. The cross-spectral density S_{xy} need not be real. Under these assumptions X_t, Y_t are not only individually wide-sense stationary processes, but are in fact jointly wide-sense stationary in the sense that for arbitrary complex constants a and b, $Z_t = aX_t + bY_t$ is again a wide-sense stationary process. For such a process Z_t, we have

$$EZ_{t+\tau}\bar{Z}_t = \int_{-\infty}^{\infty} [|a|^2S_x(\nu) + |b|^2S_y(\nu) + a\bar{b}S_{xy}(\nu) + \bar{a}b\bar{S}_{xy}(\nu)]e^{i2\pi\nu\tau}\,d\nu \tag{9.5}$$

In other words, the spectral-density function of $\{Z_t, -\infty < t < \infty\}$ is given by

$$S_z(\nu) = |a|^2 S_x(\nu) + |b|^2 S_y(\nu) + a\bar{b} S_{xy}(\nu) + \bar{a} b \bar{S}_{xy}(\nu) \tag{9.6}$$

It is obvious that S_z must be real and nonnegative for arbitrary a and b. A necessary and sufficient condition for this is that for every ν,

$$S_x(\nu) S_y(\nu) \geq |S_{xy}(\nu)|^2 \tag{9.7}$$

A proof of this fact is easily constructed by noting the fact that (9.6) is a quadratic form involving the Hermitian matrix

$$\mathbf{S}(\nu) = \begin{bmatrix} S_x(\nu) & S_{xy}(\nu) \\ \bar{S}_{xy}(\nu) & S_y(\nu) \end{bmatrix} \tag{9.8}$$

Hence, for $S_z(\nu)$ to be nonnegative for every ν and a and b, it is necessary and sufficient that $\mathbf{S}(\nu)$ be nonnegative definite for every ν. This condition is equivalent to (9.7).

For such a pair of processes $\{X_t, Y_t, -\infty < t < \infty\}$, we shall consider the problem of finding the linear least-squares estimator of Y_t given $\{X_s, -\infty < s < t\}$. If we denote by $\mathcal{H}_X{}^t$ the Hilbert space spanned by $\{X_s, -\infty \leq s < t\}$, the problem is to find the projection $\hat{E}(Y_t|\mathcal{H}_X{}^t)$. We recall that $\hat{E}(Y_t|\mathcal{H}_X{}^t)$ is characterized by two properties: (1) $\hat{E}(Y_t|\mathcal{H}_X{}^t)$ belongs to $\mathcal{H}_X{}^t$ and (2) $Y_t - \hat{E}(Y_t|\mathcal{H}_X{}^t)$ is orthogonal to $\mathcal{H}_X{}^t$. A solution to this problem in the form of a filtering operation will be constructed using the ideas of white noise and white-noise integrals that we introduced in the last section. First, we state the following fundamental result of Paley and Wiener [1934, chap. 1].

Proposition 9.2. Let $S \in L_1$ be a real nonnegative function such that

$$\int_{-\infty}^{\infty} \frac{|\ln S(\nu)|}{1 + \nu^2}\, d\nu < \infty \tag{9.9}$$

Then there exists a function $\hat{h} \subset L_2$ such that

$$|\hat{h}(\nu)|^2 = S(\nu) \tag{9.10a}$$

$$h(t) = \int_{-\infty}^{\infty} e^{i2\pi\nu t} \hat{h}(\nu)\, d\nu = 0 \qquad t < 0 \tag{9.10b}$$

$$\hat{h}(u + iv) = \int_{-\infty}^{\infty} e^{i2\pi(u+iv)t} h(t)\, dt \neq 0 \qquad v > 0 \tag{9.10c}$$

Remark: Condition (9.10b) shows that $\hat{h}$ is the transfer function of a **nonanticipative filter** (physically realizable filter). Condition (9.10c) is known as the **minimum-phase condition** in circuit theory. Roughly speaking, it means that $1/\hat{h}$ is again nonanticipative. However, $1/\hat{h}$

has no inverse Fourier transform, so the nonanticipative property is a little more difficult to define.

A complete proof is somewhat complicated and will be omitted. Instead, we shall outline a procedure by which the desired $\hat{h}$ can be found. First, consider the transformation

$$\nu = -\tan\frac{\theta}{2} \tag{9.11}$$

Then, (9.9) becomes

$$\frac{1}{2}\int_{-\pi}^{\pi}\left|\ln S\left(-\tan\frac{\theta}{2}\right)\right| d\theta < \infty$$

so that $\ln S(-\tan\theta/2)$ has a Fourier series

$$\ln S\left(-\tan\frac{\theta}{2}\right) = \sum_{n=-\infty}^{\infty} a_n e^{in\theta} \tag{9.12}$$

where a_n are given by

$$a_n = \frac{1}{2\pi}\int_{-\pi}^{\pi} e^{-in\theta} \ln S\left(-\tan\frac{\theta}{2}\right) d\theta \tag{9.13}$$

With the correspondence (9.11), we have

$$e^{i\theta} = \frac{1-i\nu}{1+i\nu} \tag{9.14}$$

so that (9.12) and (9.13) become

$$\ln S(\nu) = \sum_{n=-\infty}^{\infty} a_n \left(\frac{1-i\nu}{1+i\nu}\right)^n \tag{9.15}$$

and

$$a_n = \frac{1}{\pi}\int_{-\infty}^{\infty} \frac{1}{1+\nu^2}\left(\frac{1+i\nu}{1-i\nu}\right)^n \ln S(\nu)\, d\nu \tag{9.16}$$

If $|\hat{h}|^2 = S$, then

$$\ln S(\nu) = \ln \hat{h} + \ln \bar{\hat{h}} \tag{9.17}$$

We now identify

$$\hat{h}(\nu) = \exp\left[\frac{a_0}{2} + \sum_{n=1}^{\infty} a_n \left(\frac{1-i\nu}{1+i\nu}\right)^n\right] \tag{9.18}$$

Because $a_{-n} = \bar{a}_n$, (9.17) is satisfied, hence also (9.10a). Conditions (9.10b)

and (9.10c) follow from the fact that

$$f(u + iv) = \frac{a_0}{2} + \sum_{n=1}^{\infty} a_n \left[\frac{1 - i(u + iv)}{1 + i(u + iv)}\right]^n \tag{9.19}$$

is analytic for $v < 0$. Hence, $\hat{h}(u + iv)$ is analytic for $v < 0$, and (9.10b) follows (more or less) from contour integration closing the contour in the lower half plane. Condition (9.10c) follows from the fact that if $f(z)$ is analytic, then $e^{f(z)}$ has no zero. We call the above procedure the **spectral-factorization** procedure, since S is usually identified with a spectral-density function.

Let $\{X_t, -\infty < t < \infty\}$ have a spectral-density function S_x, and let $\hat{h}$ be obtained by factoring S_x so that (9.10) is satisfied with S_x replacing S in (9.10a). Then, in view of our discussion on (8.15), $\{X_t, -\infty < t < \infty\}$ can be regarded as the output of filtering a white noise $\{\zeta_t, -\infty < t < \infty\}$ with a nonanticipative filter having transfer function $\hat{h}$. Condition (9.10c) means that the white noise $\{\zeta_t, -\infty < t < \infty\}$ can in turn be obtained by filtering $\{X_t, -\infty < t < \infty\}$ by $1/\hat{h}$, which is also nonanticipative. A more precise statement of these results can be made in terms of the process with orthogonal increments $\{Z_t, -\infty < t < \infty\}$ corresponding to $\{\zeta_t, -\infty < t < \infty\}$. Condition (9.10$b$) then implies that there exists a process $\{Z_t, -\infty < t < \infty\}$ with $Z_0 = 0$ and $E\, dZ_t\, d\overline{Z}_s = \delta_{ts}\, dt$ such that

$$\begin{aligned} X_t &= \int_{-\infty}^{\infty} h(t - s)\, dZ_s \\ &= \int_{-\infty}^{\infty} e^{i2\pi\nu t}\hat{h}(\nu)\, d\hat{Z}_\nu \end{aligned} \tag{9.20}$$

Condition (9.10c) implies that for each t

$$Z_t \in \mathcal{H}_X{}^t \tag{9.21}$$

We can express Z_t more explicitly in terms of the spectral process $\{\hat{X}_\nu, -\infty < \nu < \infty\}$ as

$$Z_t = \int_{-\infty}^{\infty} \frac{1}{\hat{h}(\nu)} \left(\int_0^t e^{i2\pi\nu s}\, ds\right) d\hat{X}_\nu \tag{9.22}$$

which points out even more clearly that ζ_t $(=\dot{Z}_t)$ is obtained by filtering X_t by $1/\hat{h}$.

We shall now give the main result of the Wiener theory of filtering as follows [Wiener, 1949].

Proposition 9.3. Let $\{X_t, Y_t, -\infty < t < \infty\}$ be a pair of wide-sense stationary processes satisfying (9.4). Let $\hat{h}$ be obtained by factoring S_x so that (9.10a) to (9.10c) are satisfied. Let $\{Z_t, -\infty < t < \infty\}$ be as in (9.20) to (9.22). Then,

$$\hat{E}(Y_t|\mathcal{H}_X{}^t) = \int_{-\infty}^{t} g(t-s)\, dZ_s \tag{9.23}$$

where g is given by

$$g(t) = \int_{-\infty}^{\infty} e^{i2\pi\nu t}\, \overline{\frac{S_{xy}(\nu)}{\hat{h}(\nu)}}\, d\nu \qquad -\infty < t < \infty \tag{9.24}$$

Remark: Because $|S_{xy}/\hat{h}|^2 = |S_{xy}|^2/S_x$ is bounded by S_y (from (9.7)), the integral in (9.24) is well defined as a limit in L_2 mean.

Proof: To prove (9.23), we need to verify the two conditions:

$$\int_{-\infty}^{t} g(t-s)\, dZ_s \in \mathcal{H}_X{}^t \tag{9.25a}$$

$$EY_t\bar{X}_\tau = E\left[\int_{-\infty}^{t} g(t-s)\, dZ_s\right]\bar{X}_\tau \qquad \tau \le t \tag{9.25b}$$

To verify (9.25*a*), we note that $\int_{-\infty}^{t} g(t-s)\, dZ_s$ is in $\mathcal{H}_Z{}^t$, where $\mathcal{H}_Z{}^t$ denotes the Hilbert space spanned by $\{Z_s, -\infty < s \le t\}$. Since for each t, $Z_t \in \mathcal{H}_X{}^t$, so $\mathcal{H}_Z{}^t \subset \mathcal{H}_X{}^t$, verifying (9.25*a*). To verify (9.25*b*), we note that

$$\begin{aligned}\int_{-\infty}^{\infty} g(t-s)\, dZ_s &= \int_{-\infty}^{t} g(t-s)\, dZ_s + \int_{t}^{\infty} g(t-s)\, dZ_s \\ &= \int_{-\infty}^{\infty} \overline{[S_{xy}(\nu)/\hat{h}(\nu)]e^{i2\pi\nu t}}\, d\hat{Z}_\nu \end{aligned}\tag{9.26}$$

Hence, from (9.20) we have

$$\begin{aligned}E\left[\int_{-\infty}^{\infty} g(t-s)\, dZ_s\right]\bar{X}_\tau &= \int_{-\infty}^{\infty} \bar{S}_{xy}(\nu)e^{i2\pi\nu(t-\tau)}\, d\nu \\ &= EY_t\bar{X}_\tau\end{aligned}\tag{9.27}$$

On the other hand, since

$$X_\tau = \int_{-\infty}^{\tau} h(\tau - s)\, dZ_s$$

and $E\, dZ_t\, \overline{dZ_s} = \delta_{ts}\, dt$, we have

$$E\left[\int_{t}^{\infty} g(t-s)\, dZ_s\right]\bar{X}_\tau = 0 \qquad \text{whenever } \tau \le t$$

Therefore,

$$\begin{aligned}E\left[\int_{-\infty}^{t} g(t-s)\, dZ_s\right]\bar{X}_\tau &= E\left[\int_{-\infty}^{\infty} g(t-s)\, dZ_s\right]\bar{X}_\tau \\ &= EY_t\bar{X}_\tau \qquad \tau \le t\end{aligned}$$

which verifies (9.25*b*) and completes the proof. ▮

The solution (9.23) can be put in a more useful form by using (9.22). If we define

$$\hat{\gamma}(\nu) = \int_0^\infty g(t)e^{-i2\pi\nu t}\,dt \tag{9.28}$$

then we can write

$$E(Y_t|\mathcal{H}_X{}^t) = \int_{-\infty}^{\infty} \frac{\hat{\gamma}(\nu)}{\hat{h}(\nu)} e^{i2\pi\nu t}\,d\hat{X}_\nu \tag{9.29}$$

which is the output to a time-invariant linear filter with $\{X_t, -\infty < t < \infty\}$ as input and $\hat{\gamma}/\hat{h}$ as its transfer function. The filter is called the **Wiener filter.** We should note that $\hat{\gamma}$ is *not* the Fourier integral of g, since the lower limit in the integral in (9.28) is 0 and not $-\infty$, and $g(t)$ in general is not zero for $t < 0$.

For a general S_x, the spectral-factorization procedure that we outlined earlier (9.16 and 9.18) may not lead to a closed-form solution for $\hat{h}$. It is also somewhat complicated. The factorization problem becomes trivial if S_x is a rational function. If S_x is rational, then we can always write

$$S_x(\nu) = K^2 \frac{\prod_{k=1}^{m} (\nu - z_k)(\nu - \bar{z}_k)}{\prod_{k=1}^{n} (\nu - p_k)(\nu - \bar{p}_k)} \tag{9.30}$$

where every z_k and p_k have positive imaginary parts. Since $|\hat{h}|^2 = S_x$ and $\hat{h}(u + iv)$ has neither poles nor zeros for $v < 0$, $\hat{h}$ must be of the form

$$\hat{h}(\nu) = A \frac{\prod_{k=1}^{m} (\nu - z_k)}{\prod_{k=1}^{n} (\nu - p_k)} \tag{9.31}$$

where $|A|^2 = K^2$. The constant A can be determined from (9.18),

$$\hat{h}(-i) = e^{a_0/2} = A \frac{\prod_{k=1}^{m} (-i - z_k)}{\prod_{k=1}^{n} (-i - p_k)}$$

and a_0 is given by

$$a_0 = \frac{1}{\pi} \int_{-\infty}^{\infty} \frac{1}{1 + \nu^2} \ln S(\nu)\, d\nu$$

However, we note that any choice of A would yield an $\hat{h}$ satisfying (9.10). Equation (9.18) merely gives a specific one. When $\{X_t, -\infty < t < \infty\}$ is real, then $S_x(\nu)$ is an even function of ν, and the constant A in (9.31) can always be chosen so that the impulse response

$$h(t) = \int_{-\infty}^{\infty} e^{i2\pi\nu t} \hat{h}(\nu)\, d\nu$$

is real valued. This is a convenient choice. However, as far as the Wiener filtering problem is concerned, the choice of A is unimportant. It is fairly obvious that the solution (9.23) or (9.29) is independent of this choice.

For our first example, consider the following pure prediction problem: $S_x(\nu) = [1 + (2\pi\nu)^2]^{-1}$ and $Y_t = X_{t+\alpha}$, $\alpha > 0$. The factorization is trivial and we can take

$$\hat{h}(\nu) = \frac{1}{1 + i2\pi\nu}$$

From (9.24) $g(t)$ is found to be

$$\begin{aligned} g(t) &= \int_{-\infty}^{\infty} e^{i2\pi\nu t} e^{i2\pi\nu\alpha} \frac{1}{1 + i2\pi\nu}\, d\nu \\ &= \begin{cases} e^{-(t+\alpha)} & t > -\alpha \\ 0 & t < -\alpha \end{cases} \end{aligned}$$

From (9.28) we find

$$\hat{\gamma}(\nu) = e^{-\alpha} \frac{1}{1 + i2\pi\nu}$$

so that

$$\hat{E}(Y_t|\mathcal{H}_{X^t}) = e^{-\alpha} \int_{-\infty}^{\infty} e^{i2\pi\nu t}\, d\hat{X}_\nu = e^{-\alpha} X_t$$

For this simple example, the predictor is nothing more than an attenuator.

As a second example, suppose

$$\mathbf{S}(\nu) = \begin{bmatrix} \dfrac{1}{1 + (2\pi\nu)^2} + \dfrac{1}{4 + (2\pi\nu)^2} & \dfrac{1}{1 + (2\pi\nu)^2} \\ \dfrac{1}{1 + (2\pi\nu)^2} & \dfrac{1}{1 + (2\pi\nu)^2} \end{bmatrix}$$

Factorization of $S_x(\nu) = [1/1 + (2\pi\nu)^2] + [1/4 + (2\pi\nu)^2]$ yields

$$\hat{h}(\nu) = \frac{\sqrt{5} + \sqrt{2}\, i2\pi\nu}{(1 + i2\pi\nu)(2 + i2\pi\nu)}$$

This gives

$$\frac{S_{xy}(\nu)}{\hat{h}(\nu)} = \frac{2 + i2\pi\nu}{[\sqrt{5} + \sqrt{2}\,(i2\pi\nu)](1 - i2\pi\nu)}$$

Therefore, (9.24) can be computed to yield

$$\begin{aligned} g(t) &= \int_{-\infty}^{\infty} e^{i2\pi\nu t} \frac{2 - i2\pi\nu}{(1 + i2\pi\nu)[\sqrt{5} - \sqrt{2}\,(i2\pi\nu)]}\, d\nu \\ &= \int_{-\infty}^{\infty} \left(\frac{3}{\sqrt{5} + \sqrt{2}} \frac{1}{1 + i2\pi\nu} \right. \\ &\qquad \left. + \frac{2 + \sqrt{5/2}}{1 + \sqrt{5/2}} \frac{1}{\sqrt{5} - \sqrt{2}\, i2\pi\nu} \right) e^{i2\pi\nu t}\, d\nu \\ &= \begin{cases} \dfrac{3}{\sqrt{5} + \sqrt{2}}\, e^{-t} & t > 0 \\ \dfrac{2 - \sqrt{5/2}}{\sqrt{5} + \sqrt{2}} \exp\left(\sqrt{\dfrac{5}{2}}\, t\right) & t < 0 \end{cases} \end{aligned}$$

and (9.28) gives

$$\hat{\gamma}(\nu) = \frac{3}{\sqrt{5} + \sqrt{2}} \frac{1}{1 + i2\pi\nu}$$

Finally, from (9.28) we get

$$\begin{aligned} \hat{E}(Y_t | \mathcal{H}_X{}^t) &= \int_{-\infty}^{\infty} \frac{3}{\sqrt{5} + \sqrt{2}} \frac{2 + i2\pi\nu}{\sqrt{5} + \sqrt{2}\, i2\pi\nu} e^{i2\pi\nu t}\, d\hat{X}_\nu \\ &= \frac{3}{2 + \sqrt{10}} \left[X_t + (2 - \sqrt{5/2}) \int_{-\infty}^{t} \exp \right. \\ &\qquad \left. [-\sqrt{5/2}\,(t - s)] X_s\, ds \right] \end{aligned} \tag{9.31a}$$

It is quite obvious that our formulation of the filtering problem can be extended to cover cases where both the process to be estimated and the observation process are vector valued. The general case involves a pair of vector-valued processes $\{\mathbf{X}_t, \mathbf{Y}_t, t \in T\}$, not necessarily of the same dimension, whose second-order properties are completely known. The problem is then to estimate $\mathbf{Y}_t$ given $\{\mathbf{X}_s, s \leq t\}$. In wide-sense stationary cases where spectral-density functions exist, one would expect that the Wiener theory can again be developed. If we denote by $\mathbf{A}^+$ the Hermitian adjoint of a matrix $\mathbf{A}$, we can define the matrix spectral-density function $\mathbf{S}_x$ by

$$E\mathbf{X}_{t+\tau}\mathbf{X}_\tau^+ = \int_{-\infty}^{\infty} \mathbf{S}_x(\nu)e^{i2\pi\nu t}\,d\nu$$

The key step in obtaining a solution to the Wiener filtering problem is again a spectral-factorization problem. But now, we have to obtain a factorization of the matrix-valued function $\mathbf{S}_x(\nu)$ into the form

$$\mathbf{S}_x(\nu) = \hat{\mathbf{h}}(\nu)\hat{\mathbf{h}}^+(\nu)$$

such that the matrix $\hat{\mathbf{h}}(\nu)$ satisfies conditions similar to (9.10*b*) and (9.10*c*). Except that instead of (9.10*c*), the determinant of $\hat{\mathbf{h}}(u + iv)$ is to have no zero for $v > 0$. The final solution can be expressed in a form which generalizes (9.29) as

$$\hat{E}(\mathbf{Y}_t|\mathcal{H}_X{}^t) = \int_{-\infty}^{\infty} e^{i2\pi\nu t}[\hat{\mathbf{h}}^{-1}(\nu)\boldsymbol{\gamma}(\nu)]\,d\hat{\mathbf{X}}_\nu$$

where the matrix $\boldsymbol{\gamma}(\nu)$ is similarly defined, as in the scalar case. The matrix spectral-factorization problem is considerably more difficult than the scalar problem. For the rational case, a number of finite algorithms to achieve the factorization have been derived [Wiener and Masani, 1958; Wong and Thomas, 1961; Youla, 1961].

In some areas of application, the Wiener formulation of the filtering problem is not appropriate because of some of its inherent assumptions. Among these are the following: (1) wide-sense stationarity and existence of spectral densities; (2) the second-order properties of the processes $\{X_t, Y_t, -\infty < t < \infty\}$ are known, and no other information is known; (3) the estimator is to be based on the infinite past of the observation process. These limitations are removed in the formulation of the filtering problem due to Kalman and Bucy. Instead, they made other assumptions which are more natural in a great variety of applications. The form of the solution is also different. While in the Wiener theory, the final solution is in the form of a time-invariant linear and nonanticipative filter, the Kalman-Bucy theory yields a differential equation which is satisfied by the estimator. Implementation of the "filter" in feedback form is thus immediate [Kalman and Bucy, 1961].

The Kalman-Bucy filter problem is usually stated in vector form as follows: Let $\{\mathbf{X}_t, \mathbf{Y}_t, t \geq t_0\}$ be a pair of vector-valued second-order processes. The X process will be the observation process, and the Y process is to be estimated. While this notational convention is consistent with our earlier discussion, it is not universal. Often in the literature, the two letters X and Y are used in just the opposite way. The basic assumptions are the following: Throughout, **boldface** will be used to denote vectors and matrices, prime denotes transpose, + denotes Hermitian adjoint, and **I** denotes identity matrix.

1. The process to be estimated satisfies

$$\dot{\mathbf{Y}}_t = \mathbf{F}(t)\mathbf{Y}_t + \mathbf{A}(t)\boldsymbol{\zeta}_t \qquad t > t_0 \tag{9.32}$$

where $\boldsymbol{\zeta}_t$ has components that are orthogonal white-noise processes with

$$E\boldsymbol{\zeta}_t\boldsymbol{\zeta}_s^+ = \delta(t-s)\,\mathbf{I} \tag{9.33}$$

2. The observation process satisfies

$$\dot{\mathbf{X}}_t = \mathbf{H}(t)\mathbf{Y}_t + \mathbf{B}(t)\boldsymbol{\zeta}_t \qquad t > t_0 \tag{9.34}$$

Remarks:

(*a*) Both (9.32) and (9.34) are to be interpreted along the lines discussed in Sec. 8. We shall denote the process with orthogonal components which correspond to $\boldsymbol{\zeta}_t$ by $\mathbf{Z}_t$, so that formally $\dot{\mathbf{Z}}_t = \boldsymbol{\zeta}_t$. If $t_0 > -\infty$, it is convenient to set $\mathbf{Z}_{t_0} = 0$.

(*b*) We note that (9.34) is not really a differential equation, since $\mathbf{X}_t$ can be immediately expressed explicitly in terms of the **Y** and $\boldsymbol{\zeta}$ process by integrating (9.34). However, the problem would be no more general if we replace (9.34) by a linear differential equation in $\mathbf{X}_t$. Such an equation can always be changed into (9.34) by redefining the observation process.

(*c*) It is necessary to assume that the initial values $\mathbf{X}_{t_0}$ and $\mathbf{Y}_{t_0}$ are random variables orthogonal to $\{\mathbf{Z}_t, t \geq t_0\}$, in particular they can simply be constants.

As usual, let $\mathcal{H}_X{}^t$ denote the smallest Hilbert space generated by $\{\mathbf{X}_\tau, t_0 \leq \tau \leq t\}$, and let $\hat{E}$ denote projection. The Kalman-Bucy filtering problem is to find $\hat{E}(\mathbf{Y}_s|\mathcal{H}_X{}^t)$, and the main results can be summarized as follows.

Proposition 9.4. Let $\{\mathbf{X}_t, \mathbf{Y}_t, t \geq t_0\}$ satisfy (9.32) and (9.34). Let $\boldsymbol{\Phi}(s|t)$ be the unique solution of

$$\frac{d}{ds}\boldsymbol{\Phi}(s|t) = \mathbf{F}(s)\boldsymbol{\Phi}(s|t) \qquad s > t \tag{9.35}$$

with initial condition $\boldsymbol{\Phi}(t|t) = \mathbf{I}$. Let $\mathbf{A}(t)$ and $\mathbf{B}(t)$ in (9.32) and (9.33) be continuous functions on $[t_0, \infty)$.

(*a*) For $s \geq t$,

$$\hat{E}(\mathbf{Y}_s|\mathcal{H}_X{}^t) = \mathbf{\Phi}(s|t)\hat{E}(\mathbf{Y}^t|\mathcal{H}_X{}^t) \tag{9.36}$$

(*b*) Let $\tilde{\mathbf{Y}}_t$ denote $\hat{E}(\mathbf{Y}_t|\mathcal{H}_X{}^t)$, then

$$\dot{\tilde{\mathbf{Y}}}_t = \mathbf{F}(t)\tilde{\mathbf{Y}}_t + \mathbf{K}(t)[\dot{\mathbf{X}}_t - \mathbf{H}(t)\tilde{\mathbf{Y}}_t] \tag{9.37}$$

(*c*) Let $\boldsymbol{\varepsilon}_t = \mathbf{Y}_t - \tilde{\mathbf{Y}}_t$ be the error vector, and let $\mathbf{\Sigma}(t)$ denote the covariance matrix

$$\mathbf{\Sigma}(t) = E\boldsymbol{\varepsilon}_t\boldsymbol{\varepsilon}_t{}^+ \tag{9.38}$$

Then,

$$\dot{\mathbf{\Sigma}}(t) = [\mathbf{A}(t) - \mathbf{K}(t)\mathbf{B}(t)][\mathbf{A}(t) - \mathbf{K}(t)\mathbf{B}(t)]^+ + [\mathbf{F}(t) - \mathbf{K}(t)\mathbf{H}(t)]\mathbf{\Sigma}(t) + \mathbf{\Sigma}(t)[\mathbf{F}(t) - \mathbf{K}(t)\mathbf{H}(t)]^+ \tag{9.39}$$

and

$$\mathbf{K}(t)\mathbf{B}(t)\mathbf{B}^+(t) = \mathbf{A}(t)\mathbf{B}^+(t) + \mathbf{\Sigma}(t)\mathbf{H}^+(t) \tag{9.40}$$

Remarks:

(*a*) A complete proof is rather complicated, and will not be presented. Instead, we shall give a heuristic derivation.

(*b*) The continuity conditions on $\mathbf{A}(t)$ and $\mathbf{B}(t)$ are sufficient, but not necessary. However, some smoothness condition is needed. Unfortunately, this point is largely lost in our formal derivation.

(*c*) Equation (9.39) can be simplified somewhat by using (9.40). If $\mathbf{B}(t)\mathbf{B}^+(t)$ is invertible, these two equations can be combined to give a single equation in $\mathbf{\Sigma}(t)$, which is a nonlinear differential equation of the Riccati type.

(*d*) Once $\mathbf{K}$ is determined from (9.39) and (9.40), implementation of (9.37) in feedback form is immediate and yields a continuous estimate. Feedback implementation of (9.37) is often referred to as the **Kalman-Bucy filter.**

First, we derive (9.36) as follows: From (9.32) we can write for $s \geq t$,

$$\mathbf{Y}_s = \mathbf{\Phi}(s|t)\mathbf{Y}_t + \int_t^s \mathbf{\Phi}(s|t)\mathbf{A}(\tau)\, d\mathbf{Z}_\tau$$

Therefore,

$$\hat{E}(\mathbf{Y}_s|\mathcal{H}_X{}^t) = \mathbf{\Phi}(s|t)\tilde{\mathbf{Y}}_t + \int_t^s \mathbf{\Phi}(s|\tau)\mathbf{A}(\tau)\, \hat{E}(d\mathbf{Z}_\tau|\mathcal{H}_X{}^t)$$

Now, let $\mathcal{H}^t$ denote the smallest Hilbert space containing $\mathbf{X}_{t_0}$, $\mathbf{Y}_{t_0}$, and $\{\mathbf{Z}_\tau,\ \tau \leq t\}$. Because of (9.32) and (9.34), $\mathcal{H}_X{}^t$ is contained in $\mathcal{H}^t$. Because

$\mathbf{Z}_t$ is a process with orthogonal increments, and $\mathbf{X}_{t_0}$, $\mathbf{Y}_t$, are orthogonal to $\{\mathbf{Z}_t, t \geq t_0\}$,

$$\hat{E}(d\mathbf{Z}_\tau|\mathcal{H}^t) = 0 \qquad \tau \geq t$$

Hence,

$$\hat{E}(d\mathbf{Z}_\tau|\mathcal{H}_X{}^t) = \hat{E}[\hat{E}(d\mathbf{Z}_\tau|\mathcal{H}^t)|\mathcal{H}_X{}^t] = 0 \qquad \tau \geq t$$

It follows that

$$E(\mathbf{Y}_s|\mathcal{H}_X{}^t) = \Phi(s|t)\tilde{\mathbf{Y}}_t$$

which was to be derived.

To derive (9.37), we first note that every element in $\mathcal{H}_X{}^t$ can be written in the form of

$$\xi = \boldsymbol{\alpha}\mathbf{X}_{t_0} + \int_{t_0}^t \boldsymbol{\beta}(\tau)\, d\mathbf{X}_\tau$$

Since $\tilde{\mathbf{Y}}_t \in \mathcal{H}_X{}^t$ for each t, we can write

$$\tilde{\mathbf{Y}}_t = \boldsymbol{\alpha}(t)\mathbf{X}_{t_0} + \int_{t_0}^t \mathbf{K}(t|\tau)\, d\mathbf{X}_\tau \tag{9.41}$$

Thus,

$$d\tilde{\mathbf{Y}}_t = \mathbf{K}(t|t)\, d\mathbf{X}_t + dt\left[\dot{\boldsymbol{\alpha}}(t)\mathbf{X}_{t_0} + \int_{t_0}^t \frac{\partial}{\partial t}\mathbf{K}(t|\tau)\, d\mathbf{X}_\tau\right] \tag{9.42}$$

Since the bracketed terms in (9.42) are in $\mathcal{H}_X{}^t$, we can rewrite (9.42) as

$$d\tilde{\mathbf{Y}}_t = \mathbf{K}(t|t)\, d\mathbf{X}_t + \hat{E}[d\tilde{\mathbf{Y}}_t - \mathbf{K}(t|t)\, d\mathbf{X}_t|\mathcal{H}_X{}^t] \tag{9.43}$$

Now, from (9.34)

$$\hat{E}(d\mathbf{X}_t|\mathcal{H}_X{}^t) = \mathbf{H}(t)\tilde{\mathbf{Y}}_t\, dt \tag{9.44}$$

From (9.32), we have

$$\begin{aligned}\hat{E}(d\tilde{\mathbf{Y}}_t|\mathcal{H}_X{}^t) &= \hat{E}(\hat{E}(\mathbf{Y}_{t+dt}|\mathcal{H}_X{}^{t+dt}) - \hat{E}(\mathbf{Y}_t|\mathcal{H}_X{}^t)|\mathcal{H}_X{}^t)\\ &= \hat{E}(d\mathbf{Y}_t|\mathcal{H}_X{}^t) = \mathbf{F}(t)\tilde{\mathbf{Y}}_t\, dt\end{aligned} \tag{9.45}$$

Using (9.44) and (9.45) in (9.42) yields

$$d\tilde{\mathbf{Y}}_t = \mathbf{F}(t)\tilde{\mathbf{Y}}_t\, dt + \mathbf{K}(t|t)[d\mathbf{X}_t - \mathbf{H}(t)\tilde{\mathbf{Y}}_t\, dt]$$

which is just (9.37) if we set $\mathbf{K}(t|t) = \mathbf{K}(t)$.

To derive (9.39), we combine (9.32), (9.34) and (9.37) and obtain

$$d\boldsymbol{\varepsilon}_t = [\mathbf{F}(t) - \mathbf{K}(t)\mathbf{H}(t)]\boldsymbol{\varepsilon}_t + [\mathbf{A}(t) - \mathbf{K}(t)\mathbf{B}(t)]\, d\mathbf{Z}_t \tag{9.46}$$

Now,

$$\begin{aligned} d\boldsymbol{\Sigma}(t) &= \boldsymbol{\Sigma}(t+dt) - \boldsymbol{\Sigma}(t) \\ &= E\boldsymbol{\varepsilon}_{t+dt}\boldsymbol{\varepsilon}_{t+dt}^{+} - E\boldsymbol{\varepsilon}_t\boldsymbol{\varepsilon}_t^{+} \\ &= E(\boldsymbol{\varepsilon}_{t+dt} - \boldsymbol{\varepsilon}_t)(\boldsymbol{\varepsilon}_{t+dt} - \boldsymbol{\varepsilon}_t)^{+} + E(\boldsymbol{\varepsilon}_{t+dt} - \boldsymbol{\varepsilon}_t)\boldsymbol{\varepsilon}_t^{+} + E\boldsymbol{\varepsilon}_t(\boldsymbol{\varepsilon}_{t+dt} - \boldsymbol{\varepsilon}_t)^{+} \\ &= E\, d\boldsymbol{\varepsilon}_t\, d\boldsymbol{\varepsilon}_t^{+} + E\, d\boldsymbol{\varepsilon}_t\boldsymbol{\varepsilon}_t^{+} + E\boldsymbol{\varepsilon}_t\, d\boldsymbol{\varepsilon}_t^{+} \end{aligned}$$

Using (9.46) and the fact that $d\mathbf{Z}_t$ is orthogonal to $\mathcal{H}^t$, we find

$$\begin{aligned} d\boldsymbol{\Sigma}(t) = {} & [\mathbf{A}(t) - \mathbf{K}(t)\mathbf{B}(t)][\mathbf{A}(t) - \mathbf{K}(t)\mathbf{B}(t)]^{+}\, dt \\ & + [\mathbf{F}(t) - \mathbf{K}(t)\mathbf{H}(t)]\boldsymbol{\Sigma}(t)\, dt + \boldsymbol{\Sigma}(t)[\mathbf{F}(t) - \mathbf{K}(t)\mathbf{H}(t)]^{+}\, dt \end{aligned}$$

which is just (9.39).

Finally, to derive (9.40), we begin by writing $\boldsymbol{\varepsilon}_t$ as the solution of the differential equation (9.46) in the form

$$\boldsymbol{\varepsilon}_t = \boldsymbol{\psi}(t|t_0)\boldsymbol{\varepsilon}_{t_0} + \int_{t_0}^{t} \boldsymbol{\psi}(t|\tau)[\mathbf{A}(\tau) - \mathbf{K}(\tau)\mathbf{B}(\tau)]\, d\mathbf{Z}_t \tag{9.47}$$

where $\boldsymbol{\psi}$ satisfies

$$\begin{aligned} \frac{d}{dt}\boldsymbol{\psi}(t|t_0) &= [\mathbf{F}(t) - \mathbf{K}(t)\mathbf{H}(t)]\boldsymbol{\psi}(t|t_0) \qquad t > t_0 \\ \boldsymbol{\psi}(t|t_0) &= \mathbf{I} \end{aligned} \tag{9.48}$$

Since $\boldsymbol{\varepsilon}_t$ is orthogonal to $\mathcal{H}_X{}^t$, we have for $s \le t$,

$$E\boldsymbol{\varepsilon}_t\mathbf{X}_s{}^{+} = 0 = E\boldsymbol{\varepsilon}_t\left[\int_{t_0}^{s} \mathbf{H}(\tau)\mathbf{Y}_\tau\, d\tau + \int_{t_0}^{s} \mathbf{B}(\tau)\, d\mathbf{Z}_\tau\right]^{+} \tag{9.49}$$

Furthermore, for $\tau \le t$, $\tilde{\mathbf{Y}}_t \in \mathcal{H}_X{}^t$, hence

$$E\boldsymbol{\varepsilon}_t\mathbf{Y}_\tau{}^{+} = E\boldsymbol{\epsilon}_t(\mathbf{Y}_\tau - \tilde{\mathbf{Y}}_\tau)^{+} = E\boldsymbol{\varepsilon}_t\boldsymbol{\varepsilon}_\tau{}^{+} \tag{9.50}$$

Therefore, (9.49) becomes

$$\begin{aligned} 0 = \int_{t_0}^{s} (E\boldsymbol{\varepsilon}_t\boldsymbol{\varepsilon}_\tau{}^{+})\mathbf{H}^{+}(\tau)\, d\tau + \int_{t_0}^{s} \boldsymbol{\psi}(t|\tau)[\mathbf{A}(\tau) \\ - \mathbf{K}(\tau)\mathbf{B}(\tau)]\mathbf{B}^{+}(\tau)\, d\tau \qquad s \le t \end{aligned} \tag{9.51}$$

whence for $s < t$,

$$(E\boldsymbol{\varepsilon}_t\boldsymbol{\varepsilon}_s{}^{+})\mathbf{H}^{+}(s) + \boldsymbol{\psi}(t|s)\mathbf{A}(s)\mathbf{B}^{+}(s) = \boldsymbol{\psi}(t|s)\mathbf{K}(s)\mathbf{B}(s)\mathbf{B}^{+}(s) \tag{9.52}$$

Letting $s \uparrow t$ and noting the continuity assumptions, we get

$$\mathbf{K}(t)\mathbf{B}(t)\mathbf{B}^{+}(t) = \boldsymbol{\Sigma}(t)\mathbf{H}^{+}(t) + \mathbf{A}(t)\mathbf{B}^{+}(t)$$

which is (9.40). If we use (9.40) in (9.39), it is not hard to show that (9.39) can be rewritten as

$$\begin{aligned} \dot{\boldsymbol{\Sigma}}(t) = {} & \mathbf{A}(t)\mathbf{A}^{+}(t) - \mathbf{K}(t)\mathbf{B}(t)\mathbf{B}^{+}(t)\mathbf{K}^{+}(t) \\ & + \mathbf{F}(t)\boldsymbol{\Sigma}(t) + \boldsymbol{\Sigma}(t)\mathbf{F}^{+}(t) \end{aligned} \tag{9.53}$$

which is a useful alternative form to (9.39). ▮

For an example which illustrates these procedures and illustrates the difference between the Wiener filter and the Kalman-Bucy filter, consider the following problem:

We want to estimate Y_t using data $\{Y_\tau + N_\tau,\ t_0 < \tau \leq t\}$ where $\{Y_t,\ N_t,\ -\infty < t < \infty\}$ are two uncorrelated wide-sense stationary processes with spectral densities

$$S_Y(\nu) = \frac{1}{1 + (2\pi\nu)^2} \tag{9.54}$$

and

$$S_N(\nu) = \frac{1}{4 + (2\pi\nu)^2} \tag{9.55}$$

In order to use the Kalman-Bucy procedure, we first have to convert the problem into a standard form. From our earlier discussions concerning spectral factorization, we know that Y_t and N_t can be represented as

$$Y_t = \int_{-\infty}^{t} e^{-(t-\tau)}\, dZ_t \tag{9.56}$$

and

$$N_t = \frac{1}{2}\int_{-\infty}^{t} e^{-2(t-\tau)}\, dV_\tau \tag{9.57}$$

where Z_τ, V_τ are mutually uncorrelated processes with orthogonal increments. This means that Y_t and N_t satisfy

$$dY_t = -Y_t\, dt + dZ_t \tag{9.58}$$
$$dN_t = -2N_t\, dt + dV_t \tag{9.59}$$

Now, let

$$X_t = e^{2t}(Y_t + N_t) \tag{9.60}$$

Then, X_t satisfies

$$\begin{aligned} dX_t &= e^{2t}(dY_t + 2Y_t\, dt + dN_t + 2N_t\, dt) \\ &= e^{2t}Y_t\, dt + e^{2t}\, dZ_t + e^{2t}\, dV_t \end{aligned} \tag{9.61}$$

We can now identify the quantities F, H, A, B of (9.32) and (9.34) as follows:

$$\begin{aligned} F(t) &= -1 \\ H(t) &= e^{2t} \\ \mathbf{A}(t) &= [1\ 0] \\ \mathbf{B}(t) &= [e^{2t} e^{2t}] \end{aligned} \tag{9.62}$$

Equation (9.40) becomes

$$2e^{4t}K(t) = e^{2t} + e^{2t}\Sigma(t) \tag{9.63}$$

and (9.53) becomes

$$\begin{aligned}\dot{\Sigma}(t) &= 1 - 2e^{4t}K^2(t) - 2\Sigma(t) \\ &= \tfrac{1}{2} - 3\Sigma(t) - \tfrac{1}{2}\Sigma^2(t)\end{aligned} \tag{9.64}$$

Equation (9.64) can be transformed into a linear equation by the substitution $\Sigma(t) = 2\dot{\sigma}(t)/\sigma(t)$, giving us

$$\ddot{\sigma}(t) + 3\dot{\sigma}(t) - \tfrac{1}{4}\sigma(t) = 0 \tag{9.65}$$

Solving (9.65) with initial condition $2\dot{\sigma}(t_0)/\sigma(t_0) = \Sigma(t_0)$, we find

$$\begin{aligned}\Sigma(t) &= \frac{2\dot{\sigma}(t)}{\sigma(t)} \\ &= \Sigma(t_0)\left\{\frac{1 - [3/\sqrt{10} - 1/\sqrt{10}\,\Sigma(t_0)]\tanh(\sqrt{10}/2)(t-t_0)}{1 + [3/\sqrt{10} + \Sigma(t_0)/\sqrt{10}]\tanh(\sqrt{10}/2)(t-t_0)}\right\}\end{aligned} \tag{9.66}$$

The initial value $\Sigma(t_0)$ can be evaluated as follows: First, we note that the linear least-squares estimator of Y_{t_0} given X_{t_0} has the form αX_{t_0}, where α is determined by

$$E(Y_{t_0} - \alpha X_{t_0})\bar{X}_{t_0} = 0 \tag{9.67}$$

This yields

$$\begin{aligned}\alpha &= \frac{EY_{t_0}\bar{X}_{t_0}}{E|X_{t_0}|^2} \\ &= e^{-2t_0}\frac{E|Y_{t_0}|^2}{E|Y_{t_0}|^2 + E|N_{t_0}|^2} \\ &= \tfrac{1}{2}e^{-2t_0}\end{aligned} \tag{9.68}$$

Finally,

$$\begin{aligned}\Sigma(t_0) &= E|Y_{t_0} - \alpha X_{t_0}|^2 = E|\tfrac{1}{2}Y_{t_0} - \tfrac{1}{2}N_{t_0}|^2 \\ &= \tfrac{1}{4}(E|Y_{t_0}|^2 + E|N_{t_0}|^2) = \tfrac{1}{4}\end{aligned} \tag{9.69}$$

This completes the solution for $\Sigma(t)$, and via (9.63), also completes the solution for $K(t)$, and hence the Kalman-Bucy filter.

If we let $t_0 \to -\infty$ in (9.66), we get

$$\Sigma(t) \xrightarrow[t_0 \to -\infty]{} (\sqrt{10} - 3) \tag{9.70}$$

which yields $K(t) = (\sqrt{5/2} - 1)e^{-2t}$ and (9.37) becomes

$$d\tilde{Y}_t = -\sqrt{\tfrac{5}{2}}\,\tilde{Y}_t\,dt + (\sqrt{\tfrac{5}{2}} - 1)e^{-2t}\,dX_t \tag{9.71}$$

This gives us

$$\tilde{Y}_t = \int_{-\infty}^{t} (\sqrt{\tfrac{5}{2}} - 1) \exp\left[-\sqrt{\tfrac{5}{2}}\,(t-\tau)\right] e^{-2\tau}\, dX_\tau$$
$$= \int_{-\infty}^{t} (\sqrt{\tfrac{5}{2}} - 1) \exp\left[-\sqrt{\tfrac{5}{2}}\,(t-\tau)\right] (d\xi_\tau + 2\xi_\tau\, d\tau) \qquad (9.72)$$

where we have set $\xi_\tau = e^{-2t}X_\tau = Y_\tau + N_\tau$. The final expression in (9.72) represents the output of a linear time-invariant filter with $Y_t + N_t$ as input and with a transfer function given by

$$\left(\sqrt{\frac{5}{2}} - 1\right)\frac{(2\pi i\nu) + 2}{(2\pi i\nu) + \sqrt{\frac{5}{2}}} = \frac{3}{\sqrt{5} + \sqrt{2}}\,\frac{2 + i2\pi\nu}{\sqrt{5} + \sqrt{2}\, i2\pi\nu}$$

which should be compared with (9.31a).

EXERCISES

1. Test whether each of the following functions is nonnegative definite.

(a) $R(t,s) = \begin{cases} 1 - |t-s| & |t-s| \le 1 \\ 0 & |t-s| > 1 \end{cases}$

(b) $R(t,s) = e^{-|t-s|} \qquad -\infty < t, s < \infty$

(c) $R(t,s) = e^{|t-s|}$

(d) $R(t,s) = \begin{cases} 1 & |t-s| \le 1 \\ 0 & |t-s| > 1 \end{cases}$

2. Let $\lambda_0, \lambda_1, \ldots$ be the eigenvalues of (4.12).

(a) Show that

$$R(t,t) - \sum_{n=0}^{N} \lambda_n |\varphi_n(t)|^2 = E\left| X_t - \sum_{n=0}^{N} \varphi_n(t) \int_a^b \bar{\varphi}_n(s) X_s\, ds \right|^2 \ge 0$$

(b) Show that

$$\sum_{n=0}^{N} \lambda_n \le \int_a^b R(t,t)\, dt < \infty$$

Hence, $\sum_{n=0}^{N} \lambda_n$ must converge and $\lambda_n \xrightarrow[n\to\infty]{} 0$.

3. Suppose that a q.m. continuous and wide-sense stationary process $\{X_t, -\infty < t < \infty\}$ has a covariance function $R(\cdot)$ which is periodic with period T, that is,

$$R(\tau + T) = R(\tau) \qquad \text{for all } \tau \text{ in } (-\infty, \infty)$$

Define for $n = 0, \pm1, \pm2, \ldots$

$$Z_n = \frac{1}{T}\int_0^T X_t e^{-in(2\pi/T)t}$$

(*a*) Show that $\{Z_n\}$ are mutually orthogonal, that is,

$$EZ_m\bar{Z}_n = 0 \qquad \text{whenever } m \neq n$$

(*b*) Show that for each t

$$X_t = \sum_{n=-\infty}^{\infty} Z_n e^{in(2\pi/T)t}$$

(*c*) Suppose that

$$R(\tau) = \sum_{n=0}^{\infty} \frac{1}{1+n^2} \cos n\left(\frac{2\pi}{T}\right)\tau$$

find the eigenvalues and a complete set of orthonormal eigenfunctions to

$$\int_0^T R(t-s)\varphi(s)\,ds = \lambda\varphi(t) \qquad 0 \le t \le T$$

4. Let $R(t,s)$ be given by

$$R(t,s) = \begin{cases} 1 - |t-s| & 0 \le |t-s| \le 1 \\ 0 & \text{elsewhere} \end{cases}$$

Find the eigenvalues and a complete set of O-N eigenfunctions for

$$\int_0^{\frac{1}{2}} R(t,s)\varphi(s)\,ds = \lambda\varphi(t) \qquad 0 \le t \le \tfrac{1}{2}$$

Hint: $\dfrac{\partial^2 R(t,s)}{\partial t^2} = -2\delta(t-s) \qquad \text{for } 0 < t,\ s < \tfrac{1}{2}$

5. (*a*) Suppose that $R(t,s)$ satisfies

$$\sum_{k=0}^{n} a_k \frac{\partial^{2k}}{\partial t^{2k}} R(t,s) = \sum_{k=0}^{m} b_k \frac{\partial^{2k}}{\partial t^{2k}} \delta(t-s) \qquad a < t,\ s < b$$

Show that the integral equation

$$\int_a^b R(t,s)\varphi(s)\,ds = \lambda\varphi(t) \qquad a \le t \le b$$

can be reduced to the differential equation

$$\lambda \sum_{k=0}^{n} a_k \frac{\partial^{2k}\varphi(t)}{\partial t^{2k}} = \sum_{k=0}^{m} b_k \frac{\partial^{2k}\varphi(t)}{\partial t^{2k}} \qquad a < t < b$$

together with appropriate boundary conditions.

(b) Verify that, formally at least, if $R(t,s)$ satisfies

$$R(t,s) = \int_{-\infty}^{\infty} \frac{\sum_{k=0}^{m} b_k (2\pi i \nu)^{2k}}{\sum_{k=0}^{n} a_k (2\pi i \nu)^{2k}} e^{i2\pi\nu(t-s)}\, d\nu$$

then it also satisfies

$$\sum_{k=0}^{n} a_k \frac{\partial^{2k}}{\partial t^{2k}} R(t,s) = \sum_{k=0}^{m} b_k \frac{\partial^{2k}}{\partial t^{2k}} \delta(t-s)$$

(c) Reduce the equation

$$\int_a^b e^{-|t-s|}\varphi(s)\, ds = \lambda\varphi(t) \qquad a \le t \le b$$

to a differential equation and two linearly independent boundary conditions on $\varphi(a)$, $\varphi(b)$, $\dot\varphi(a)$, and $\dot\varphi(b)$. [Hint: Each of these four quantities can be expressed as a linear combination of $\int_a^b e^{-s}\varphi(s)\, ds$ and $\int_a^b e^{s}\varphi(s)\, ds$.]

6. Suppose that $\{W_t, t \ge 0\}$ is a standard Brownian motion and

$$X_t(\omega) = f(t) W_{\tau(t)}(\omega) \qquad 0 \le t \le T$$

where $f(t)$ is a continuous function and $\tau(t)$ is a continuous nondecreasing function with $\tau(0) = 0$. Using the results on the solutions of (4.31), find an expansion of this form

$$X_t(\omega) = \sum_{n=0}^{\infty} \alpha_n(t) Z_n(\omega)$$

where $\{Z_n, n = 0, 1, 2, \ldots\}$ are independent Gaussian random variables with $EZ_mZ_n = \delta_{mn}$. What conditions do we need, if any, in order that each Z_n belongs to $\mathcal{H}_X$?

7. Suppose that $\{X_t, -\infty < t < \infty\}$ has a covariance function

$$R(\tau) = \tfrac{1}{8}e^{-|\tau|}(3\cos\tau + \sin|\tau|)$$

Find its spectral-density function $S(\nu)$, $-\infty < \nu < \infty$.

8. Suppose that $\{X_t, -\infty < t < \infty\}$ has the spectral density given in Exercise 7, and $\{Y_t, -\infty < t < \infty\}$ is such that:

(a) $Y_t \in \mathcal{H}_X$ for every t

(b) $EY_t\bar{X}_s = \frac{1}{8}e^{-|t-s|}[\cos(t-s) + \sin(t-s)]$. Find $EY_t\bar{Y}_s$.

9. Suppose that $\{X_t, -\infty < t < \infty\}$ is wide-sense stationary. Show that for a fixed constant W, $\{e^{i2\pi Wt}X_t, -\infty < t < \infty\}$ is again wide-sense stationary. Is $\{\cos 2\pi Wt X_t, -\infty < t < \infty\}$ wide-sense stationary?

10. Suppose that $X_t = \sum_{k=1}^{N} X_{kt}$, $-\infty < t < \infty$, is the sum of N wide-sense stationary and q.m. continuous processes $X_{1t}, X_{2t}, \ldots, X_{Nt}$. Show that we have a representation

$$X_t = \int_{-\infty}^{\infty} e^{i2\pi\nu t}\, d\hat{X}_\nu$$

Is the process $\{\hat{X}_\nu, -\infty < \nu < \infty\}$ always a process with orthogonal increments? (Note: $X_t = \cos 2\pi W t Z_t$, with Z_t stationary, is an example of such a process.)

11. For a process of the type given in Exercise 10, show that

$$\int_{-\infty}^{\infty} h(t-\tau)X_\tau\, d\tau = \int_{-\infty}^{\infty} \hat{h}(\nu)e^{i2\pi\nu t}\, d\hat{X}_\nu$$

12. Let $\{X_t, -\infty < t < \infty\}$ be a wide-sense stationary process with a spectral representation

$$X_t = \int_{-\infty}^{\infty} e^{i2\pi\nu t}\, d\hat{X}_\nu$$

Let

$$Y_t = \int_{-\infty}^{\infty} \eta(\nu)e^{i\varphi(\nu)t}\, d\hat{X}_\nu$$

where $\varphi(\nu)$, $-\infty < \nu < \infty$, is a real-valued function. Show that $\{Y_t, -\infty < t < \infty\}$ is wide-sense stationary.

13. Suppose that $\{X_t, -\infty < t < \infty\}$ is a real-valued wide-sense stationary process, and let $\tilde{X}_t$ denote its Hilbert transform

$$\tilde{X}_t = \int_{-\infty}^{\infty} (-i \operatorname{sgn} \nu)e^{i2\pi\nu t}\, d\hat{X}_\nu$$

Let $EX_tX_s = e^{-|t-s|}$.

(*a*) Show that $\tilde{X}_t$ is real valued

(*b*) Find $E\tilde{X}_t\tilde{X}_s$ and $E\tilde{X}_tX_s$

(*c*) Let $Z_t = \cos 2\pi\nu_0 t X_t + \sin 2\pi\nu_0 t\tilde{X}_t$. Show that Z_t is of the form

$$Z_t = \int_{-\infty}^{\infty} e^{i\varphi(\nu)t}\, d\hat{X}_\nu$$

with $\varphi(\nu) = 2\pi[\nu - \nu_0 \operatorname{sgn} \nu]$. [Hence, it must be wide-sense stationary (see Exercise 12).]

14. Let $\{Z_t, -\infty < t < \infty\}$ be a process with orthogonal increments such that $EZ_t = 0$, $E|Z_t - Z_s|^2 = |t - s|$ and $Z_0 = 0$.

(*a*) Show that

$$EZ_t\bar{Z}_s = \tfrac{1}{2}(|t| + |s| - |t - s|)$$

(b) Let $\{\zeta_{nt}, -\infty < t < \infty\}$ be defined by

$$\zeta_{nt} = n(Z_{t+1/n} - Z_t) \qquad n = 1, 2, \ldots$$

Show that for each n, $\{\zeta_{nt}, -\infty < t < \infty\}$ is wide-sense stationary and find its spectral-density function $S_n(\nu)$, $-\infty < \nu < \infty$.

(c) Show that ζ_{nt} converges to a white noise in the sense of (8.2) and (8.3).

15. Let f be a differentiable function such that its derivative $\dot{f}$ is continuous on $[a,b]$. Show that

$$\int_a^b f(t)\, dZ_t + \int_a^b \dot{f}(t) Z_t\, dt = f(b)Z_b - f(a)Z_a$$

where $\{Z_t, -\infty < t < \infty\}$ is the process described in Exercise 14. (Hint: Make use of the sequence $\{\zeta_{nt}\}$ defined in Exercise 14 and show that

$$\int_a^b \dot{f}(t) \int_0^t \zeta_{ns}\, ds\, dt \xrightarrow[n\to\infty]{\text{q.m.}} \int_a^b \dot{f}(t) Z_t\, dt$$

The rest is easy.)

16. Use the results of Exercise 15 and show that the solution of the integral equation

$$Y_t = Y_a - \int_a^t Y_s\, ds + Z_t - Z_a$$

is given by $Y_t = Y_a e^{-(t-a)} + \int_a^t e^{-(t-s)}\, dZ_s$.

17. Suppose that $\{X_t, a \le t \le b\}$ is a q.m. continuous process such that

$$\frac{\partial^2}{\partial t\, \partial s} EX_t\bar{X}_s = \delta(t - s) + \rho(t,s) \qquad a < t < b$$

where ρ is a continuous nonnegative definite function on $[a,b]$.

(a) Show that $X(f) \equiv \int_a^b f(t)\, dX_t$ is well defined for all f satisfying $\int_a^b |f(t)|^2\, dt$ such that

(1) $X(\cdot)$ is linear, that is, $X(\alpha f + \beta g) = \alpha X(f) + \beta X(g)$
(2) If $a \le c < d \le b$ and $f = I_{cd}$ is the indicator function of $[c,d)$, then $X(f) = X_d - X_c$

(b) Find an expression for $EX(f)\bar{X}(g)$

(c) Show that every element Y in $\mathcal{H}_X$ can be represented as

$$Y = \int_a^b \eta(t) dX_t + kX_a$$

18. Let $\{W_t, t \ge 0\}$ be a standard Brownian motion process, and let A be a real-valued second-order random variable ($EA = 0$, $EA^2 = 1$) independent of the W process. Suppose that the process

$$X_t = At + W_t \qquad t \ge 0$$

is observed. Let $\tilde{A}_t$ denote the linear least-squares estimator of A given X_s, $0 \leq s \leq t$. Find an explicit expression for $\tilde{A}_t$ in the form of

$$\tilde{A}_t = \int_0^t h(t,s)\, dX_s$$

19. Let $X_t = Y \cos 2\pi Wt + N_t$, where $\{N_t, -\infty < t < \infty\}$ is a second-order process with zero mean and $EN_t\bar{N}_s = e^{-\nu_0|t-s|}$, and Y is a random variable with $EY = 0$, $E|Y|^2 = 1$ and $EY\bar{N}_t = 0$ for all t. Let $\hat{Y}_t$ denote the linear least-squares estimator of Y given X_s, $0 \leq s \leq t$.

(*a*) Find $\hat{Y}_t$ in the form of

$$\hat{Y}_t = \int_0^t h(t,s)\, dX_s + X_0$$

(*b*) Find $\hat{Y}_t$ as the solution of a Kalman filtering problem with observation process $Z_t = e^{\nu_0 t}X_t$. That is, show that $\hat{Y}_t$ satisfies

$$\dot{\hat{Y}}_t = \alpha(t)\hat{Y}_t + K(t)[\dot{Z}_t - \beta(t)\hat{Y}_t]$$

and find $\alpha(\cdot)$, $K(\cdot)$, and $\beta(\cdot)$.

20. Suppose that $\{X_t, -\infty < t < \infty\}$ has a spectral density given by

$$S_x(\nu) = \frac{1}{4 + (2\pi\nu)^4} \qquad -\infty < \nu < \infty$$

Find the predictor $\hat{E}(X_{t+\alpha}|\mathcal{H}_X{}^t)$, and express it in the form

$$\hat{E}(X_{t+\alpha}|\mathcal{H}_X{}^t) = \int_{-\infty}^{\infty} e^{t2\pi\nu t}H_\alpha(\nu)\, d\hat{X}_\nu$$

21. Let $\{X_t, -\infty < t < \infty\}$ be as in Exercise 20, and let $\{N_t, -\infty < t < \infty\}$ have spectral density $S_N(\nu) = 1/[1 + (2\pi\nu)^2]$. Assume that $EX_t\bar{N}_s = 0$ for all t and s and define

$$Y_t = X_t + N_t \qquad -\infty < t < \infty$$

Find $\hat{E}(X_t|\mathcal{H}_Y{}^t)$ and express it in the form

$$\hat{E}(X_t|\mathcal{H}_Y{}^t) = \int_{-\infty}^{\infty} e^{i2\pi\nu t}H(\nu)\, d\hat{Y}_\nu$$

22. Suppose that the quantities **A**, **B**, **F**, and **H** appearing in (9.32) and (9.34) are given as follows:

$$\mathbf{A}(t) = \begin{bmatrix} 1 \\ 0 \end{bmatrix}$$
$$\mathbf{B}(t) = 1$$
$$\mathbf{F}(t) = \begin{bmatrix} 0 & -1 \\ 1 & 0 \end{bmatrix}$$
$$\mathbf{H}(t) = [0 \quad 1]$$

Find the matrix $\mathbf{K}(t)$ appearing in (9.37) for $t \geq 0$, assuming $\mathbf{\Sigma}(0) = \mathbf{0}$.

23. Suppose that Y_t satisfies a differential equation

$$\ddot{Y}_t = Y_t + \eta_t$$

where $E\eta_t\bar{\eta}_s = \delta(t - s)$. Let X_t satisfy

$$\dot{X}_t = Y_t + \xi_t$$

with $E\eta_t\bar{\xi}_s = 0$ and $E\xi_t\bar{\xi}_s = \delta(t - s)$. Reexpress these equations in the form of (4.32) and (4.34) with

$$\mathbf{Y}_t = \begin{bmatrix} \dot{Y}_t \\ Y_t \end{bmatrix} \qquad \text{and} \qquad \zeta_t = \begin{bmatrix} \eta_t \\ \xi_t \end{bmatrix}$$

24. Suppose that in Exercise 23 the process ξ_t, instead of being white, satisfies $E\xi_t\bar{\xi}_s = e^{-|t-s|}$. Reexpress the two differential equations in the form of (4.32) and (4.34) with suitable choices for $\mathbf{X}_t$ and ζ_t. (Hint: Now $(d/dt)\xi_t + \xi_t$ is a white noise.)

4 Stochastic Integrals and Stochastic Differential Equations

1. INTRODUCTION

Roughly speaking, stochastic differential equations are differential equations driven by Gaussian white noise. Here, we are using the term "stochastic differential equations" in a restricted sense and not merely to denote differential equations with some probabilistic aspects. The importance of stochastic differential equations is largely due to the fact that the solution of such an equation is a sample-continuous Markov process, and conversely, a large and important class of sample-continuous Markov processes can be modeled by the solutions of stochastic differential equations. From the point of view of applications, this is a direct benefit of using white noise as a noise model, and this fact accounts for its popularity. After all, white noise is, at best, a tolerable abstraction and is never a completely faithful representation of a physical noise source. Its *raison d'être* is the simplicity of analysis that it brings about. We have seen this in connection with Kalman filtering, and we shall see it again in the Markovian nature of solutions to stochastic differential equations.

Stochastic differential equations are defined in terms of stochastic integrals, which in turn need to be defined. The generally accepted definition for a stochastic integral is the one due to Ito and is often referred to as the Ito integral. We shall call it simply the stochastic integral. In addition to its role in stochastic differential equations, the stochastic integral is of tremendous importance in its own right. The main source of its importance is due to the fact that an important class of martingales can be represented as stochastic integrals. We shall explore this aspect of stochastic integrals a little later.

It turns out that a calculus based on Ito's definition of stochastic integrals is not compatible with rules of ordinary calculus. Since the differential equations that we encounter in practice must usually be interpreted in terms of ordinary calculus, it raises the question whether stochastic differential equations can really be used to model differential equations driven by "approximately white" noise. This is an important question, since the uniqueness, existence, and Markov property of the solution to a stochastic differential equation are all based on the interpretation of the equation in terms of stochastic integrals. We shall pose this question in a more precise way and give an answer. The answer is roughly that a differential equation driven by "nearly white" Gaussian noise can indeed be modeled by a stochastic differential equation, but in general, a "correction term" has to be added to the equation.

The solution of a stochastic differential equation not only is a sample-continuous Markov process, but under quite general conditions, its transition-probability density function can be shown to satisfy a pair of partial differential equations called the backward and forward equations of Kolmogorov, or the diffusion equations. These equations were derived by Kolmogorov some time before Ito's work on stochastic differential equations, and they took on new light in view of Ito's results.

Both Kolmogorov and Ito were motivated in their work by the goal of discovering conditions under which "local" properties of a Markov process completely determine its probability law. This problem is brought into a sharper focus in the case of Markov processes with stationary transition probabilities. In that case, the results of semigroup theory can be used to give a more complete answer to the question, "When do local properties determine completely the probability law?" Theory of Markov semigroups is sufficiently different to require a separate treatment. It will be treated in the next chapter.

Finally, we should mention that by and large we shall restrict ourselves to scalar-valued processes in this chapter. For applications, extension to vector-valued processes is of great importance. This extension is not difficult in spirit, but is rather complicated in detail and involves cumbersome notations.

2. STOCHASTIC INTEGRALS

Let $(\Omega,\mathcal{A},\mathcal{P})$ be a fixed probability space. Let $\{\mathcal{A}_t, -\infty < t < \infty\}$ be an increasing family of sub-σ algebras of $\mathcal{A}$, and let $\{W_t, -\infty < t < \infty\}$ be a Brownian motion process such that for each s, the aggregate $\{W_t - W_s, t \geq s\}$ is independent of $\mathcal{A}_s$ and W_t is $\mathcal{A}_t$ measurable for each t. It follows that for $s \geq 0$

$$
\begin{aligned}
&E^{\mathcal{A}_t}W_{t+s} = W_t \\
&E^{\mathcal{A}_t}(W_{t+s} - W_t)^2 = s \qquad \text{a.s.}
\end{aligned}
\tag{2.1}
$$

We recall that we refer to this situation by saying that $\{W_t, \mathcal{A}_t, -\infty < t < \infty\}$ is a Brownian motion. By a **stochastic integral** we mean a quantity of the form

$$
I(\varphi) = \int_a^b \varphi(\omega,t)\, dW(\omega,t) \tag{2.2}
$$

Because W_t is neither differentiable nor of bounded variation, (2.2) does not have a well-defined interpretation as an integral in the ordinary sense. If the integrand φ does not depend on ω, then (2.2) can be treated as a second-order stochastic integral as introduced in Sec. 3.6. Interpreted in that way, only the property of a Brownian motion as a process with orthogonal increments is made use of, and not its other properties. If φ is random, i.e., it depends on ω, then (2.2) has to be defined anew. It turns out that its definition now depends in a crucial way on the martingale property (2.1) of W_t.

We make the following assumptions on the integrand φ:

The function φ is jointly measurable in (ω,t) (with respect to $\mathcal{A}$ in ω and the Lebesgue measure in t). For each t, φ_t is measurable with respect to $\mathcal{A}_t$ (2.3)

$$
\varphi \text{ satisfies } \int_a^b E|\varphi_t|^2\, dt < \infty \tag{2.4}
$$

The stochastic integral is now defined in the following way:

1. If there exist times $t_0, t_1, \ldots, t_n$ independent of ω, such that $a = t_0 < t_1 < \cdots < t_n = b$ and

$$
\varphi(\omega,t) = \varphi_\nu(\omega) \qquad \begin{aligned} &t_\nu \leq t < t_{\nu+1} \\ &\nu = 0, \ldots, n-1 \end{aligned} \tag{2.5}
$$

and if φ satisfies (2.3) and (2.4), then we call φ an (ω,t)-**step function** and define the stochastic integral by

$$
\int_a^b \varphi(\omega,t)\, dW(\omega,t) = \sum_{\nu=0}^{n-1} \varphi_\nu(\omega)[W(\omega,t_{\nu+1}) - W(\omega,t_\nu)] \tag{2.6}
$$

2. If φ satisfies (2.3) and (2.4), then we shall show that there exists a sequence of (ω,t)-step functions $\{\varphi_n(\omega,t)\}$ satisfying (2.3) and (2.4) such that

$$\|\varphi - \varphi_n\|^2 = \int_a^b E|\varphi(\cdot,t) - \varphi_n(\cdot,t)|^2 \, dt \xrightarrow[n\to\infty]{} 0 \tag{2.7}$$

It will then follow that $\int_a^b \varphi_n(\omega,t) \, dW(\omega,t)$ converges in q.m. as $n \to \infty$, and this q.m. limit is the same for any sequence of (ω,t)-step functions which satisfy (2.7). Therefore, we can define

$$\int_a^b \varphi(\omega,t) \, dW(\omega,t) = \lim_{n\to\infty} \text{ in q.m.} \int_a^b \varphi_n(\omega,t) \, dW(\omega,t) \tag{2.8}$$

where $\{\varphi_n\}$ is any sequence of steps satisfying (2.7). We assume that both a and b are finite. If not, the stochastic integral is defined as the limit in q.m. as $a \to -\infty$, or $b \to \infty$, or both.

Proposition 2.1. Let $\{W_t, -\infty < t < \infty\}$ be a Brownian motion, and let $\varphi(\omega,t)$ satisfy (2.3) and (2.4). Then,

(*a*) There exists a sequence of (ω,t) steps $\{\varphi_n\}$ satisfying (2.3) and (2.4) such that

$$\|\varphi - \varphi_n\|^2 = \int_a^b E|\varphi(\cdot,t) - \varphi_n(\cdot,t)|^2 \, dt \xrightarrow[n\to\infty]{} 0 \tag{2.9}$$

(*b*) For each n, $I(\varphi_n) = \int_a^b \varphi_n(\omega,t) \, dW(\omega,t)$ is well defined by (2.6), and $\{I(\varphi_n)\}$ converges in q.m. as $n \to \infty$.

(*c*) If $\{\varphi_n\}$ and $\{\varphi_n'\}$ are two sequences of (ω,t) steps satisfying (2.3) and (2.4) such that $\|\varphi - \varphi_n\|$ and $\|\varphi - \varphi_n'\|$ both go to zero as $n \to \infty$, then

$$\lim_{n\to\infty} \text{ in q.m. } I(\varphi_n) = \lim_{n\to\infty} \text{ in q.m. } I(\varphi_n') \tag{2.10}$$

Proof:

(*a*) Suppose $E\varphi(\cdot,t)\bar{\varphi}(\cdot,s)$ is continuous on $[a,b] \times [a,b]$, that is, φ_t is q.m. continuous on $[a,b]$, then an approximating sequence of (ω,t) step functions $\{\varphi_n\}$ can be constructed by partitioning $[a,b]$, sampling $\varphi(\omega,t)$ at partition points $t_\nu^{(n)}$, defining $\varphi_n(\omega,t) = \varphi(\omega,t_\nu^{(n)})$, $t_\nu^{(n)} \le t < t_{\nu+1}^{(n)}$, and refining the partitions to zero $[\max_\nu (t_{\nu+1}^{(n)} - t_\nu^{(n)}) \xrightarrow[n\to\infty]{} 0]$. Since φ_t is q.m. continuous, $E|\varphi(\cdot,t) - \varphi_n(\cdot,t)|^2 \xrightarrow[n\to\infty]{} 0$ for every t in $[a,b]$. By the dominated convergence theorem, we have

$$\int_a^b E|\varphi(\cdot,t) - \varphi_n(\cdot,t)|^2 \, dt \xrightarrow[n\to\infty]{} 0$$

More generally, if φ merely satisfies (2.3) and (2.4) and is not necessarily q.m. continuous on $[a,b]$, we construct a sequence of approximating (ω,t)-

step functions in the following manner: First, let f_n be φ with its real and imaginary parts truncated to $\pm n$, then

$$\int_a^b E|\varphi(\cdot,t) - f_n(\cdot,t)|^2\,dt \xrightarrow[n\to\infty]{} 0 \tag{2.11}$$

Thus, we can always assume φ to be bounded. If φ is bounded, define

$$g_n(\omega,t) = \int_0^\infty e^{-\tau}\varphi\left(\omega,\, t - \frac{\tau}{n}\right) d\tau \tag{2.12}$$

Then, $g_n(\cdot,t)$ is q.m. continuous on $[a,b]$ and satisfies (2.3) and (2.4). Now,

$$\int_a^b E|\varphi(\cdot,t) - g_n(\cdot,t)|^2\,dt \le \int_a^b \int_0^\infty e^{-\tau}E\left|\varphi(\cdot,t) - \varphi\left(\cdot,\, t - \frac{\tau}{n}\right)\right|^2 dt\,d\tau \tag{2.13}$$

Since $\int_a^b \left|\varphi(\cdot,t) - \varphi\left(\cdot,\, t - \frac{\tau}{n}\right)\right|^2 dt \xrightarrow[n\to\infty]{} 0$ whenever φ is bounded and Lebesgue measurable, we have

$$\int_a^b E|\varphi(\cdot,t) - g_n(\cdot,t)|^2\,dt \xrightarrow[n\to\infty]{} 0 \tag{2.14}$$

It is now clear that we can construct a sequence of (ω,t)-step functions approximating φ by sampling g_n at the partition points of a sequence of partitions refining to zero. This completes the proof for (a).

(b) We assume that $\{\varphi_n\}$ is a sequence of (ω,t)-step functions such that $\|\varphi - \varphi_n\| \xrightarrow[n\to\infty]{} 0$ and define $I(\varphi_n)$ by (2.6) as

$$I(\varphi_n) = \sum_\nu \varphi_{n\nu}(\omega)\,[W(\omega,t_{\nu+1}^{(n)}) - W(\omega,t_\nu^{(n)})] \tag{2.15}$$

Hence, if we write $\Delta_\nu{}^n W = W(t_{\nu+1}^{(n)}) - W(t_\nu{}^{(n)})$, we have

$$E|I(\varphi_n)|^2 = \sum_\nu \sum_\mu E(\varphi_{n\nu}\bar{\varphi}_{n\mu}\,\Delta_\nu{}^n W\,\Delta_\mu{}^n W)$$

If $\mu > \nu$, then $\Delta_\mu{}^n W$ is independent of $\mathcal{A}_{t_\nu{}^{(n)}}$ while $(\varphi_{n\nu}\varphi_{n\mu}\,\Delta_\nu{}^n W)$ is $\mathcal{A}_{t_\nu{}^{(n)}}$ measurable. Hence

$$E(\varphi_{n\nu}\bar{\varphi}_{n\mu}\,\Delta_\nu{}^n W\,\Delta_\mu{}^n W) = 0 \qquad \text{if } \mu \ne \nu$$

and

$$\begin{aligned} E|I(\varphi_n)|^2 &= \sum_\nu E[|\varphi_{n\nu}|^2(\Delta_\nu{}^n W)^2] \\ &= \sum_\nu E|\varphi_{n\nu}|^2 E^{\mathcal{A}_{t_\nu^{(n)}}}(\Delta_\nu{}^n W)^2 \\ &= \sum_\nu E|\varphi_{n\nu}|^2(t_{\nu+1}^{(n)} - t_\nu{}^{(n)}) \\ &= \int_a^b E|\varphi_n(\cdot,t)|^2\,dt \end{aligned} \tag{2.16}$$

Now, $I(\varphi_{m+n}) - I(\varphi_n) = I(\varphi_{m+n} - \varphi_n)$ and $\varphi_{m+n} - \varphi_n$ is again a step. Therefore,

$$\begin{aligned} E|I(\varphi_{m+n}) - I(\varphi_n)|^2 &= \int_a^b E|\varphi_{m+n}(\cdot,t) - \varphi_n(\cdot,t)|^2\, dt \\ &\leq 2\int_a^b E|\varphi_{m+n}(\cdot,t) - \varphi(\cdot,t)|^2\, dt \\ &\quad + 2\int_a^b E|\varphi_n(\cdot,t) - \varphi(\cdot,t)|^2\, dt \underset{n\to\infty}{\longrightarrow} 0 \end{aligned} \tag{2.17}$$

Hence, $\{I(\varphi_n)\}$ is a mutually q.m. convergent sequence, and there exists a second-order random variable $I(\varphi)$ such that

$$E|I(\varphi_n) - I(\varphi)|^2 \underset{n\to\infty}{\longrightarrow} 0$$

and (*b*) is proved.

(*c*) Suppose $\{\varphi_n\}$ and $\{\varphi_n'\}$ are (ω,t) steps such that $\|\varphi - \varphi_n\| \underset{n\to\infty}{\longrightarrow} 0$ and $\|\varphi - \varphi_n'\| \underset{n\to\infty}{\longrightarrow} 0$. Then,

$$\|\varphi_n - \varphi_n'\| \leq \sqrt{2}\,(\|\varphi_n - \varphi\|^2 + \|\varphi_n' - \varphi\|^2)^{\frac{1}{2}} \underset{n\to\infty}{\longrightarrow} 0$$

Therefore, $E|I(\varphi_n) - I(\varphi_n')|^2 \underset{n\to\infty}{\longrightarrow} 0$ and

$$\underset{n\to\infty}{\text{lim in q.m.}}\, I(\varphi_n) = \underset{n\to\infty}{\text{lim in q.m.}}\, I(\varphi_n')$$

This proves (*c*). ∎

Proposition 2.2. Let $\varphi(\omega,t)$ and $\psi(\omega,t)$ satisfy (2.3) and (2.4). Then

$$E\left[\int_a^b \varphi(\cdot,t)\, dW_t \cdot \int_a^b \psi(\cdot,t)\, dW_t\right] = \int_a^b E\varphi_t\psi_t\, dt \tag{2.18}$$

Proof: It is enough to prove (2.18) for $\varphi = \psi$, because

$$\begin{aligned} EI(\varphi)\overline{I(\psi)} &= \tfrac{1}{4}[E|I(\varphi) + I(\psi)|^2 - E|I(\varphi) - I(\psi)|^2] \\ &\quad + i\tfrac{1}{4}[E|I(-i\varphi) + I(\psi)|^2 - E|I(-i\varphi) - I(\psi)|^2] \\ &= \tfrac{1}{4}[E|I(\varphi + \psi)|^2 - E|I(\varphi - \psi)|^2] \\ &\quad + \frac{i}{4}[E|I(-i\varphi + \psi)|^2 - E|I(-i\varphi - \psi)|^2] \end{aligned} \tag{2.19}$$

which means that (2.18) follows from the seemingly more special case $E|I(\varphi)|^2 = \int_a^b E|\varphi_t|^2\, dt$. Now if φ is an (ω,t)-step function, we have already proved in (2.16) that

$$E|I(\varphi)|^2 = \int_a^b E|\varphi(\cdot,t)|^2\, dt$$

If φ is not a step, let $\{\varphi_n\}$ be an approximating sequence of steps, then

$$\begin{aligned} E|I(\varphi)|^2 &= E|I(\varphi - \varphi_n) + I(\varphi_n)|^2 \\ &= E|I(\varphi_n)|^2 + 2 \operatorname{Re} EI(\varphi - \varphi_n)\overline{I(\varphi_n)} + E|I(\varphi - \varphi_n)|^2 \end{aligned}$$

Since $E|I(\varphi - \varphi_n)|^2 \xrightarrow[n\to\infty]{} 0$, we have

$$\begin{aligned} E|I(\varphi)|^2 &= \lim_{n\to\infty} E|I(\varphi_n)|^2 \\ &= \lim_{n\to\infty} \int_a^b E|\varphi_n(\cdot,t)|^2\,dt = \int_a^b E|\varphi(\cdot,t)|^2\,dt \quad \blacksquare \end{aligned}$$

3 PROCESSES DEFINED BY STOCHASTIC INTEGRALS

Proposition 3.1. Let $\{W_t,\mathcal{A}_t\}$ be a Brownian motion and let φ satisfy (2.3) and (2.4). Define a process $\{X_t,\ a \le t \le b\}$ by

$$X(\omega,t) = \int_a^t \varphi(\omega,s)\,dW(\omega,s) \tag{3.1}$$

Then, $\{X_t,\ \mathcal{A}_t,\ a \le t \le b\}$ is a martingale, that is,

$$t > s \Rightarrow E^{\mathcal{A}_s}X_t = X_s \quad \text{a.s.} \tag{3.2}$$

Remark: This important martingale property is intimately connected with the fact that a stochastic integral is inherently defined by forward difference approximations. We recall that if φ is an (ω,t)-step function, then

$$\int_a^b \varphi(\omega,t)\,dW(\omega,t) = \sum_\nu \varphi_\nu(\omega)[W(\omega,t_{\nu+1}) - W(\omega,t_\nu)]$$

where each summand consists of a term φ_ν measurable with respect to $\mathcal{A}_{t_\nu}$ and a forward increment $\Delta_\nu W$ which is independent of $\mathcal{A}_{t_\nu}$. For a general φ, the stochastic integral is defined by approximating φ with steps, and it involves forward difference approximation once again.

Proof: First, suppose that the integrand φ in (3.1) is an (ω,t)-step function. Let $t > s$, and let $t_1, t_2, \ldots, t_n$ be jump points of φ between s and t. We can write

$$\begin{aligned} X(\omega,t) - X(\omega,s) &= \int_s^t \varphi(\omega,\tau)dW(\omega,\tau) \\ &= \varphi_0(\omega)[W(\omega,t_1) - W(\omega,s)] + \varphi_1(\omega) \\ &\quad [W(\omega,t_2) - W(\omega,t_1)] + \cdots + \varphi_n(\omega)[W(\omega,t) - W(\omega,t_n)] \end{aligned} \tag{3.3}$$

where φ_0 is $\mathcal{A}_s$ measurable and φ_k is $\mathcal{A}_{t_k}$ measurable, $k = 1, \ldots, n$. By successively taking conditional expectation with respect to $\mathcal{A}_{t_n}$, $\mathcal{A}_{t_{n-1}}$,

. . . , $\mathcal{A}_s$, we find

$$E^{\mathcal{A}_s}(X_t - X_s) = E^{\mathcal{A}_s}E^{\mathcal{A}_{t_1}} \cdots E^{\mathcal{A}_{t_n}}(X_t - X_s)$$
$$= 0 \quad \text{a.s.}$$

Since X_s is obviously $\mathcal{A}_s$ measurable, this proves the proposition for φ equal to a step. If φ is not a step, let $\{\varphi_n\}$ be step approximations to φ, and define

$$X_n(\omega,t) = \int_a^t \varphi_n(\omega,t)\, dW(\omega,\tau) \tag{3.4}$$

For each n, $\{X_n(\omega,t), \mathcal{A}_t, a \le t \le b\}$ is a martingale, and

$$E^{\mathcal{A}_s}(X_t - X_s) = E^{\mathcal{A}_s}[X(\cdot,t) - X_n(\cdot,t)] - E^{\mathcal{A}_s}[X(\cdot,s) - X_n(\cdot,s)]$$

Since $E|X(\cdot,t) - X_n(\cdot,t)|^2 \xrightarrow[n\to\infty]{} 0$, we have

$$E^{\mathcal{A}_s}(X_t - X_s) = E^{\mathcal{A}_s}[X(\cdot,t) - X_n(\cdot,t)] - E^{\mathcal{A}_s}[X(\cdot,s) - X_n(\cdot,s)] \xrightarrow[n\to\infty]{\text{q.m.}} 0 \tag{3.5}$$

Hence, $E^{\mathcal{A}_s}(X_t - X_s) = 0$, a.s., and the proof is complete. ∎

A process $\{X_t, a \le t \le b\}$ as defined by (3.1) is obviously q.m. continuous. Thus, we can choose a version of $\{X_t, a \le t \le b\}$ which is separable and measurable. If we choose such a version and if we assume that the Brownian process $\{W_t, a \le t \le b\}$ in (3.1) is also separable, then $\{X_t, a \le 1 \le b\}$ is sample continuous with probability 1. When φ is an (ω,t)-step function, sample continuity is obvious since $\{X_t, a \le t \le b\}$ is then a separable Brownian motion pieced together in a continuous way. If φ is not a step, let $\{\varphi_n\}$ be a sequence of (ω,t)-step functions satisfying (2.3) and (2.4) such that

$$\|\varphi - \varphi_n\|^2 = \int_a^b E|\varphi(\cdot,s) - \varphi_n(\cdot,s)|^2\, ds \le \frac{1}{n^4}$$

Such a sequence can always be obtained by choosing a subsequence of any sequence $\{\psi_m\}$ such that $\|\psi_m - \varphi\| \xrightarrow[m\to\infty]{} 0$.

If we set $X_{nt} = \int_a^t \varphi_{ns}\, dW_s$ and choose it to be separable, then for each n, $\{X_{nt}, a \le t \le b\}$ is sample continuous with probability 1. For each n, $\{X_{nt} - X_t, a \le t \le b\}$ is a separable second-order martingale. If we apply the version of Proposition 2.3.2 for complex-valued martingales, we get

$$\mathcal{P}\left(\sup_{a \le t \le b} |X_{nt} - X_t| \ge \frac{1}{n}\right) \le n^2 E|X_{nb} - X_b|^2 = n^2\|\varphi_n - \varphi\|^2 \le \frac{1}{n^2}$$

It follows that $\sum_n \mathcal{P}(\sup_{a \le t \le b} |X_{nt} - X_t| \ge 1/n) < \infty$ and the Borel-Cantelli lemma implies that

$$\Lambda = \limsup_n \left\{ \omega: \sup_{a \leq t \leq b} |X_{nt}(\omega) - X_t(\omega)| \geq \frac{1}{n} \right\}$$

is a null set, that is, for every $\omega \notin \Lambda$, $\sup_t |X_{nt}(\omega) - X_t(\omega)| \geq \frac{1}{n}$ for, at most, a finite number of n. Therefore, for $\omega \notin \Lambda$,

$$\lim_{n\to\infty} \sup_{a \leq t \leq b} |X_{nt}(\omega) - X_t(\varphi)| = 0$$

and $X_t(\omega)$, $a \leq t \leq b$, being the uniform limit of a sequence of continuous functions, is itself continuous. This proves the sample continuity of $\{X_t, a \leq t \leq b\}$.

One immediate consequence of the martingale property is that a stochastic integral does not behave like an ordinary integral. Consider, for example, the stochastic integral $\int_0^t W_s \, dW_s$. If the integral is like an ordinary integral, surely it must be equal to $\frac{1}{2}(W_t^2 - W_0^2) = \frac{1}{2}W_t^2$. However, $\frac{1}{2}W_t^2$ is not a martingale, as is seen from the relationship

$$E^{\mathfrak{A}_s}(\tfrac{1}{2}W_t^2) = \tfrac{1}{2}W_s^2 + \tfrac{1}{2}(t - s)$$

Therefore, $\int_0^t W_s \, dW_s$ cannot be equal to $\frac{1}{2}W_t^2$. What $\int_0^t W_s \, dW_s$ is will be clarified by the so-called **Ito's differentiation rule,** which will be stated below.

To state the differentiation rule for stochastic integrals under its natural hypotheses, we need to generalize the definition of stochastic integrals to include integrands φ which satisfy (2.3), but instead of (2.4), the weaker condition

$$\int_a^b |\varphi(\omega,t)|^2 \, dt < \infty \qquad \text{almost surely} \tag{3.6}$$

This generalization is discussed in detail in Sec. 6 of this chapter (see, in particular, Proposition 6.1). For now, we merely note that if φ satisfies (2.3) and (3.6), then the stochastic integral $\int_a^b \varphi_t \, dW_t$ is defined by

$$\int_a^b \varphi(\omega,t) \, dW(\omega,t) = \lim_{n\to\infty} \text{in p.} \int_a^b \varphi^n(\omega,t) \, dW(\omega,t)$$

where φ^n is defined by

$$\varphi^n(\omega,t) = \begin{cases} \varphi(\omega,t) & \text{if } \int_a^t |\varphi(\omega,s)|^2 \, ds \leq n \\ 0 & \text{otherwise} \end{cases}$$

Stochastic integrals appearing in the following proposition will be assumed to be defined in this way if the integrands satisfy (3.6) rather than (2.4).

Proposition 3.2. Let $X_1(\omega,t), X_2(\omega,t), \ldots, X_n(\omega,t)$ be processes defined in terms of a single Brownian motion $W(\omega,t)$ as follows:

$$X_k(\omega,t) = \int_a^t f_k(\omega,t')\,dt' + \int_a^t \varphi_k(\omega,t')\,dW(\omega,t') \qquad k = 1, \ldots, n \tag{3.7}$$

Let $Y(\omega,t) = \psi(X_{1t}, X_{2t}, \ldots, X_{nt}, t)$, where ψ is once continuously differentiable with respect to t and has continuous second partials with respect to the X's. Then, with probability 1,

$$Y(\omega,t) = Y(\omega,a) + \int_a^t \dot{\psi}(\mathbf{X}(\omega,t'), t')\,dt' + \sum_{k=1}^{n} \int_a^b \psi_k(\mathbf{X}(\omega,t'), t')\,dX_k(\omega,t') + \sum_{j=1}^{n}\sum_{k=1}^{n} \frac{1}{2}\int_a^b \psi_{jk}(\mathbf{X}(\omega,t'), t')\varphi_j(\omega,t')\varphi_k(\omega t')\,dt' \tag{3.8}$$

where $\dot{\psi} = \partial\psi/\partial t$, $\psi_k = \partial\psi/\partial x_k$, and $\psi_{jk} = \partial^2\psi/\partial x_j\,\partial x_k$.

Remark: The surprising thing about (3.8) is the last term. It comes about in roughly the following way. We recall that a Brownian motion W_t has the curious property $(dW_t)^2 \cong dt$. Therefore, $dX_j(t)\,dX_k(t) \cong \varphi_j\varphi_k\,dt$. Now,

$$\begin{aligned} dY_t &= Y_{t+dt} - Y_t = \psi(\mathbf{X}_{t+dt}, t+dt) - \psi(\mathbf{X}_t,t) \\ &= \dot{\psi}\,dt + \sum_k \psi_k\,dX_k(t) + \frac{1}{2}\sum_j\sum_k \psi_{jk}\,dX_j(t)\,dX_k(t) + \cdots \end{aligned} \tag{3.9}$$

Both the first and the third term in (3.9) are of order dt, hence

$$dY_t = \dot{\psi}\,dt + \sum_k \psi_k\,dX_k(t) + \frac{1}{2}\sum_{j,k} \psi_{jk}\varphi_j\varphi_k\,dt + o(dt) \tag{3.10}$$

which is nothing but a symbolic way of writing (3.8). We note that $dX_k(t)$ can be replaced by $f_k\,dt + \varphi_k\,dW_t$ in both (3.8) and (3.10), permitting Y_t to be written in the form of

$$Y_t = Y_a = \int_a^t g(\omega,t')\,dt' + \int_a^t \gamma(\omega,t')\,dW(\omega,t')$$

If we apply (3.8) to $Y_t = \frac{1}{2}W_t^2$, we find immediately

$$dY_t = W_t\,dW_t + \tfrac{1}{2}dt$$

or

$$Y_t = \int_0^t W_s\,dW_s + \tfrac{1}{2}t$$

or

$$\int_0^t W_s\,dW_s = Y_t - \tfrac{1}{2}t = \tfrac{1}{2}W_t^2 - \tfrac{1}{2}t \tag{3.11}$$

which is indeed a martingale.

It might be useful to isolate two special cases of Proposition 3.2. First, suppose $Y_t = \psi(X_t,t)$ depends only on a single X process, and $dX_t = f_t\,dt + \varphi_t\,dW_t$. Then, (3.8) becomes

$$Y_t = Y_a + \int_a^t \dot{\psi}(X_s,s)\,ds + \int_a^t \psi'(X_s,s)\,dX_s + \frac{1}{2}\int_a^t \psi''(X_s,s)\varphi^2(\omega,s)\,ds \tag{3.12}$$

where prime denotes differentiation with respect to the first variable. For the second special case, consider the product $Y(\omega,t) = X_1(\omega,t)X_2(\omega,t)$, where X_1 and X_2 satisfy (3.7) with $k = 1, 2$. Then (3.8) becomes

$$Y_t = Y_a + \int_a^t X_2(\omega,t')\,dX_1(\omega,t') + \int_a^t X_1(\omega,t')\,dX_2(\omega,t') + \frac{1}{2}\int_a^t \varphi_1(\omega,t')\varphi_2(\omega,t')\,dt' \tag{3.13}$$

4. STOCHASTIC DIFFERENTIAL EQUATIONS

From the point of applications, a major motivation for studying stochastic differential equations is to give meaning to an equation of the form

$$\frac{d}{dt}X_t = m(X_t,t) + \sigma(X_t,t)\zeta_t \tag{4.1}$$

where ζ_t is a Gaussian white noise. At least formally, we know that $\int_0^t \zeta_s\,ds$ has all the attributes of a Brownian motion W_t. Hence, formally again, (4.1) appears to be equivalent to

$$X_t = X_a + \int_a^t m(X_s,s)\,ds + \int_a^t \sigma(X_s,s)\,dW_s \tag{4.2}$$

With stochastic integrals having been defined, (4.2) is capable of taking on a precise interpretation. Whether the interpretation is the one that we really want to give to (4.1) is something else again. We postpone examination of this question until Sec. 5. For the time being, we confine ourselves to a study of (4.2) as an equation in the unknown X_t and with the last integral interpreted as a stochastic integral.

By a **stochastic differential equation,** we mean an equation of the form

$$dX(\omega,t) = m(X(\omega,t),\,t)\,dt + \sigma(X(\omega,t),\,t)\,dW(\omega,t) \tag{4.3}$$

which is nothing more or less than a symbolic way of writing (4.2). A process $\{X_t,\, t \geq a\}$ is said to satisfy (4.3) with initial condition $X_a = X$ if (1) for each t

$$\int_a^t \sigma(X_s,s)\,dW_s$$

is capable of being interpreted as a stochastic integral and (2) for each t, X_t is almost surely equal to the random variable defined by

$$X + \int_a^t m(X_s,s)\,ds + \int_a^t \sigma(X_s,s)\,dW_s$$

Under the conditions that we shall assume, we can in fact assert a stronger result than a.s. equality of the two random variables, viz., q.m. difference between the two is zero.

We shall first state and prove an existence and uniqueness theorem following Ito.

Proposition 4.1. Let $\{W_t, \mathcal{A}_t, a \le t \le T < \infty\}$ be a separable Brownian motion. Let X be a random variable measurable with respect to $\mathcal{A}_a$ and satisfy $EX^2 < \infty$. Let $m(x,t)$ and $\sigma(x,t)$, $-\infty < x < \infty$, $a \le t \le T$, be Borel measurable functions in the pair (x,t). Let m and σ satisfy the following conditions:

$$|m(x,t) - m(y,t)| + |\sigma(x,t) - \sigma(y,t)| \le K|x - y| \tag{4.4}$$

$$|m(x,t)| + |\sigma(x,t)| \le K\sqrt{1 + x^2} \tag{4.5}$$

Under these hypotheses, there exists a separable and measurable process $\{X_t, a \le t \le T\}$ with the following properties:

P_1: For each t in $[a,T]$ X_t is $\mathcal{A}_t$-measurable

P_2: $\int_a^T EX_t^2\,dt < \infty$

P_3: $\{X_t, a \le t \le T\}$ satisfies (4.2) with $X_a = X$

P_4: With probability 1, $\{X_t, a \le t \le T\}$ is sample continuous

P_5: $\{X_t, a \le t \le T\}$ is unique with probability 1

P_6: $\{X_t, a \le t \le T\}$ is a Markov process

Remark: Condition (4.4) is known as the **uniform Lipschitz condition.** Without loss of generality, the constants K in (4.4) and (4.5) can be assumed to be the same.

Proof: We shall give a proof by constructing a solution. Since we shall be dealing with a sequence of stochastic process, we shall write $X(\omega,t)$ or $X(\cdot,t)$ rather than X_t, because the subscript will be used to index terms in the sequence. First, define a sequence of processes $\{X_n(\cdot,t), a \le t \le T\}$ as follows:

$$X_0(\omega,t) = X(\omega)$$

$$X_{n+1}(\omega,t) = X(\omega) + \int_a^t m(X_n(\omega,s), s)\,ds + \int_a^t \sigma(X_n(\omega,s), s)\,dW(\omega,s) \tag{4.6}$$

We need to show that the last integral is well defined as a stochastic integral for each n. That is, we need to show that for each n

$$\sigma(X_n(\omega,t), t) \text{ is jointly measurable in } (\omega,t), \text{ and for each } t \text{ is } \mathcal{A}_t \text{ measurable} \tag{4.7}$$

and

$$\int_{a]}^{T} E\sigma^2(X_n(\cdot,t), t)\, dt < \infty \tag{4.8}$$

This can be done by induction. First, we verify (4.7) and (4.8) for $n = 0$. Since $X_0(\omega,t) = X(\omega)$, (4.7) is satisfied, because σ is a Borel measurable function, and $\sigma(X,t)$ is not only $\mathcal{A}_t$ measurable, it is $\mathcal{A}_a$ measurable. Using (4.5), we have

$$\sigma^2(X,t) \leq K^2(1 + X^2)$$

so that

$$\int_a^T E\sigma^2(X_0(\cdot,t), t)\, dt \leq K^2(1 + EX^2)(T - a) < \infty$$

and (4.8) is verified for $n = 0$. Now, assume that (4.7) and (4.8) are both satisfied for $n = 0, 1, 2, \ldots, k$. Then from (4.6),

$$X_{k+1}(\omega,t) = X + \int_a^t m(X_k(\omega,s), s)\, ds + \int_a^t \sigma(X_k(\omega,s), s)\, dW(\omega,s)$$

each of the three terms on the right is $\mathcal{A}_t$ measurable, because $\{X_k(\cdot,s), a \leq s \leq t\}$ is $\mathcal{A}_t$ measurable. Next, we note that for $a \leq t_0 \leq t \leq T$,

$$[X_{k+1}(\omega,t) - X_{k+1}(\omega,t_0)]^2 \leq 2\left\{\left[\int_{t_0}^t m(X_k(\omega,s), s)\, ds\right]^2 + \left[\int_{t_0}^t \sigma(X_k(\omega,s), s)\, dW(\omega,s)\right]^2\right\} \tag{4.9}$$

By using the Schwarz inequality on the second term and (2.18) on the last term, we get

$$E[X_{k+1}(\cdot,t) - X_{k+1}(\cdot,t_0)]^2 \leq 2\left[(t - t_0)\int_{t_0}^t Em^2(X_k(\cdot,s), s)\, ds + \int_{t_0}^t E\sigma^2(X_k(\cdot,s), s)\, ds\right] \tag{4.10}$$

Now, (4.5) can be applied again, and we get

$$E[X_{k+1}(\cdot,t) - X_{k+1}(\cdot,t_0)]^2 \leq 2\left\{K^2[1 + (t - t_0)]\int_{t_0}^t [1 + EX_k^2(\cdot,s)]\, ds\right\} \tag{4.11}$$

Therefore, $\{X_{k+1}(\cdot,t), a \leq t \leq T\}$ is q.m. continuous and a measurable version can be chosen. Furthermore,

$$\begin{aligned}\int_a^T EX_{k+1}^2(\cdot,t)\, dt &\leq 2\int_a^T E[X_{k+1}(\cdot,t) - X]^2\, dt + 2\int_a^T EX^2\, dt \\ &\leq 4K^2[1 + (T - a)](T - a)\int_a^T [1 + EX_k^2(\cdot,s)]\, ds \\ &\quad + 2(T - a)EX^2 < \infty\end{aligned} \tag{4.12}$$

The induction is complete, and we have verified (4.7) and (4.8) for every n. Therefore, the sequence of processes $\{X_n(\cdot,t),\ a \le t \le T,\ n = 0, 1, \ldots\}$ is well defined.

Next we prove that for each t, $\{X_n(\cdot,t),\ n = 0, 1, \ldots\}$ converges in quadratic mean. To do this, define

$$\begin{aligned} \Delta_0(\omega,t) &= X(\omega) \\ \Delta_n(\omega,t) &= X_n(\omega,t) - X_{n-1}(\omega,t) \qquad n = 1, 2, \ldots \end{aligned} \tag{4.13}$$

Using (4.6), we get

$$\begin{aligned} \Delta_{n+1}(\omega,t) = &\int_a^t [m(X_n(\omega,s), s) - m(X_{n-1}(\omega,s), s)]\, ds \\ &+ \int_a^t [\sigma(X_n(\omega,s), s) - \sigma(X_{n-1}(\omega,s), s)]\, dW(\omega,s) \end{aligned} \tag{4.14}$$

If we make use of the inequality $(A + B)^2 \le 2A^2 + 2B^2$, the Schwarz inequality, (2.18) on the stochastic integral, and the uniform Lipschitz condition, we find

$$E\Delta_{n+1}^2(\cdot,t) \le 2K^2[1 + (T - a)] \int_a^t E\Delta_n{}^2(\cdot,s)\, ds \tag{4.15}$$

The inequality (4.15) can be iterated starting from $E\Delta_0{}^2(\cdot,t) = EX^2$, and we get

$$E\Delta_n{}^2(\cdot,t) \le \{2K^2[1 + (T - a)]\}^n \frac{(t-a)^n}{n!} EX^2 \tag{4.16}$$

Now,

$$X_{n+m}(\omega,t) - X_n(\omega,t) = \sum_{k=1}^{m} \Delta_{n+k}(\omega,t)$$

and by the Cauchy-Schwarz inequality

$$[X_{n+m}(\omega,t) - X_n(\omega,t)]^2 \le \left(\sum_{k=1}^{m} \frac{1}{2^{n+k}}\right)\left[\sum_{k=1}^{m} 2^{n+k}\Delta_{n+k}^2(\omega,t)\right] \tag{4.17}$$

Therefore, from (4.16) we get

$$\begin{aligned} \sup_{m\ge 0} E[X_{n+m}(\cdot,t) - X_n(\cdot,t)]^2 &\le \sum_{k=1}^{\infty} 2^{n+k} E\Delta_{n+k}^2(\cdot,t) \\ &\le EX^2 \sum_{k=n+1}^{\infty} \frac{\{4K^2[1 + (T-a)](t-a)\}^k}{k!} \end{aligned} \tag{4.18}$$

Since $\sum_{k=0}^{\infty} \alpha^k/k!$ converges to e^α for every finite α,

$$\sup_{m\ge 0} E[X_{n+m}(\cdot,t) - X_n(\cdot,t)]^2 \xrightarrow[n\to\infty]{} 0 \tag{4.19}$$

uniformly in t. Therefore, for every $t \in [a,T]$, $\{X_n(\cdot,t)\}$ is a q.m. convergent sequence. Let the q.m. limit be denoted by $X(\cdot,t)$.

Thus, we have obtained a process $\{X(\omega,t),\ a \le t \le T\}$ such that

$$\sup_{a\le t\le T} E[X_n(\cdot,t) - X(\cdot,t)]^2 \xrightarrow[n\to\infty]{} 0 \tag{4.20}$$

Because for each n, $\{X_n(\cdot,t),\ a \le t \le T\}$ is q.m. continuous, the limit process $\{X_t,\ a \le t \le T\}$ is also q.m. continuous, hence continuous in probability. It follows from Proposition 2.2.3 that a separable and measurable version can be chosen for $\{X_t,\ a \le t \le T\}$. We shall now show that $\{X_t,\ a \le t \le T\}$ so constructed satisfies $P_1 - P_5$.

First, for each t, $X_n(\cdot,t)$ is $\mathcal{A}_t$ measurable for every n. Therefore, $X(\cdot,t)$ is also $\mathcal{A}_t$ measurable for each t, and P_1 is proved. Next,

$$\int_a^T EX_t^2\,dt \le 2\left\{\int_a^T E[X(\cdot,t) - X_n(\cdot,t)]^2\,dt + \int_a^T EX_n^2(\cdot,t)\,dt\right\} \tag{4.21}$$

From (4.18) we have for some constant α,

$$\sup_n \int_a^T EX_n^2(\cdot,t)\,dt \le 2\left[\int_a^T e^{\alpha t}\,dt + (T-a)\right] EX^2 = A < \infty \tag{4.22}$$

Hence, using (4.20) on (4.21) we get

$$\int_a^T EX_t^2\,dt \le 2A < \infty$$

which proves P_2. Together, P_1 and P_2 ensure that $\int_a^t \sigma(X_s,s)\,dW_s$ is well defined as a stochastic integral.

To prove that the process $\{X_t,\ a \le t \le T\}$ is indeed a solution to (4.2) with $X_a = X$, we define

$$D_t = X_t - X - \int_a^t m(X_s,s)\,ds - \int_a^t \sigma(X_s,s)\,dW_s$$

Using (4.6), we can rewrite D_t as

$$D_t = [X(\cdot,t) - X_{n+1}(\cdot,t)] - \int_a [m(X(\cdot,s),\ s) - m(X_n(\cdot,s),\ s)]\,ds$$
$$- \int_a^t [\sigma(X(\cdot,s),\ s) - \sigma(X_n(\cdot,s),\ s)]\,ds$$

It is now easy to show that each of the three terms on the right-hand side goes to zero in quadratic mean as $n \to \infty$. Therefore

$$ED_t^2 = 0$$

and for each $t \in [a,T]$,

$$X_t = X + \int_a^t m(X_s,s)\,ds + \int_a^t \sigma(X_s,s)\,dW_s \tag{4.23}$$

with probability 1. Further, both sides of (4.23) represent separable processes if we choose a separable version for the stochastic integral. Hence,

$$\mathcal{P}\left(X_t = X + \int_a^t m(X_s,s)\,ds + \int_a^t \sigma(X_s,s)\,dW_s,\ a \le t \le T\right) = 1 \tag{4.24}$$

and P_3 has been proved. Since the right-hand side of (4.23) represents a sample-continuous process, $\{X_t,\ a \le t \le T\}$ is also sample continuous, and P_4 is proved. To prove uniqueness, suppose that X_t and $\tilde{X}_t$ are both solutions of (4.2) with $X_a = \tilde{X}_a = X$. Then, we can write

$$(X_t - \tilde{X}_t) = \int_a^t [m(X_s,s) - m(\tilde{X}_s,s)]\,ds + \int_a^t [\sigma(X_s,s) - \sigma(\tilde{X}_s,s)]\,dW_s \tag{4.25}$$

Equation (4.25) has the same form as (4.14), and the same arguments yield

$$E(X_t - \tilde{X}_t)^2 \le 2K^2[1 + (T - a)] \int_a^t E(X_s - \tilde{X}_s)^2\,ds \tag{4.26}$$

Inequality (4.26) has the form

$$\frac{d}{dt} f(t) \le cf(t) \qquad t > a \tag{4.27}$$

with $f(t) \ge 0$ and $f(a) = 0$. Rewriting, we get

$$\frac{d}{dt}(e^{-ct}f(t)) \le 0 \qquad t > a$$

Therefore, by integrating we get $e^{-ct}f(t) \le f(a)e^{-ca}$, and it follows that

$$0 \le f(t) \le f(a)e^{c(t-a)} = 0 \qquad a \le t \le T$$

This proves that $f(t) = \int_a^t E(X_s - \tilde{X}_s)^2\,ds = 0$, $a \le t \le T$. From (4.26) we have

$$E(X_t - \tilde{X}_t)^2 = 0 \qquad a \le t \le T$$

Therefore, for each t in $[a,T]$,

$$\mathcal{P}(X_t \ne \tilde{X}_t) = 0$$

and by σ additivity,

$$\mathcal{P}(X_t \ne \tilde{X}_t \text{ at one or more rational points in } [a,T]) = 0$$

If X_t and $\tilde{X}_t$ are both chosen to be separable, then they are both sample continuous and

$$\mathcal{P}(X_t = \tilde{X}_t \text{ for all } t \in [a,T]) = 1$$

This proves that with probability 1 there is only one sample-continuous solution to (4.2) with the same initial condition.

Finally, we prove that $\{X_t,\ a \le t \le T\}$ is a Markov process. Using (4.2) we write

$$X_t = X_s + \int_s^t m(X_\tau,\tau)\, d\tau + \int_s^t \sigma(X_\tau,\tau)\, dW_\tau \qquad a \le s < t \le T$$

which can be regarded as a stochastic integral equation on the interval $s \le t \le T$ with X_s as the initial condition. Thus, for each $t \in [s,T]$, X_t can be obtained as a function of X_s and $\{W_\tau - W_s,\ s \le \tau \le t\}$, that is, X_t is measurable with respect to the σ algebra generated by X_s and $\{W_\tau - W_s,\ s \le \tau \le t\}$. Since X_s is $\mathcal{A}_s$ measurable, and $\{W_\tau - W_s,\ s \le \tau \le t\}$ is independent of $\mathcal{A}_s$, X_t is conditionally independent of $\mathcal{A}_s$ given X_s. *A fortiori*, X_t is conditionally independent of $\{X_\tau,\ a \le \tau \le s\}$ given X_s, and this proves the Markov property. ∎

Summarizing the preceding results, we find that, under the conditions on X, m, and σ given in Proposition 4.1, the stochastic integral equation

$$X_t = X + \int_a^t m(X_s,s)\, ds + \int_a^t \sigma(X_s,s)\, dW_s \tag{4.28}$$

has a unique sample-continuous solution which is Markov. We emphasize again that, by definition, the last integral $\int_a^t \sigma(X_s,s)\, dW_s$ in the integral equation is to be interpreted as a stochastic integral. The question whether this stochastic integral equation adequately models a differential equation driven by Gaussian white noise

$$\begin{aligned} \frac{d}{dt} X_t &= m(X_t,t) + \sigma(X_t,t)\zeta_t \\ X_a &= X \end{aligned} \tag{4.29}$$

remains unanswered. Indeed, this question cannot be answered without more being said about what we want (4.29) to mean. As it stands, (4.29) is merely a string of symbols, nothing more. We shall take up this question in the next section.

Finally, we note that the existence of a solution to (4.2) is ensured even without the Lipschitz condition (4.4), but then the uniqueness is no longer guaranteed [Skorokhod, 1965, p. 59].

5. WHITE NOISE AND STOCHASTIC CALCULUS

In this section, we offer an interpretation of differential equations driven by white noise, and examine its relationship with stochastic differential equations. The equation that we would like to give a precise meaning to is the following:

$$\frac{d}{dt} X(\omega,t) = m(X(\omega,t), t) + \sigma(X(\omega,t), t)\zeta(\omega,t) \tag{5.1}$$

where ζ_t is a Gaussian white noise. Since white noise is an abstraction and not a physical process, what one really means by (5.1) in practice is probably an equation driven by a stationary Gaussian process with a spectral density that is flat over a very wide range of frequencies. If we take ζ_t to be such a process in (5.1), then there is no difficulty interpreting (5.1) as an ordinary differential equation for each sample function, provided that the spectral density of ζ_t eventually goes to zero sufficiently rapidly so that the sample functions are well behaved. While this is probably what we want (5.1) to mean, this is not how we want (5.1) to be handled mathematically. If we take ζ_t to be a process with well-behaved sample functions, we lose some of the simple statistical properties of X_t, the primary one being the Markov property. In practice, the interpretation of (5.1) that we really want is probably the following. Take a sequence of Gaussian processes $\{\zeta_n(\cdot,t)\}$ which "converges" in some suitable sense to a white Gaussian noise, and yet for each n $\zeta_n(\cdot,t)$ has well-behaved sample functions. Now, for each n the equation

$$\frac{d}{dt} X_n(\omega,t) = m(X_n(\omega,t), t) + \sigma(X_n(\omega,t), t)\zeta_n(\omega,t) \qquad a \le t \le T \tag{5.2}$$

together with the initial condition $X_n(\omega,a) = X(\omega)$ can be solved. We assume that m and σ are such that the solution exists and is unique for almost all sample functions. Thus, we obtain a sequence of processes $\{X_n(\cdot,t),\ a \le t \le T\}$. Suppose that as $n \to \infty$, $\{\zeta_n(\cdot,t)\}$ converges in a suitable sense to white noise, and the sequence $\{X_n(\cdot,t),\ a \le t \le T\}$ converges almost surely, or in quadratic mean, or even merely in probability, to a process $\{X(\cdot,t),\ a \le t \le T\}$. Then it is natural to say that X_t is the solution of

$$\dot{X}_t = m(X_t,t) + \sigma(X_t,t)\zeta_t$$

where ζ_t is a Gaussian white noise. This makes precise the interpretation of (5.1). We still have to determine whether (5.1) can be modeled by a stochastic differential equation as defined in the last section.

In order to make precise the notion of $\{\zeta_n(\cdot,t)\}$ converging to a white noise, we define

$$W_n(\omega,t) = \int_a^t \zeta_n(\omega,s)\, ds \tag{5.3}$$

and rewrite (5.2) as an integral equation

$$X_n(\omega,t) = X_n(\omega,a) + \int_a^t m(X_n(\omega,s), s)\, ds + \int_a^t \sigma(X_n(\omega,s), s)\, dW_n(\omega,s) \quad (5.4)$$

Since a Gaussian white noise ζ_t is the formal derivative of a Brownian motion, we make precise the notion of $\{\zeta_n(\cdot,t)\}$ converging to a Gaussian white noise by requiring that

$$W_n(\cdot,t) \xrightarrow[n\to\infty]{\text{a.s.}} K[W(\cdot,t) - W(\cdot,a)] \quad (5.5)$$

where K is a constant and $\{W(\cdot,t),\ a \le t \le T\}$ is a Brownian motion process. Since the constant K can always be absorbed into σ in (5.4), we shall assume it to be 1. We want to resolve the following two questions: First, under what conditions will $\{X_n(\cdot,t), a \le t \le T\}$ converge? Secondly, if $\{X_n(\cdot,t),\ a \le t \le T\}$ converges, does the limit $\{X(\cdot,t),\ a \le t \le T\}$ satisfy a stochastic differential equation, and if so, what stochastic differential equation?

Before stating the precise results that can be proved concerning these questions, we shall give a preliminary and heuristic discussion of what we can expect. This is especially important since what can be proved precisely at present is a little complicated and undoubtedly falls far short of what is in fact true. To begin with, consider a sequence of processes $\{Y_n(\cdot,t)\}$ defined by

$$Y_n(\omega,t) = \int_a^t \varphi(W_n(\omega,t), t)\, dW_n(\omega,t) \quad (5.6)$$

where φ is a known function and $\{W_n(\cdot,t)\}$ converges to a Brownian motion, and we want to determine what $\{Y_n(\cdot,t)\}$ converges to. Suppose we define a function $\psi(x,t)$ by

$$\psi(x,t) = \int_0^x \varphi(z,t)\, dz \quad (5.7)$$

If we denote $(\partial/\partial t)\psi(x,t)$ by $\dot{\psi}(x,t)$, we find

$$d\psi(W_n(\omega,t), t) = \varphi(W_n(\omega,t), t)\, dW_n(\omega,t) + \dot{\psi}(W_n(\omega,t), t)\, dt \quad (5.8)$$

In other words, we have

$$\psi(W_n(\omega,t), t) - \psi(W_n(\omega,a), a) - \int_a^t \dot{\psi}(W_n(\omega,s), s)\, ds = Y_n(\omega,t) \quad (5.9)$$

Now, if ψ and $\dot{\psi}$ are reasonable functions, we would certainly expect that as $W_n(\omega,t) \xrightarrow[n\to\infty]{} W(\omega,t)$,

$$\psi(W_n(\omega,t), t) \xrightarrow[n\to\infty]{} \psi(W(\omega,t), t)$$
$$\dot{\psi}(W_n(\omega,t), t) \xrightarrow[n\to\infty]{} \dot{\psi}(W(\omega,t), t)$$

Therefore, if all this is true, then

$$Y_n(\omega,t) \xrightarrow[n\to\infty]{} Y(\omega,t) = \psi(W(\omega,t), t) - \psi(W(\omega,a), a) - \int_a^t \dot{\psi}(W(\omega,s), s)\, ds \quad (5.10)$$

Now, by the Ito's differentiation rule (Proposition 3.2),

$$\psi(W(\omega,t), t) = \psi(W(\omega,a), a) + \int_a^t \dot{\psi}(W(\omega,s), s)\, ds + \int_a^t \psi'(W(\omega,s), s)\, dW(\omega,s) + \frac{1}{2}\int_a^t \psi''(W(\omega,s), s)\, ds \quad (5.11)$$

Noting $\psi'(x,t) = \varphi(x,t)$, we get

$$Y(\omega,t) = \int_a^t \varphi(W(\omega,s), s)\, dW(\omega,s) + \frac{1}{2}\int_a^t \varphi'(W(\omega,s), s)\, ds \quad (5.12)$$

Comparing (5.12) against (5.6), we get the interesting result

$$\int_a^t \varphi(W_n(\omega,s), s)\, dW_n(\omega,s) \xrightarrow[n\to\infty]{} \int_a^t \varphi(W(\omega,s), s)\, dW(\omega,s) + \frac{1}{2}\int_a^t \varphi'(W(\omega,s), s)\, ds \quad (5.13)$$

where the first term on the right-hand side in (5.13) is a stochastic integral. The reason for the extra term is the same as the reason for the extra term in the Ito's differentiation formula (3.8). As we discussed it at that time, roughly speaking, the extra term is due to the fact that $(dW)_t^2$ is approximately dt.

In light of (5.13), we should expect a similar development for (5.4) as $n \to \infty$, namely, there will be an extra term. To find out what this extra term is, we first rewrite (5.4) as

$$dX_n(\omega,t) = m(X_n(\omega,t), t)\, dt + \sigma(X_n(\omega,t), t)\, dW_n(\omega,t) \quad (5.14)$$

Now define

$$\psi(x,t) = \int_0^x \frac{1}{\sigma(z,t)}\, dz \quad (5.15)$$

so that

$$d\psi(X_n(\omega,t), t) = \dot{\psi}(X_n(\omega,t), t)\, dt + \psi'(X_n(\omega,t), t)\, dX_n(\omega,t) = \dot{\psi}(X_n(\omega,t), t) + \frac{m(X_n(\omega,t), t)}{\sigma(X_n(\omega,t), t)}\, dt + dW_n(\omega,t) \quad (5.16)$$

or

$$\psi(X_n(\omega,t), t) - \psi(X_n(\omega,a), a) = \int_a^t \mu(X_n(\omega,s), s)\, ds + W_n(\omega,t) - W_n(\omega,a) \quad (5.17)$$

where we have set $\mu = (m/\sigma) + \dot{\psi}$. Suppose $\{W_n(\cdot,t)\}$ converges to a Brownian motion $W(\cdot,t)$ and suppose that $\{X_n(\cdot,t)\}$ converges to a process $X(\cdot,t)$. Then, under reasonable conditions, we would expect

$$\psi(X(\omega,t), t) - \psi(X(\omega,a), a) = \int_a^t \mu(X(\omega,s), s)\, ds + W(\omega,t) - W(\omega,a) \quad (5.18)$$

If we assume that $X(\omega,t)$ can be written 'n the form of

$$X(\omega,t) = X(\omega,a) + \int_a^t f(\omega,s)\, ds + \int_a^t \varphi(\omega,s)\, dW(\omega,s) \quad (5.19)$$

then we can apply Ito's differentiation formula (3.8) to $\psi(X_t,t)$ and get

$$\psi(X(\omega,t), t) - \psi(X(\omega,a), a) = \int_a^t \psi'(X(\omega,s), s) f(\omega,s)\, ds + \int_a^t \dot{\psi}(X(\omega,s), s)\, ds + \int_a^t \psi'(X(\omega,s), s)\varphi(\omega,s)\, dW(\omega,s) + \frac{1}{2}\int_a^t \psi''(X(\omega,s), s)\varphi^2(\omega,s)\, ds \quad (5.20)$$

We can now equate (5.18) with (5.20) and get

$$\varphi(\omega,s)\psi'(X(\omega,s), s) = 1$$

Therefore, by noting that $\psi' = 1/\sigma$, we get

$$\varphi(\omega,t) = \sigma(X(\omega,t), t) \quad (5.21)$$

Further,

$$\dot{\psi} + \frac{1}{\sigma} f + \tfrac{1}{2}\sigma^2 \left(\frac{1}{\sigma}\right)' = \mu = \frac{m}{\sigma} + \dot{\psi}$$

Hence,

$$f(\omega,t) = m(X(\omega,t), t) + \tfrac{1}{2}\sigma(X(\omega,t), t)\sigma'(X(\omega,t), t) \quad (5.22)$$

Putting (5.21) and (5.22) into (5.19), we get

$$X_t = X_a + \int_a^t [m(X_s,s) + \tfrac{1}{2}\sigma(X_s,s)\sigma'(X_s,s)]\, ds + \int_a^t \sigma(X_s,s)\, dW_s \quad (5.23)$$

What we have shown, at least formally, is that if we interpret a white-noise-driven equation

$$\frac{d}{dt} X_t = m(X_t,t) + \sigma(X_t,t)\zeta_t \quad (5.24)$$

by a sequence of equations like (5.2), then the white-noise-driven differential equation is equivalent to a stochastic differential equation given by

$$dX_t = m(X_t,t)\, dt + \tfrac{1}{2}\sigma(X_t,t)\sigma'(X_t,t)\, dt + \sigma(X_t,t)\, dW_t \quad (5.25)$$

Again, we note the presence of an extra term $\frac{1}{2}\sigma\sigma'$, which will be referred to as the **correction term.**

We shall now state some convergence results concerning (5.13) and (5.23) [Wong and Zakai, 1965*a* and *b*, 1966]. We need to define some types of approximations $\{W_n(\omega,t)\}$ to a Brownian motion $W(\omega,t)$ as follows:

A_1: For each t, $W_n(\cdot,t) \xrightarrow[n\to\infty]{\text{a.s.}} W(\cdot,t)$. For each n and almost all ω, $W_n(\omega,\cdot)$ is sample continuous and of bounded variation on $[a,T]$.

A_2: A_1 and also for almost all ω, $W_n(\omega,\cdot)$ uniformly bounded, i.e., for almost all ω,

$$\sup_n \sup_{t\in[a,b]} |W_n(\omega,t)| < \infty$$

A_3: A_2 and for each n and almost all ω, $W_n(\omega,t)$ has a continuous derivative $\dot{W}_n(\omega,t)$.

A_4: For each n, $W_n(\omega,t)$ is a polygonal approximation of $W(\omega,t)$ defined by

$$W_n(\omega,t) = W(\omega,t_j^{(n)}) + [W(\omega,t_{j+1}^{(n)}) - W(\omega,t_j^{(n)})]\frac{t - t_j^{(n)}}{t_{j+1}^{(n)} - t_j^{(n)}}$$
$$t_j^{(n)} \le t \le t_{j+1}^{(n)} \quad (5.26)$$

where $a = t_0^{(n)} < t_1^{(n)} < \cdots < t_n^{(n)} = T$ and

$$\max_j (t_{j+1}^{(n)} - t_j^{(n)}) \xrightarrow[n\to\infty]{} 0$$

Proposition 5.1. Let $\varphi(x,t)$ have continuous partial derivatives $\varphi'(x,t) = (\partial/\partial x)\varphi(x,t)$ and $(\partial/\partial t)\varphi(x,t)$ in $-\infty < x < \infty$, $a \le t \le b$. Let $\{W_n(\omega,t)\}$ satisfy A_2, then

$$\int_a^b \varphi(W_n(\omega,t), t)\, dW_n(\omega,t) \xrightarrow[n\to\infty]{\text{a.s.}} \int_a^b \varphi(W(\omega,t), t)\, dW(\omega,t) + \frac{1}{2}\int_a^b \varphi'(W(\omega,t), t)\, dt \quad (5.27)$$

Further, if $\varphi(x,t)$ does not depend on t, then the conclusion holds with A_1 replacing A_2.

Proposition 5.2. Let $m(x,t)$, $\sigma(x,t)$, $\sigma'(x,t) = (\partial/\partial x)\sigma(x,t)$, and $\dot{\sigma}(x,t) = (\partial/\partial t)\sigma(x,t)$ be continuous in $-\infty < x < \infty$, $a \le t \le b$. Let $m(x,t)$, $\sigma(x,t)$, and $\sigma(x,t)\sigma'(x,t)$ satisfy a uniform Lipschitz condition, i.e., if f denotes any of three quantities m, σ, $\sigma\sigma'$, then

$$|f(x,t) - f(y,t)| \le K|x - y| \quad (5.28)$$

Let $\{X_n(\omega,t), t \ge a\}$ satisfy (5.2), and let $\{X(\omega,t), t \ge a\}$ satisfy the stochastic differential equation

$$dX(\omega,t) = m(X(\omega,t), t)\,dt + \sigma(X(\omega,t), t)\,dW(\omega,t) + \tfrac{1}{2}\sigma(X(\omega,t), t)\sigma'(X(\omega,t), t)\,dt \tag{5.29}$$

Let $X_n(\omega,a) = X(\omega) = X(\omega,a)$, where X is independent of the aggregate of differences $\{W_t - W_a, t \geq a\}$ and $EX^2 < \infty$.
(*a*) If in addition, $|\sigma(x,t)| \geq \beta > 0$ and $|\sigma(x,t)| < K\sigma^2(x,t)$, then with $\{W_n(\omega,t)\}$ satisfying A_3

$$X_n(\omega,t) \xrightarrow[n\to\infty]{\text{a.s.}} X(\omega,t) \qquad a \leq t \leq b \tag{5.30}$$

(*b*) If $\{W_n(\omega,t)\}$ satisfies A_4 and $EX^4 < \infty$, then

$$X_n(\omega,t) \xrightarrow[n\to\infty]{\text{q.m.}} X(\omega,t) \qquad a \leq t \leq b \tag{5.31}$$

It should be mentioned that a symmetrized definition for stochastic integrals has been proposed [Fisk, 1963; Stratonovich, 1966] for which rules of ordinary calculus apply. Rewritten in terms of the Fisk-Stratonovich's integral, neither (5.13) nor (5.23) would contain an extra term. However, this approach has the disadvantage that conditions which guarantee the convergence of Fisk-Stratonovich's integral are less natural and more difficult to verify than those of the stochastic integral. Furthermore, the martingale property of Ito's integral would be lost. As we shall see in Chap. 6, an important application of the stochastic integral is in the representation of likelihood ratios and filtering operations, and this application depends on the martingale property. While these representations, under suitable restrictions, can be reexpressed in terms of the Fisk-Stratonovich integral, the resulting formulas will be considerably more complicated.

Equations (5.27) and (5.29) can be interpreted as expressions relating a white-noise integral $\int_a^t \varphi(\omega,s)\zeta(\omega,s)\,ds$ to the stochastic integral $\int_a^t \varphi(\omega,s)\,dW(\omega,s)$ for the following two special cases:

1. $\varphi(\omega,s) = \psi(W(\omega,s), s)$
2. $\varphi(\omega,s) = \psi(X(\omega,s), s)$, and X_s is related to W_s via a stochastic differential equation.

In general, $\varphi(\omega,t)$ may depend in a much more complicated way on $\{W_s, a \leq s \leq t\}$. The question arises as to whether it is possible to relate the white-noise integral to the corresponding stochastic integral in the general situation. This question has been resolved [Wong and Zakai, 1969]. Roughly speaking, the white-noise integral is equal to the corresponding stochastic integral plus a correction term. If $\varphi(\omega,t)$ is viewed as a functional on $\{W(\omega,s), a \leq s \leq t\}$, then the correction term can be expressed in terms of the Frechét differential of this functional.

In applications, differential equations driven by white noise frequently appear in a vector form as follows:

$$\dot{\mathbf{X}}_t = \mathbf{m}(\mathbf{X}_t,t) + \boldsymbol{\sigma}(\mathbf{X}_t,t)\boldsymbol{\zeta}_t \tag{5.32}$$

where $\mathbf{X}_t$ and $\mathbf{m}$ are n vectors, $\boldsymbol{\zeta}_t$ is a p vector of independent Gaussian white noise, and $\boldsymbol{\sigma}$ is a matrix of appropriate dimensions. There is no difficulty in extending the definition of stochastic integral to the form

$$\int_a^b \boldsymbol{\varphi}(\omega,t)\, d\mathbf{W}(\omega,t)$$

where $\mathbf{W}_t$ is a vector of independent Brownian motions, and $\boldsymbol{\varphi}_t$ is a matrix provided that

$$\int_a^b \sum_{i,j} E|\varphi_{ij}(\cdot,t)|^2\, dt < \infty$$

In terms of the extended definition of stochastic integral, stochastic differential equations in the vector form can be treated. Intuitively, it is highly plausible that the white-noise equation (5.32) is equivalent to a stochastic differential equation

$$d\mathbf{X}_t = \mathbf{f}(\mathbf{X}_t,t)\, dt + \mathbf{g}(\mathbf{X}_t,t)\, d\mathbf{W}_t \tag{5.33}$$

The problem is to determine $\mathbf{f}$ and $\mathbf{g}$. It was conjectured by Wong and Zakai [1965*a*] that $\mathbf{g} = \boldsymbol{\sigma}$, but

$$f_k(\mathbf{x},t) = m_k(\mathbf{x},t) + \frac{1}{2}\sum_{l,m} \frac{\partial\sigma_{km}(\mathbf{x},t)}{\partial x_l}\,\sigma_{lm}(\mathbf{x},t) \tag{5.34}$$

This has since been verified [McShane, 1969, 1970] under suitable conditions.

As the final topic in this section, we briefly consider problems arising in simulation. Suppose that we want to simulate a white-noise differential equation

$$\dot{X}_t = m(X_t,t) + \sigma(X_t,t)\zeta_t \tag{5.35}$$

Roughly speaking, there is a time constant or a bandwidth associated with the equation. While it is not clear how a bandwidth should be defined, it clearly should be related to the maximum rate of change of X_t in some way. The following definition may be useful:

$$B = \sup_{x,t}\left[\frac{|m(x,t)|}{1+|x|} + \frac{\sigma^2(x,t)}{1+|x|^2}\right]$$

Under assumption (4.5), this quantity is always finite. If z_t is a stationary Gaussian process with a spectral density that is constant over a bandwidth much greater than B, then it is intuitively clear that (5.35) can be simulated by replacing ζ_t by z_t. Hence, an analog and continuous-time simulation of (5.35) can be achieved by implementing

$$\dot{X}_t = m(X_t,t) + \sigma(X_t,t)z_t \tag{5.36}$$

with a wide-band noise source z_t. Of course, this also simulates the stochastic differential equation

$$dX_t = m(X_t,t)\,dt + \sigma(X_t,t)\,dW_t + \tfrac{1}{2}\sigma\sigma'(X_t,t)\,dt \tag{5.37}$$

The situation is less clear in discrete-time simulation. All depends on the noise source. If one uses a random-number generator which produces a sequence of *independent* Gaussian random variables $z_1, z_2, \ldots,$ then the difference equation

$$X_{k+1} = X_k + \Delta m(X_k,t_k) + \Delta\tfrac{1}{2}\sigma\sigma'(X_k,t_k) + \sigma(X_k,t_k)z_k \tag{5.38}$$

simulates (5.37) well, provided that we choose $Ez_k^2 = \Delta$ and $\Delta \ll 1/B$. Hence, (5.35) can be simulated by implementing (5.38). On the other hand, suppose that the noise source is a wide-band noise generator with bandwidth $B_0 \gg B$. If we sample this noise source at a rate to permit a faithful reproduction of this noise, we would have to sample at $2B_0$ or more. If we do this and produce a sequence $z_1, z_2, \ldots,$ then the difference equation

$$X_{k+1} = X_k + \frac{1}{2B_0}\,m(X_k,t_k) + \sigma(X_k,t_k)z_k \tag{5.39}$$

is a good approximation to (5.35). The difference here is that $z_1, z_2, \ldots,$ are no longer independent.

6. GENERALIZATIONS OF THE STOCHASTIC INTEGRAL

For a Brownian motion $\{W_t,\mathcal{A}_t\}$, we have defined the stochastic integral

$$I(\varphi,\omega) = \int_a^b \varphi(\omega,t)\,dW(\omega,t) \tag{6.1}$$

for integrands satisfying (1) φ jointly measurable in (ω,t), (2) for each t φ_t is $\mathcal{A}_t$ measurable, and (3) $\int_a^b E|\varphi_t|^2\,dt < \infty$. The stochastic integral (6.1) can be generalized in two important directions. First, it can be defined for integrands satisfying (1), (2), and instead of (3), the weaker condition

$$\int_a^b |\varphi(\omega,t)|^2\,dt < \infty \qquad \text{a.s.} \tag{6.2}$$

Secondly, the Brownian motion $\{W_t,\mathcal{A}_t\}$ in (6.1) can be replaced by a class of martingales $\{X_t,\mathcal{A}_t\}$. In this section, we shall consider both these generalizations and their applications.

Proposition 6.1. Let $\{W_t, \mathcal{A}_t\}$ be a Brownian motion and let $\varphi(\omega,t)$ satisfy:

(*a*) φ is jointly measurable in (ω,t).

(*b*) For each t, φ_t is $\mathcal{A}_t$ measurable,

(*c*) $\int_a^b |\varphi(\omega,t)|^2 \, dt < \infty$ almost surely.

Let φ^m be defined by

$$\varphi^m(\omega,t) = \begin{cases} \varphi(\omega,t) & \text{if } \int_a^t |\varphi(\omega,t)|^2 \, dt \leq m \\ 0 & \text{otherwise} \end{cases} \tag{6.3}$$

and let $I(\varphi^m)$ denote the stochastic integral

$$I(\varphi^m) = \int_a^b \varphi^m(\omega,t) \, dW(\omega,t)$$

Then, $\{I(\varphi^m),\ m = 1, \ldots\}$ converges in probability as $m \to \infty$, and we define

$$I(\varphi) = \int_a^b \varphi(\omega,t) \, dW(\omega,t) = \lim_{m\to\infty} \text{in p. } I(\varphi^m) \tag{6.4}$$

Proof: Let φ^m be defined by (6.3). For each m, φ^m satisfies (2.3) and (2.4) so that $I(\varphi^m)$ is well defined. Now, for any ω such that

$$\int_a^b |\varphi(\omega,t)|^2 \, dt \leq \min(m,n)$$

we have from (6.3)

$$\sup_t |\varphi^m(\omega,t) - \varphi^n(\omega,t)| = 0$$

which in turn implies that $\int_a^b \varphi^m(\omega,t) \, dW(\omega,t) = \int_a^b \varphi^n(\omega,t) \, dW(\omega,t)$. It follows that for every $\epsilon > 0$,

$$\mathcal{P}(|I(\varphi^m) - I(\varphi^n)| \geq \epsilon) \leq \mathcal{P}\left(\int_a^b |\varphi_t|^2 \, dt > \min(m,n)\right) \xrightarrow[m,n\to\infty]{} 0$$

which proves that $\{I(\varphi^n)\}$ converges in probability so that (6.4) is an adequate definition for $I(\varphi)$. ∎

Remarks.

(*a*) If $\{\varphi^n\}$ is a sequence of functions satisfying conditions (2.3) and (6.2), if $|\varphi_m(\omega,t)| \leq |\varphi(\omega,t)|$, and if

$$\varphi_m \xrightarrow[m\to\infty]{} \varphi \text{ in } \mathcal{P} \times \mathcal{L} \text{ measure} \tag{6.5}$$

then

$$I(\varphi_m) \xrightarrow[m\to\infty]{\text{in p.}} I(\varphi) \tag{6.6}$$

(*b*) Now the process

$$X_t = \int_a^t \varphi(\omega,s)\, dW(\omega,s) \tag{6.7}$$

is no longer necessarily a martingale. Of course, a sufficient condition for X_t to be a martingale is precisely

$$\int_a^b E|\varphi_s|^2\, ds < \infty \tag{6.8}$$

However, this is not a necessary condition. If we define $\tau_n(\omega) = \min t\colon \int_a^t \varphi^2(\omega,s)\, ds \geq n$ and set $\tau_n(\omega) = \infty$ if $\int_a^b \varphi^2(\omega,s)\, ds < n$, then for each n, $X_{nt} = X_{\min (t,\tau_n)}$ is a martingale. By definition X_t is said to be a **local martingale** [see, e.g., Kunita and Watanabe, 1967].

Next, we shall consider generalizations of the stochastic integral by replacing the Brownian motion W_t by a more general process. As a first step in this direction, we shall replace W_t by a process Z_t satisfying the following properties. Throughout, $\{\mathcal{A}_t\}$ again denotes an increasing family of σ algebras.

$$\{Z_t, \mathcal{A}_t, a \leq t \leq b\} \text{ is a martingale and } EZ_t^2 < \infty \tag{6.9}$$

$$E(Z_t - Z_s)^2 = E^{\mathcal{A}_s}(Z_t - Z_s)^2 \quad \text{a.s.} \tag{6.10}$$

Let $F(t)$ be a nondecreasing function so that

$$E(Z_t - Z_s)^2 = F(t) - F(s) \qquad t \geq s \tag{6.11}$$

Then, the stochastic integral

$$\int_a^b \varphi(\omega,s)\, dZ(\omega,s) \tag{6.12}$$

is well defined for any φ satisfying

$$\varphi \text{ is jointly measurable in } (\omega,t) \text{ and for each } t,\ \varphi_t \text{ is } \mathcal{A}_t \text{ measurable} \tag{6.13}$$

$$\int_a^b |\varphi(\omega,s)|^2\, dF(s) < \infty \qquad \text{a.s.} \tag{6.14}$$

The procedure for defining (6.12) is exactly the same as before and will not be repeated.

The class of processes satisfying both (6.9) and (6.10) is still quite restricted. In particular, if Z_t is almost surely sample continuous, then $F(t)$ is necessarily continuous [for convenience, we set $F(0) = 0$] and Z_t can be expressed as

$$Z_t = W_{F(t)} \tag{6.15}$$

where W_t is a Brownian motion. Therefore, if we consider only sample-continuous Z_t, then the stochastic integral (6.12) is really the same as what we already defined. The next step in generalizing the stochastic integral is to get rid of the restriction (6.10). We begin with the following result.

Proposition 6.2. Let $\{Z_t, \mathcal{A}_t, a \le t \le b\}$ be a sample-continuous second-order martingale. Then there is a unique decomposition

$$Z_t^2 = Z_{1t} + Z_{2t} \qquad a \le t \le b \tag{6.16}$$

where $\{Z_{2t}, \mathcal{A}_t, a \le t \le b\}$ is a sample-continuous first-order martingale, and $\{Z_{1t}, a \le t \le b\}$ is sample continuous, nondecreasing, with $Z_{1a} = 0$.

Remark: This proposition is a special case of the well-known supermartingale-decomposition theorem of Meyer [1966, Chap. 7]. We note that if Z_t is a Brownian motion, then Z_{1t} is simply t.

Proposition 6.3. Let $\{Z_t, \mathcal{A}_t, a \le t \le b\}$ be a sample-continuous second-order martingale. Let $\{Z_{1t}, a \le t \le b\}$ be defined as in (6.16). Suppose that $\varphi(\omega,t)$, $\omega \in \Omega$, $t \in [a,b]$, is a jointly measurable function[1] such that for each t, φ_t is $\mathcal{A}_t$ measurable and

$$\int_a^b \varphi_t^2 \, dZ_{1t} < \infty \tag{6.17}$$

with probability 1. Then the stochastic integral

$$I(\varphi,\omega) = \int_a^b \varphi(\omega,t) \, dZ(\omega,t) \tag{6.18}$$

is well defined by the following two properties:

(*a*) If φ is an (ω,t)-step function, then

$$I(\varphi,\omega) = \sum_\nu \varphi_\nu(\omega)[Z(\omega,t_{\nu+1}) - Z(\omega,t_\nu)]$$

(*b*) If $\int_a^b |\varphi(\omega,t) - \varphi_k(\omega,t)| \, dZ_1(\omega,t) \xrightarrow[k\to\infty]{\text{a.s.}} 0$, then $I(\varphi_k,\omega) \xrightarrow[k\to\infty]{\text{in p.}} I(\varphi,\omega)$

Remark: It is clear that Z_{1t} now plays the role played by t in the original definition of stochastic integral.

If X_t is of the form

$$X_t = \int_a^t f(\omega,s) \, ds + \int_a^t \varphi(\omega,s) \, dZ(\omega,s) \tag{6.19}$$

where the last integral is defined as in (6.18), then a transformation rule similar to Proposition 3.2 holds once again. Let $\psi(x,t)$ be twice continuously

[1] Here measurability in t refers to Borel measurability.

differentiable in x and once in t. Then

$$\psi(X_t,t) = \psi(X_a,a) + \int_a^t \psi'(X_s,s)\,dX_s + \int_a^t \dot{\psi}(X_s,s)\,ds + \frac{1}{2}\int_a^t \psi''(X_s,s)\varphi_s^2\,dZ_{1s} \tag{6.20}$$

with probability 1 [Kunita and Watanabe, 1967].

Suppose that there exists a continuous and nondecreasing function $F(t)$, $a \le t \le b$, such that for almost all ω, $Z_1(\omega,t)$ as a function of t is absolutely continuous with respect to the Borel measure generated by F. That is, there exists an a.s. nonnegative function $z(\omega,t)$ such that

$$Z_1(\omega,t) = \int_a^t z(\omega,s)\,dF(s) \qquad \text{a.s.} \tag{6.21}$$

If such an F exists it can always be taken to be EZ_{1t}, because (6.21) implies that

$$Z_1(\omega,t) = \int_a^t \frac{z(\omega,s)}{Ez_s}\,d(EZ_{1s}) \tag{6.22}$$

If Z_{1t} has the representation (6.21), then Z_t can be represented as a stochastic integral

$$Z(\omega,t) = \int_a^t \sqrt{z(\omega,s)}\,dW(\omega,F(s)) \tag{6.23}$$

where $\{W_s,\ 0 \le s \le F(b)\}$ is a Brownian motion. We note that (6.23) is a stochastic integral of the type given by (6.12). Now, with (6.23) we can rewrite any stochastic integral in terms of Z_t,

$$I(\varphi,\omega) = \int_a^b \varphi_s\,dZ_s = \int_a^t \varphi_s \sqrt{z_s}\,dW_{F(s)} \tag{6.24}$$

Once again, we return to the basic definition of a stochastic integral in terms of a Brownian motion.

As the final step in generalizing the stochastic integral, consider a sample-continuous process $\{X_t,\ a \le t \le b\}$ which has a decomposition

$$X_t = Y_t + Z_t \tag{6.25}$$

where Y_t is almost surely of bounded variation, and Z_t is a second-order sample-continuous martingale. Clearly, we can define

$$\int_a^b \varphi_t\,dX_t = \int_a^b \varphi_t\,dY_t + \int_a^b \varphi_t\,dZ_t \tag{6.26}$$

provided that the first integral exists almost surely as a Stieltjes integral and the second as a stochastic integral. A process that can be decomposed as in (6.25) was termed a quasi-martingale by Fisk [1965] who also gave necessary and sufficient conditions for the existence of such a decomposition. Unfortunately, these conditions are not always easily verified.

As an example of applications of the generalized definition of stochastic integral, we consider the following important representation theorem due to Doob [1953, pp. 287–291].

Proposition 6.4. Let $\{X_t, a \le t \le b\}$ be a sample-continuous second-order process. Let $m(x,t)$ and $\sigma(x,t)$ be Borel functions of (x,t) satisfying

$$|m(x,t)| \le K\sqrt{1 + x^2} \tag{6.27}$$

$$0 \le \sigma(x,t) \le K\sqrt{1 + x^2} \tag{6.28}$$

Let $\mathcal{A}_t$ denote the smallest σ algebra such that X_s, $s \le t$, are all measurable, and suppose that $\{X_t, a \le t \le b\}$ satisfies the following conditions:

(*a*) There exists $\{Z(\omega,t), a \le t \le b\}$ such that $Z_t \ge 0, EZ_t < \infty$, and

$$\sup_{t>s} E^{\mathcal{A}_s} X_t^2 \le Z_s \tag{6.29}$$

(*b*) There exists a nondecreasing function f with $\lim_{h \downarrow 0} f(h) = 0$ such that whenever $a \le t < t + h \le b$, we have with probability 1,

$$\left| E^{\mathcal{A}_t}(X_{t+h} - X_t) - \int_t^{t+h} m(X_s,s)\, ds \right| \le hf(h)(1 + X_t^2) \tag{6.30}$$

$$\left| E^{\mathcal{A}_t}(X_{t+h} - X_t)^2 - \int_t^{t+h} \sigma^2(X_s,s)\, ds \right| \le hf(h)(1 + X_t^2) \tag{6.31}$$

Under these conditions, $\{X_t, a \le t \le b\}$ is a Markov process and satisfies a stochastic differential equation

$$X_t = X_a + \int_a^t m(X_s,s)\, ds + \int_a^t \sigma(X_s,s)\, dW_s \tag{6.32}$$

where $\{W_t, a \le t \le b\}$ is a Brownian motion.

Remark: We have made no assumption that m and σ satisfy a Lipschitz condition. Without such an assumption, we cannot be sure that there is a unique solution to (6.32). One possible consequence of this is that the finite-dimensional distributions of $\{X_t, a \le t \le b\}$ may not be completely determined by m, σ, and the distribution of X_a.

Proof: We shall give an outline of the proof. Let $\{Z_t, a \le t \le b\}$ be defined by

$$Z_t = X_t - X_a - \int_a^t m(X_s,s)\, ds \tag{6.33}$$

Because of (6.30), we can show that $\{Z_t, \mathcal{A}_t, a \le t \le b\}$ is a sample-continuous martingale. Because of (6.27), it is also second order. Furthermore, if we define

$$Y_t = Z_t^2 - \int_a^t \sigma^2(X_s,s)\, ds \tag{6.34}$$

then because of (6.31), $\{Y_t, \mathcal{A}_t, a \le t \le b\}$ is also a sample-continuous martingale. Therefore, the process $\{Z_{1t}, a \le t \le b\}$ defined by the decomposition (6.16) is simply

$$Z_{1t} = \int_a^t \sigma^2(X_s,s)\, ds \tag{6.35}$$

Clearly, Z_{1t} has the form of (6.21) with

$$F(t) = t \tag{6.36}$$

and

$$z(\omega,t) = \sigma^2(X(\omega,t), t) \tag{6.37}$$

From (6.23) we get

$$Z_t = \int_a^t \sigma(X_s,s)\, dW_s \tag{6.38}$$

Equation (6.38) combined with (6.33) yields (6.32). ▮

7. DIFFUSION EQUATIONS

In this section, we shall try to show that the transition probabilities of a process satisfying a stochastic differential equation can be obtained by solving either of a pair of partial differential equations. These equations are called the backward and forward equations of Kolmogorov or, alternatively, diffusion equations. The forward equation is also sometimes called the Fokker-Planck equation. The situation, however, is not completely satisfactory. As we shall see, the original derivation of Kolmogorov involved assumptions that cannot be directly verified. Attempts in circumventing these assumptions involve other difficulties. We begin with a derivation of the diffusion equations following the lines of Kolmogorov [1931].

Let $\{X_t, a \le t \le b\}$ be a Markov process, and denote

$$P(x,t|x_0,t_0) = \mathcal{P}(X_t < x | X_{t_0} = x_0) \tag{7.1}$$

We call $P(x,t|x_0,t_0)$ the **transition function** of the process. If there is a function $p(x,t|x_0,t_0)$ so that

$$P(x,t|x_0,t_0) = \int_{-\infty}^{x} p(u,t|x_0,t_0)\, du \tag{7.2}$$

then we call $p(x,t|x_0,t_0)$ the **transition density function.** Since $\{X_t, a \le t \le b\}$ is a Markov process, $P(x,t|x_0,t_0)$ satisfies the Chapman-Kolmogorov equation

$$P(x,t|x_0,t_0) = \int_{-\infty}^{\infty} P(x,t|z,s)\, dP(z,s|x_0,t_0) \tag{7.3}$$

We now assume the crucial conditions on $\{X_t, a \le t \le b\}$ which make the derivation of the diffusion equations possible. These conditions are very similar to conditions (6.30) and (6.31) which made it possible to represent a process as the solution of a stochastic differential equation.

Define for a positive ϵ,

$$M_k(x,t;\epsilon,\Delta) = \int_{|y-x|\le\epsilon} (y-x)^k \, dP(y, t+\Delta|x,t) \qquad k = 0, 1, 2$$

$$M_3(x,t;\epsilon,\Delta) = \int_{|y-x|\le\epsilon} |y-x|^3 \, dP(y, t+\Delta|x,t) \tag{7.4}$$

We assume that the Markov process $\{X_t,\ a \le t \le b\}$ satisfies the following conditions:

$$\frac{1}{\Delta}[1 - M_0(x,t;\epsilon,\Delta)] \xrightarrow[\Delta\downarrow 0]{} 0 \tag{7.5}$$

$$\frac{1}{\Delta} M_1(x,t;\epsilon,\Delta) \xrightarrow[\Delta\downarrow 0]{} m(x,t) \tag{7.6}$$

$$\frac{1}{\Delta} M_2(x,t;\epsilon,\Delta) \xrightarrow[\Delta\downarrow 0]{} \sigma^2(x,t) \tag{7.7}$$

$$\frac{1}{\Delta} M_3(x,t;\epsilon,\Delta) \xrightarrow[\Delta\downarrow 0]{} 0 \tag{7.8}$$

It is clear that if $1 - M_0(x,t;\epsilon,\Delta) \xrightarrow[\Delta\downarrow 0]{} 0$, then by dominated convergence,

$$\mathcal{P}(|X_{t+\Delta} - X_t| > \epsilon) = \int_{-\infty}^{\infty} [1 - M_0(x,t;\epsilon,\Delta)] \, dP(x,t) \xrightarrow[\Delta\downarrow 0]{} 0$$

Therefore, (7.5) is considerably stronger than continuity in probability. In addition, suppose that the transition function $P(x,t|x_0,t_0)$ satisfies the condition:

For each (x,t), $P(x,t|x_0,t_0)$ is once differentiable in t_0 and three-times differentiable in x_0, and the derivatives are continuous and bounded in (x_0,t_0) (7.9)

Now we can derive the background equation as follows. Write the Chapman-Kolmogorov equation in the form

$$P(x,t|x_0,t_0) = \int_{-\infty}^{\infty} P(x,t|z, t_0+\Delta) \, dP(z, t_0+\Delta|x_0,t_0) \tag{7.10}$$

Because of (7.9), we can write, by virtue of the Taylor's theorem,

$$\begin{aligned} P(x,t|z, t_0+\Delta) &= P(x,t|x_0, t_0+\Delta) + \frac{\partial P(x,t|x_0, t_0+\Delta)}{\partial x_0}(z - x_0) \\ &\quad + \frac{1}{2}\frac{\partial^2 P(x,t|x_0, t_0+\Delta)}{\partial x_0^2}(z-x_0)^2 \\ &\quad + \frac{1}{6}\frac{\partial^3 P(x,t|z, t_0+\Delta)}{\partial z^3}\bigg|_{z=\theta}(z-x_0)^3 \qquad |\theta - x_0| \le |z - x_0| \end{aligned} \tag{7.11}$$

Using (7.11) in (7.10) and using (7.4), we can write

$$P(x,t|x_0,t_0) = \int_{|z-x_0|>\epsilon} P(x,t|z,\, t_0+\Delta)\, dP(z,\, t_0+\Delta|x_0,t_0) + \sum_{k=0}^{2} \frac{1}{k!} M_k(x_0,t_0;\,\epsilon,\Delta) \frac{\partial^k}{\partial x_0{}^k} P(x,t|x_0,\, t_0+\Delta) + \frac{1}{6}\int_{|z-x_0|\le\epsilon} \left.\frac{\partial^3 P(x,t|z,\, t_0+\Delta)}{\partial z^3}\right|_{z=\theta} (z-x_0)^3\, dP(z,\, t_0+\Delta|x_0,t_0) \tag{7.12}$$

This means that

$$\left|\left[\frac{P(x,t|x_0,t_0) - P(x,t|x_0,\, t_0+\Delta)}{\Delta}\right] - \frac{1}{\Delta} M_1(x_0,t_0;\,\epsilon,\Delta) \frac{\partial P(x,t|x_0,\, t_0+\Delta)}{\partial x_0} - \frac{1}{2}\frac{1}{\Delta} M_2(x_0,t_0;\,\epsilon,\Delta) \frac{\partial^2 P(x,t|x_0,\, t_0+\Delta)}{\partial x_0{}^2}\right| \le \left|\left[\frac{1 - M_0(x_0,t_0;\,\epsilon,\Delta)}{\Delta}\right]\right| + \frac{M_3(x_0,t_0;\,\epsilon,\Delta)}{6\Delta} \sup_{|z-x_0|\le\epsilon} \left|\frac{\partial^3 P(x,t|z,\, t_0+\Delta)}{\partial z^3}\right| \tag{7.13}$$

If we let $\Delta \downarrow 0$ and use conditions (7.5) through (7.8), (7.13) becomes

$$-\frac{\partial}{\partial t_0} P(x,t|x_0,t_0) = m(x_0,t_0) \frac{\partial}{\partial x_0} P(x,t|x_0,t_0) + \tfrac{1}{2}\sigma^2(x_0,t_0) \frac{\partial^2}{\partial x_0{}^2} P(x,t|x_0,t_0) \qquad a < t_0 < t < b \tag{7.14}$$

The "terminal" condition for (7.14) is

$$\lim_{t_0 \uparrow t} P(x,t|x_0,t_0) = \begin{cases} 1 & x > x_0 \\ 0 & x < x_0 \end{cases} \tag{7.15}$$

Equation (7.14) is the **backward equation** of diffusion. The name is due to the fact that it is an equation in the pair of initial variables (x_0,t_0) moving backward from t.

The forward equation can be derived in the following indirect way: Let $f(x)$, $-\infty < x < \infty$, be a Schwartz function of rapid descent. That is, f is infinitely differentiable, and for any k and m,

$$|x|^k |f^{(m)}(x)| \xrightarrow[|x|\to\infty]{} 0 \tag{7.16}$$

As we did in Chap. 3, the space of all such functions is denoted by $\mathcal{S}$ (cf. 3.5.6c). Define the function

$$\begin{aligned} \hat{f}(t|x_0,t_0) &= E[f(X_t)|X_{t_0} = x_0] \\ &= \int_{-\infty}^{\infty} f(x)\, dP(x,t|x_0,t_0) \end{aligned} \tag{7.17}$$

Now, write

$$\begin{aligned}\hat{f}(t + \Delta|x_0,t_0) &= \int_{-\infty}^{\infty} f(x)\, dP(x, t + \Delta|x_0,t_0) \\ &= \int_{-\infty}^{\infty} \left[\int_{-\infty}^{\infty} f(x)\, dP(x, t + \Delta|z,t)\right] dP(z,t|x_0,t_0) \quad (7.18)\end{aligned}$$

Since f is infinitely differentiable and satisfies (7.16), we have

$$f(x) = \sum_{k=0}^{2} \frac{1}{k!} f^{(k)}(z)(x - z)^k + \tfrac{1}{6} f^{(3)}(\theta)(x - z)^3 \tag{7.19}$$

Repeating the arguments leading to (7.14), we find

$$\frac{\partial \hat{f}(t|x_0,t)}{\partial t} = \int_{-\infty}^{\infty} \left[m(z,t) \frac{df(z)}{dz} + \tfrac{1}{2}\sigma^2(z,t) \frac{d^2 f(z)}{dz^2} \right] dP(z,t|x_0,t_0) \tag{7.20}$$

Now, if $P(x,t|x_0,t_0)$ satisfies

For each (x_0,t_0), $P(x,t|x_0,t_0)$ is four-times differentiable in x and once in t, and the derivatives are continuous in (x,t) (7.21)

and if $\sigma^2(x,t)$ is twice continuously differentiable in x, and $m(x,t)$ is once continuously differentiable in x, then we have from (7.20) and integrations by parts,

$$\int_{-\infty}^{\infty} f(x) \left\{ \frac{\partial}{\partial t} p(x,t|x_0,t_0) - \frac{1}{2} \frac{\partial^2}{\partial x^2} [\sigma^2(x,t)p(x,t|x_0,t_0)] + \frac{\partial}{\partial x} [m(x,t)p(x,t|x_0,t_0)] \right\} dx = 0 \tag{7.22}$$

Since (7.22) holds for all $f \in \mathcal{S}$, the quantity in the brackets must be zero for almost all x, but being continuous, it must be zero for all x. Therefore,

$$\frac{\partial}{\partial t} p(x,t|x_0,t_0) = \frac{1}{2} \frac{\partial^2}{\partial x^2} [\sigma^2(x,t)p(x,t|x_0,t_0)] - \frac{\partial}{\partial x} [m(x,t)p(x,t|x_0,t_0)] \qquad b > t > t_0 > a \tag{7.23}$$

Equation (7.23) is the **forward equation** of diffusion, and is also called the **Fokker-Planck equation.** The initial condition to be imposed is

$$\int_{-\infty}^{\infty} f(x)p(x,t|x_0,t_0)\, dx \xrightarrow[t \downarrow t_0]{} f(x_0) \qquad \forall f \in \mathcal{S} \tag{7.24}$$

that is, $p(x,t_0|x_0,t_0) = \delta(x - x_0)$. A solution of (7.23), satisfying (7.24), will be called its **fundamental solution.** Our derivation of the two equations of diffusion are now complete.

If we view the two diffusion equations as possible means for determining the transition probabilities of the solution of a stochastic differential equation, then the situation is still not entirely satisfactory. This is because the diffusion equations have been derived under differentiability assumptions (7.9) and (7.21). If we don't know $P(x,t|x_0,t_0)$, how do we know whether it is differentiable the required number of times? This difficulty is in part resolved by the following proposition.

Proposition 7.1. Let $m(x,t)$ and $\sigma(x,t)$ satisfy the following conditions on $-\infty < x < \infty$, $a \le t \le b$:
There exist positive constants σ_0 and K so that

$$\begin{aligned} |m(x,t)| &\le K\sqrt{1+x^2} \\ 0 < \sigma_0 \le \sigma(x,t) &\le K\sqrt{1+x^2} \end{aligned} \tag{7.25}$$

There exist positive constants γ and K so that

$$\begin{aligned} |m(x,t) - m(y,t)| &\le K|x-y|^\gamma \\ |\sigma(x,t) - \sigma(y,t)| &\le K|x-y|^\gamma \end{aligned} \qquad \text{(Hölder condition)} \tag{7.26}$$

Then, the following conclusions are valid:
(*a*) The backward equation

$$\tfrac{1}{2}\sigma^2(x_0,t_0)\frac{\partial^2 P(x,t|x_0,t_0)}{\partial x_0^2} + m(x_0,t_0)\frac{\partial P(x,t|x_0,t_0)}{\partial x_0} = -\frac{\partial}{\partial t_0}P(x,t|x_0,t_0) \qquad t > t_0 \tag{7.27}$$

has a unique solution corresponding to condition (7.15). Further, for $t > t_0$, $P(x,t|x_0,t_0)$ is differentiable with respect to x so we have the transition density

$$p(x,t|x_0,t) = \frac{\partial}{\partial x}P(x,t|x_0,t_0) \tag{7.28}$$

(*b*) There exists a sample-continuous Markov process $\{X_t,\ a \le t \le b\}$ with transition function $P(x,t|x_0,t_0)$.
(*c*) Conditions (7.5) to (7.8) are satisfied.
(*d*) If $m'(x,t)$, $\sigma'(x,t)$, $\sigma''(x,t)$ satisfy (7.25) and (7.26), then $p(x,t|x_0,t_0)$ is the unique fundamental solution of the forward equation.
(*e*) If γ can be taken to be 1 in (7.26), then $p(x,t|x_0,t_0)$ is the transition density of the unique solution to the stochastic integral equation

$$X_t = X_a + \int_a^t m(X_s,s)\,ds + \int_a^t \sigma(X_s,s)\,dW_s \tag{7.29}$$

Example 1. Suppose that $\{X_t, t \geq 0\}$ satisfies a stochastic differential equation

$$dX_t = -X_t\,dt + \sqrt{2(1 + X_t^2)}\,dW_t \tag{7.30}$$

Here, we have $m(x,t) = -x$ and $\sigma(x,t) = \sqrt{2(1 + x^2)}$. Therefore, the forward equation is given by

$$\frac{\partial^2}{\partial x^2}[(1 + x^2)p(x,t|x_0,t_0)] + \frac{\partial}{\partial x}[xp(x,t|x_0,t_0)] = \frac{\partial}{\partial t}p(x,t|x_0,t_0) \tag{7.31}$$

Because m and σ do not depend on t, $p(x,t|x_0,t_0)$ will depend only on $t - t_0$ and not on t and t_0 separately. Furthermore, we can rewrite (7.30) as

$$X_t = e^{-(t-t_0)}X_{t_0} + \sqrt{2}\int_{t_0}^{t} e^{-(t-s)}\sqrt{1 + X_s^2}\,dW_s$$

and from this we expect that as $t \to \infty$, the conditional density $p(x,t|x_0,t_0)$ will approach the stationary density $p(x)$. We also expect that

$$\frac{\partial p(x,t|x_0,t_0)}{\partial t} \xrightarrow[t\to\infty]{} 0$$

Therefore,

$$\frac{d^2}{dx^2}[(1 + x^2)p(x)] + \frac{d}{dx}[xp(x)] = 0$$

or

$$\frac{d}{dx}[(1 + x^2)p(x)] + xp(x) = \text{constant}$$

Because $p'(x), p(x) \xrightarrow[|x|\to\infty]{} 0$, the constant must be 0, and by direct integration we get

$$p(x) = p(0)\frac{1}{(1 + x^2)^{\frac{3}{2}}} \tag{7.32}$$

By requiring $\int_{-\infty}^{\infty} p(x)\,dx = 1$, we find $p(0) = \frac{1}{2}$.

The above procedure illustrates how (7.31) is very frequently used, namely, to find the stationary density when it exists. Actually, for this example, (7.31) can be solved completely. If we set

$$u(x,t;\,x_0,t_0) = \frac{\sinh^{-1} x - \sinh^{-1} x_0}{2\sqrt{t - t_0}} \tag{7.33}$$

then it can be shown that $p(x,t|x_0,t_0)$ is given by

$$p(x,t|x_0,t_0) = \frac{1}{2(1+x^2)^{\frac{3}{2}}} \frac{1}{\sqrt{\pi(t-t_0)}} e^{-t}e^{-u^2} + \frac{1}{\sqrt{\pi}} \int_{u-\sqrt{t-t_0}}^{u+\sqrt{t-t_0}} e^{-z^2}\,dz \qquad (7.34)$$

Example 2. Suppose that $\{Y_t, t \geq 0\}$ and $\{Z_t, t \geq 0\}$ are two independent standard Brownian motion processes. Let

$$X_t = Y_t^2 + Z_t^2 \qquad (7.35)$$

We shall use Proposition 6.4 and show that X_t is Markov and derive its stochastic differential equation. Let $\mathcal{B}_t$ denote the smallest σ algebra such that Y_s, Z_s, $s \leq t$ are all measurable. Let $\mathcal{A}_t$ denote the smallest σ algebra with respect to which X_s, $s \leq t$, are all measurable. Clearly, for each t, $\mathcal{B}_t \supset \mathcal{A}_t$. Now, for $h > 0$,

$$\begin{aligned} E^{\mathcal{A}_t}(X_{t+h} - X_t) &= E^{\mathcal{A}_t}E^{\mathcal{B}_t}(X_{t+h} - X_t) \\ &= E^{\mathcal{A}_t}E^{\mathcal{B}_t}\{(Y_{t+h}{}^2 - Y_t{}^2) + (Z_{t+h}{}^2 - Z_t{}^2)\} \\ &= E^{\mathcal{A}_t}(h + h) = 2h \end{aligned}$$

Similarly, we find that

$$E^{\mathcal{A}_t}(X_{t+h} - X_t)^2 = 8h^2 + 4hX_t$$

Clearly, the hypotheses of Proposition 6.4 are satisfied with

$$m(x,t) = 2 \qquad (7.36)$$

$$\sigma^2(x,t) = 4x \qquad (7.37)$$

Therefore, $\{X_t, t \geq 0\}$ satisfies the stochastic differential equation

$$dX_t = 2dt + \sqrt{X_t}\,dW_t \qquad (7.38)$$

We note that σ does not satisfy a uniform Lipschitz condition. However, it turns out that (7.38) has a unique nonnegative solution for $X_0 = 0$ anyway [McKean, 1960, 1969].

According to Proposition 7.1 the conditional density for $\{X_t, t \geq 0\}$ must satisfy the Fokker-Planck equation

$$\frac{\partial^2}{\partial x^2}[xp(x,t|x_0,t_0)] - 2\frac{\partial}{\partial x}p(x,t|x_0,t_0) = \frac{\partial}{\partial t}p(x,t|x_0,t_0) \qquad (7.39)$$

This equation has a unique fundamental solution given by

$$p(x,t|x_0,t_0) = \frac{1}{(t-t_0)} \exp -\frac{1}{2}\left(\frac{x+x_0}{t-t_0}\right) I_0\left(\frac{\sqrt{xx_0}}{t-t_0}\right) \qquad (7.40)$$

which can be verified by direct computation. In this example $p(x,t|x_0,t_0)$ can also be obtained directly from the definition for X_t.

Finally, we shall state, but not derive, the diffusion equations in the multidimensional case. Let $\{\mathbf{W}_t,\ a \le t \le b\}$ be a vector, the components of which are independent Brownian motion processes. Let $\mathbf{X}$ be a vector random variable independent of $\{\mathbf{W}_t,\ a \le t \le b\}$. Under appropriate conditions on $\mathbf{m}$ and $\boldsymbol{\sigma}$, existence and uniqueness of the solution to the vector stochastic integral equation

$$\mathbf{X}_t = \mathbf{X} + \int_a^t \mathbf{m}(\mathbf{X}_s,s)\,ds + \int_a^t \boldsymbol{\sigma}(\mathbf{X}_s,s)\,d\mathbf{W}_s \tag{7.41}$$

can be established. The solution $\{\mathbf{X}_t,\ a \le t \le b\}$ is a vector Markov process in the sense that

$$\mathcal{P}(\mathbf{X}_t \in E|\mathbf{X}_t,\ t \le s) = \mathcal{P}(\mathbf{X}_t \in E|\mathbf{X}_s) \tag{7.42}$$

whenever E is a Borel set and $s \le t$. Furthermore, with probability 1,

$$\lim_{h \downarrow 0} \frac{1}{h} E[(\mathbf{X}_{t+h} - \mathbf{X}_t)|\mathbf{X}_t] = \mathbf{m}(\mathbf{X}_t,t) \tag{7.43}$$

$$\lim_{h \downarrow 0} \frac{1}{h} E[(\mathbf{X}_{t+h} - \mathbf{X}_t)(\mathbf{X}_{t+h} - \mathbf{X}_t)^T|\mathbf{X}_t] = \boldsymbol{\sigma}(\mathbf{X}_t,t)\boldsymbol{\sigma}^T(\mathbf{X}_t,t) = \boldsymbol{\beta}(\mathbf{X}_t,t) \tag{7.44}$$

Suppose that $\{\mathbf{X}_t,\ a \le t \le b\}$ is a vector Markov process satisfying (7.43) and (7.44). Then under some additional conditions similar to the scalar case, the conditional density $p(\mathbf{x},t|\mathbf{x}_0,t_0)$ can be shown to satisfy a pair of partial differential equations

$$\frac{1}{2}\sum_{i,j} \beta_{ij}(\mathbf{x}_0,t_0) \frac{\partial^2}{\partial x_{0i}\,\partial x_{0j}} p(\mathbf{x},t|\mathbf{x}_0,t_0) + \sum_i m_i(\mathbf{x}_0,t_0) \frac{\partial}{\partial x_{0i}} p(\mathbf{x},t|\mathbf{x}_0,t_0) = -\frac{\partial}{\partial t_0} p(\mathbf{x},t|\mathbf{x}_0,t_0) \tag{7.45}$$

and

$$\frac{1}{2}\sum_{i,j} \frac{\partial^2}{\partial x_i\,\partial x_j} [\beta_{ij}(\mathbf{x},t)p(\mathbf{x},t|x_0,t_0)] - \sum_i \frac{\partial}{\partial x_i} [m_i(\mathbf{x},t)p(\mathbf{x},t|\mathbf{x}_0,t_0)] = \frac{\partial}{\partial t} p(\mathbf{x},t|\mathbf{x}_0,t_0) \tag{7.46}$$

with initial condition $p(\mathbf{x},t_0|\mathbf{x}_0,t_0) = \delta(\mathbf{x} - \mathbf{x}_0)$. Under some additional conditions, $\{\mathbf{X}_t,\ a \le t \le b\}$ can also be shown to satisfy a stochastic differential equation of the form (7.41). Therefore, if $\mathbf{m}$ and $\boldsymbol{\sigma}$ are such that (7.41) has a unique solution, and (7.46) has a unique fundamental solution that is a density function, then the process $\{\mathbf{X}_t,\ a \le t \le b\}$ satisfying

(7.41) must have a density function which satisfies (7.46). There is a major difference in the vector case, however. The best result that is known on the uniqueness of solution for (7.46) requires that **m** and **σ** be bounded, which is a strong assumption. In practice there is an abundance of examples to show that this boundedness condition is not essential. However, the situation remains not completely clear.

EXERCISES

1. Let φ satisfy (2.3) and (2.4). In addition, assume that its derivative $\dot{\varphi}$ is continuous on $[a,b]$. Show that

$$\int_a^b \dot{\varphi}(\omega,t) W(\omega,t)\, dt + \int_a^b \varphi(\omega,t)\, dW(\omega,t) = \varphi(\omega,b) W(\omega,b) - \varphi(\omega,a) W(\omega,a)$$

almost surely. (Hint: Let $\{W_n\}$ be a sequence of step approximations to the Brownian motion W and consider

$$\varphi_b W_b - \varphi_a W_a - \int_a^b \dot{\varphi}_t W_{nt}\, dt \Big)$$

2. Suppose that Z_t satisfies

$$Z_t = 1 + \int_0^t Z_s \varphi_s\, dW_s$$

Show that Z_t is given by

$$Z_t = \exp\left(\int_0^t \varphi_s\, dW_s - \frac{1}{2} \int_0^t \varphi_s^2\, ds \right)$$

(Hint: Consider $\ln Z_t$, and use Ito's differentiation formula.)

3. Let X_t satisfy the stochastic differential equation

$$dX_t = X_t[m(t)\, dt + \sigma(t)\, dW_t] \qquad t > 0$$

where m and σ are nonrandom, i.e., independent of ω. Show that X_t, $t \geq 0$ has the form

$$X_t = X_0 \exp\left[\int_0^t \sigma(s)\, dW_s + \int_0^t f(s)\, ds \right]$$

and find f.

4. Let X_t satisfy a stochastic differential equation

$$dX_t = m(X_t,t)\, dt + \sigma(X_t,t)\, dW_t \qquad t > 0$$

where m and σ satisfy conditions given in Proposition 4.1. Show that

$$EX_t^2 \leq 3 \left[EX_0^2 + K(1 + t) \int_0^t (1 + EX_s^2)\, ds \right]$$

From this inequality, show that if $EX_0^2 < \infty$ then there exist finite constants A and α such that

$$EX_t^2 \leq A e^{\alpha t^2} \qquad 0 \leq t < \infty$$

5. Let $\varphi(x)$, $-\infty < x < \infty$, be a real-valued Borel function with bounded-continuous second derivative φ'', and let $\{W_t, t \geq 0\}$ be a standard Brownian motion. Show that

$$X_t = \varphi(W_t) - \frac{1}{2}\int_0^t \varphi''(W_s)\,ds$$

is a martingale. Thus, if $\varphi'' \geq 0$ ($\varphi'' \leq 0$) then $\varphi(W_t)$ is a submartingale (respectively, supermartingale).

6. Let φ be as in Exercise 5, and set

$$\mathbf{X}_t = \begin{bmatrix} X_{1t} \\ X_{2t} \end{bmatrix} = \begin{bmatrix} W_t \\ \varphi(W_t) \end{bmatrix}$$

Show that $\mathbf{X}_t$ is the solution of a vector stochastic differential equation

$$d\mathbf{X}_t = \mathbf{f}(\mathbf{X}_t)\,dt + \mathbf{g}(\mathbf{X}_t)\,dW_t$$

where

$$\mathbf{f}(\mathbf{x}) = \begin{bmatrix} 0 \\ \frac{1}{2}\varphi''(x_1) \end{bmatrix} \quad \text{and} \quad \mathbf{g}(\mathbf{x}) = \begin{bmatrix} 1 \\ \varphi'(x_1) \end{bmatrix}$$

Verify, both directly and by using (5.34), that $\mathbf{X}_t$ can also be considered to be the solution of a white-noise equation

$$\dot{\mathbf{X}}_t = \begin{bmatrix} 1 \\ \varphi'(X_{1t}) \end{bmatrix} \zeta_t$$

7. Let W_{kt}, $k = 1, \ldots, n$ be independent standard Brownian motion processes. Define

$$X_t = \sqrt{\sum_{k=1}^{n} W_{kt}^2}$$

Show that X_t satisfies a stochastic differential equation, and find this equation. Is X_t Markov?

8. Consider a white-noise equation

$$\dot{X}_t = \mu_t(b + \sigma\zeta_t) \qquad t > 0$$

where X_t is interpreted as the state at t, μ_t is the control at t, ζ_t is a standard Gaussian white noise, and b, σ are constants.

(*a*) For open-loop control, μ_t can depend only on the initial state X_0 and t. Find the open-loop control $\mu(X_0,t)$ which minimizes EX_T^2.

(*b*) For linear closed-loop control, we permit μ_t to take on the form $\mu_t = \alpha(t)X_t$. Find the function $\alpha(t)$ which minimizes EX_T^2.

9. Consider the forward equation (7.23) for the case where m and σ^2 are functions only of x and not t, that is,

$$\frac{\partial}{\partial t}p(x,t|x_0,t_0) = \frac{1}{2}\frac{\partial^2}{\partial x^2}[\sigma^2(x)p(x,t|x_0,t_0)] - \frac{\partial}{\partial x}[m(x)p(x,t|x_0,t_0)]$$

Let w be any positive solution of the equation

$$\frac{1}{2}\frac{d}{dx}[\sigma^2(x)w(x)] = m(x)w(x)$$

Show that $p(x,t|x_0,t_0)$ can be written in the form

$$p(x,t|x_0,t_0) = w(x)\int_\Lambda e^{-\lambda(t-t_0)}\varphi_\lambda(x)\psi_\lambda(x_0)\,\mu(d\lambda)$$

where Λ is a Borel subset of the real line, μ a Borel measure, and φ_λ, ψ_λ are both solutions of the Sturm-Liouville equation

$$\frac{1}{2}\frac{d}{dx}\left[\sigma^2(x)w(x)\frac{df(x)}{dx}\right] + \lambda w(x)f(x) = 0$$

10. For $\sigma^2(x) = 1$, $m(x) = 0$, we have the Brownian motion case and

$$p(x,t|x_0,t_0) = \frac{1}{\sqrt{2\pi(t-t_0)}}\exp\left[-\frac{1}{2}\frac{(x-x_0)^2}{(t-t_0)}\right]$$

Show that $p(x,t|x_0,t_0)$ has the form prescribed by Exercise 9. In particular,

$$p(x,t|x_0,t_0) = \frac{1}{2\pi}\int_{-\infty}^{\infty} e^{-\frac{1}{2}\nu^2(t-t_0)}e^{i\nu(x-x_0)}\,d\nu$$

11. Suppose that $\sigma^2(x,t) = \beta(t)$ and $m(x,t) = \alpha(t)x$. Show that the fundamental solution of the Fokker-Planck equation has the form

$$p(x,t|x_0,t_0) = \frac{1}{\sqrt{2\pi a^2(t,t_0)}}\exp\left\{-\frac{1}{2a^2(t,t_0)}[x-b(t,t_0)x_0]^2\right\}$$

How are $a^2(t,t_0)$ and $b(t,t_0)$ obtained?

12. Suppose that $\sigma^2(x,t) = 1$ and $m(x,t) = -\operatorname{sgn} x$ where $\operatorname{sgn} x$ is $+1$ or -1 according as $x \geq 0$ or $x < 0$.

(*a*) Find the limiting density $p(x) = \lim_{t\to\infty} p(x,t|x_0,t_0)$.

(*b*) Find $p(x,t|x_0,t_0)$.

5

One-Dimensional Diffusions

1. INTRODUCTION

This chapter is an introduction to the semigroup treatment of Markov processes with stationary transition functions. Modern theory of Markov processes is primarily a semigroup theory. Even though much of this theory has not found its way into applications in physical problems, the elucidation that is made possible with the semigroup approach makes it indispensable in any treatment of Markov processes.

Consider a Markov process $\{X(\omega,t),\ t \in [0,\infty)\}$ defined on a probability space $(\Omega,\mathcal{A},\mathcal{P})$. We assume that the transition function

$$\mathcal{P}(X_t < b|X_s = a) = P(b,t|a,s) \qquad t > s$$

depends only on $t - s$ and not on t and s separately. We call such transition functions **stationary transition functions.** A process with a stationary transition function need not be a stationary process. For example, Brownian motion has a stationary transition function, but is not a stationary process.

It is rather important that the set of values that $X(\omega,t)$ can assume be explicitly stated. We assume that this set is an *interval* S which we shall call the **state space.** The interval S can be finite or infinite, closed or open at either end. Let $\mathcal{R}$ denote the σ algebra of Borel sets in S. That is, $\mathcal{R}$ is the smallest σ algebra containing all subintervals of S. For $E \in \mathcal{R}$, we denote

$$P_a(E,t) = \mathcal{P}(X_{t+s} \in E | X_s = a) \tag{1.1}$$

and call $\{P_a(E,t),\ a \in S,\ E \in \mathcal{R},\ t \in [0, \infty)\}$ the transition function. Since $\{X_t,\ 0 \leq t \leq \infty\}$ is Markov, $P_a(E,t)$ satisfies the Chapman-Kolmogorov equation, viz.,

$$P_a(A, t + s) = \int_S P_b(E,t) P_a(db,s) \tag{1.2}$$

If a transition density $p_a(b,t)$ exists for $t > 0$, that is, if $P_a(E,t)$ can be written as

$$P_a(E,t) = \int_E p_a(b,t)\, db \qquad \begin{matrix} E \in \mathcal{R} \\ t > 0 \end{matrix} \tag{1.3}$$

then $p_a(b,t)$ satisfies

$$p_a(b, t + s) = \int_S p_a(x,s) p_x(b,t)\, dx \tag{1.4}$$

Equations (1.2) and (1.4) reveal the basic theme of this chapter. If $P_a(\cdot,t)$ is known for $0 \leq t \leq \epsilon$, then $P_a(\cdot,t)$ for any t can be determined by iterating (1.2), no matter how small ϵ is. Therefore, the knowledge of $P_a(\cdot,t)$ for $t \approx 0$ suffices to determine $P_a(\cdot,t)$ for all t. It turns out that under suitable continuity conditions and with proper interpretation, $P_a(\cdot,0)$ and $(\partial/\partial t)P_a(\cdot,t)\big|_{t=0}$ completely determine $P_a(\cdot,t)$ for all t. Furthermore, since $P_a(\cdot,0)$ is always the same, viz.,

$$P_a(A,0) = \begin{cases} 1 & \text{if } a \in A \\ 0 & \text{if } a \notin A \end{cases}$$

the behavior of Markov processes with stationary transition functions depends entirely on $(\partial/\partial t)P_a(\cdot,t)\big|_{t=0}$. The goal is to deduce everything about the behavior of a Markov process from $(\partial/\partial t)P_a(\cdot,t)\big|_{t=0}$. We note that this is very much the same motivation underlying the derivation of the diffusion equations (backward and forward equations of Kolmogorov). The only difference is that in the stationary case, the setting is much more general, and the answers are more clear cut. Of course, when we say "everything about the behavior of the process" we mean everything up to an arbitrary distribution of X_0, since we only make use of the transition functions. In this sense, we do not distinguish between processes differing only in the distribution of X_0. For example, we shall define in this chapter a Brownian motion as any process $\{X_t, t \geq 0\}$ such that $\{X_t - X_0, t \geq 0\}$ is a Brownian motion in the earlier sense. That is, here a Brownian motion is any Markov process with transition density

$$p_a(b,t) = \frac{1}{\sqrt{2\pi t}} e^{-(1/2t)(b-a)^2} \tag{1.5}$$

Incidentally, this broadening of the definition of Brownian motion also avoids the awkwardness in interpreting $P_a(E,t)$ as

$$P_a(E,t) = \mathcal{P}(X_t \in E | X_0 = a)$$

when our earlier definition specified $X_0 = 0$, a.s.

2. THE MARKOV SEMIGROUP

Let $\{X_t, 0 \leq t < \infty\}$ be a Markov process with a stationary transition function P_a and continuous in probability. We shall adopt the notations

$$E_a Z = E(Z | X_0 = a) \tag{2.1}$$

and

$$\mathcal{P}_a(B) = \mathcal{P}(B | X_0 = a) \tag{2.2}$$

Because of the continuity in probability, for *every open set E* containing a,

$$\lim_{t \downarrow 0} \mathcal{P}_a(X_t \in E) = \lim_{t \downarrow 0} P_a(E,t) = 1 \tag{2.3}$$

Let **B** denote the space of all complex-valued bounded functions defined on the state space S and measurable with respect to the Borel σ algebra $\mathcal{R}$. With the usual norm

$$\|f\| = \sup_{x \in S} |f(x)| \tag{2.4}$$

the space **B** is a Banach space. For each $t \in [0, \infty)$, we define an operator H_t by

$$(H_t f)(a) = E_a f(X_t) \tag{2.5}$$

Since

$$|E_a f(X_t)| = \left| \int_S f(x) P_a(dx,t) \right| \leq \sup_{x \in S} |f(x)| \int_S P_a(dx,t) = \|f\|$$

H_t maps **B** into **B** and is a contraction, that is,

$$\|H_t f\| \leq \|f\| \tag{2.6}$$

Because of (2.6), for each t, H_t is continuous, that is,

$$\|f_n - f\| \xrightarrow[n\to\infty]{} 0 \Rightarrow \|H_t f_n - H_t f\| \xrightarrow[n\to\infty]{} 0 \tag{2.7}$$

A sequence $\{f_n\}$ in **B** is said to **converge strongly** to f if $\|f_n - f\| \xrightarrow[n\to\infty]{} 0$, and we shall use the notation

$$s \lim_{n\to\infty} f_n = f \tag{2.8}$$

The one-parameter family of operators $\{H_t,\ 0 \le t < \infty\}$ that we have defined has some additional important properties. First and foremost, is the **semigroup property.** From the Chapman-Kolmogorov equation (1.2), we can write

$$\begin{aligned}(H_{t+s}f)(a) = E_a f(X_{t+s}) &= \int_S f(x)P_a(dx, t+s) \\ &= \int_S f(x) \int_S P_b(dx,t)P_a(db,s) \\ &= \int_S (H_t f)(b)P_a(db,s) = (H_s H_t f)(a)\end{aligned}$$

Therefore,

$$H_{t+s} = H_s H_t = H_t H_s \tag{2.9}$$

In addition, because of (2.3) we also have

$$(H_t f)(x) \xrightarrow[t\downarrow 0]{} f(x) \tag{2.10}$$

at every continuity point of f. Roughly speaking, (2.9) and (2.10) imply that H_t must be of the form

$$H_t = e^{tA}$$

where A is necessarily given by $A = (d/dt)H_t\big|_{t=0}$. Thus, the first step is to define $(d/dt)H_t\big|_{t=0}$.

Let $\mathfrak{D}_A$ denote the set of all functions f in **B** such that the limit

$$s \lim_{t\downarrow 0} \frac{1}{t}(H_t f - f)$$

exists. This limit defines a linear operator A mapping $\mathfrak{D}_A$ into **B**, that is,

$$Af = s \lim_{t\downarrow 0} \frac{1}{t}(H_t f - f) \tag{2.11}$$

The operator A is called the **generator** of the semigroup and of the Markov process. If $\mathfrak{D}_A = \mathbf{B}$ and if A is a bounded operator (i.e., there exists finite K such that $\|Af\| \le K\|f\|$ for all $f \in \mathbf{B}$), then we truly would have

$$H_t = e^{tA} \equiv \sum_{n=0}^{\infty} \frac{t^n}{n!} A^n$$

However, this happens only in a few relatively uninteresting cases, for example, when S contains only a finite number of points. In the general case $\mathfrak{D}_A \neq \mathbf{B}$, A is unbounded and e^{tA} is not well defined. The main goal of this section is to show that if a Markov process is continuous in probability and has a stationary transition function, then its transition function is uniquely determined by its generator.

We note that if f is such that $H_t f \in \mathfrak{D}_A$, then

$$\frac{d}{dt} H_t f = A H_t f \tag{2.12}$$

and this is a version of the backward equation. To see that this is the case, let I_E denote the indicator function of E and suppose that $H_t I_E \in \mathfrak{D}_A$ for $t > 0$. Then

$$(H_t I_E)(a) = P_a(E,t)$$

and

$$\frac{\partial}{\partial t} P_a(E,t) = (AP.(E,t))(a) \tag{2.13}$$

As we shall see a little later, for a Brownian motion $H_t I_E \in \mathfrak{D}_A$ for $t > 0$ and

$$(Af)(a) = \frac{1}{2} \frac{d^2 f(a)}{da^2}$$

Therefore, for a Brownian motion,

$$\frac{\partial}{\partial t} P_a(E,t) = \frac{1}{2} \frac{\partial^2}{\partial a^2} P_a(E,t) \qquad t > 0$$

which is just the backward equation for that process.

The procedure of constructing the semigroup $\{H_t, 0 \leq t < \infty\}$, or equivalently, determining the transition function, from its generator A involves in an essential way the resolvent. First, define $\mathbf{B}_0$ as the set of all functions f in $\mathbf{B}$ such that

$$\|H_t f - f\| \xrightarrow[t \downarrow 0]{} 0 \tag{2.14}$$

It is clear that $\mathbf{B}_0 \supset \mathfrak{D}_A$. It turns out that $\mathfrak{D}_A$ is dense in $\mathbf{B}_0$. That is, every $f \in \mathbf{B}_0$ is the strong limit of a sequence from $\mathfrak{D}_A$. To show this, take any $f \in \mathbf{B}_0$ and set

$$f_n = n \int_0^{1/n} H_s f \, ds \tag{2.15}$$

Then, for $n = 1, 2, \ldots,$

$$\operatorname*{s\,lim}_{t \downarrow 0} (H_t f_n - f_n) = n(H_{1/n} f - f)$$

so that $f_n \in \mathfrak{D}_A$ for each n. Furthermore,

$$\|f_n - f\| = \left\| n \int_0^{1/n} (H_t f - f)\, dt \right\| \le \sup_{0 \le t \le 1/n} \|H_t f - f\| \xrightarrow[n \to \infty]{} 0$$

Therefore, $\mathfrak{D}_A$ is dense in $\mathbf{B}_0$. For $f \in \mathbf{B}_0$ and $\lambda > 0$. We define

$$(R_\lambda f)(a) = \int_0^\infty e^{-\lambda t}(H_t f)(a)\, dt \tag{2.16}$$

The family of operators $\{R_\lambda,\ 0 < \lambda < \infty\}$ is called the **resolvent** of the semigroup $\{H_t,\ 0 \le t < \infty\}$. The importance of R_λ is that it is simply related to the generator A.

Proposition 2.1. For every $g \in \mathbf{B}_0$ and every $\lambda > 0$, $R_\lambda g \in \mathfrak{D}_A$. Furthermore, $f = R_\lambda g$ is the unique solution to the equation

$$\lambda f - Af = g \qquad f \in \mathfrak{D}_A \tag{2.17}$$

Proof: We shall sketch a proof with some details omitted. First, we verify that $R_\lambda g \in \mathfrak{D}_A$ by computing

$$\begin{aligned}
\frac{1}{t}(H_t R_\lambda g - R_\lambda g) &= \frac{1}{t}\int_0^\infty e^{-\lambda s}(H_{t+s} g - H_s g)\, ds \\
&= \frac{1}{t}\left(e^{\lambda t}\int_t^\infty e^{-\lambda s} H_s g\, ds - \int_0^\infty e^{-\lambda s} H_s g\, ds\right) \\
&= -e^{\lambda t}\frac{1}{t}\int_0^t e^{-\lambda s} H_s g\, ds + \frac{1}{t}(e^{\lambda t} - 1) R_\lambda g \xrightarrow[t \downarrow 0]{s} -g \\
&\qquad + \lambda R_\lambda g
\end{aligned}$$

where $\xrightarrow{s}$ denotes strong convergence. Therefore,

$$AR_\lambda g = -g + \lambda R_\lambda g$$

and

$$\lambda R_\lambda g - AR_\lambda g = g$$

Therefore, $f = R_\lambda g$ is a solution to (2.17). Next, we prove that it is the only solution.

Suppose that f_1 and f_2 are two solutions to (2.17). Set

$$\varphi = f_1 - f_2$$

Then $\varphi \in \mathfrak{D}_A$ and

$$\lambda\varphi - A\varphi = 0$$

Therefore,

$$\frac{d}{dt}(e^{-\lambda t} H_t \varphi) = -H_t(\lambda\varphi - A\varphi) = 0$$

and

$$e^{-\lambda t}H_t\varphi = H_0\varphi = \varphi$$

It follows that

$$0 \le \|\varphi\| = e^{-\lambda t}\|H_t\varphi\| \le e^{-\lambda t}\|\varphi\| \xrightarrow[t\to\infty]{} 0$$

so that $\|\varphi\| = 0$, which proves uniqueness. ∎

Proposition 2.1 shows that the mapping $R_\lambda: \mathbf{B}_0 \to \mathfrak{D}_A$ is one to one and onto. If we denote the identity operator by I, then we have

$$R_\lambda = (\lambda I - A)^{-1} \tag{2.18}$$

and

$$A = \lambda I - R_\lambda^{-1} \tag{2.19}$$

By using R_λ, we can now construct the semigroup $\{H_t,\ 0 \le t < \infty\}$ from A.

Proposition 2.2. Let $\{X_t,\ 0 \le t \le \infty\}$ be a Markov process with a stationary transition function $P_a(E,t)$, and let $\{X_t,\ 0 \le t < \infty\}$ be continuous in probability. Then its transition function, equivalently, its semigroup $\{H_t,\ 0 \le t < \infty\}$, is uniquely determined by its generator A.

Proof: First, we note that

$$\lambda R_\lambda f = \lambda \int_0^\infty e^{-\lambda t}H_t f\,dt = \int_0^\infty e^{-t}H_{t/\lambda}f\,dt$$

so that $\|\lambda R_\lambda f\| \le \|f\|$ and

$$\|\lambda R_\lambda f - f\| = \left\| \int_0^\infty e^{-t}(H_{t/\lambda}f - f)\,dt \right\|$$

which goes to zero as $\lambda \to \infty$ by dominated convergence. Therefore,

$$\lambda R_\lambda \xrightarrow[\lambda\to\infty]{} I \tag{2.20}$$

Next, define A_λ by

$$A_\lambda = \lambda A R_\lambda \tag{2.21}$$

From (2.19) we have

$$\|A_\lambda f\| = \lambda\|AR_\lambda f\| = \lambda\|\lambda R_\lambda f - f\| \le \lambda\|\lambda R_\lambda f\| + \lambda\|f\| \le 2\lambda\|f\| \tag{2.22}$$

so that A_λ is a bounded operator. We can define

$$e^{tA_\lambda} = \sum_{n=0}^{\infty} \frac{t^n}{n!} A_\lambda{}^n$$

Given A, we now determine $\{H_t,\ 0 \le t < \infty\}$ as follows: For $f \in \mathfrak{D}_A$ we can show that

$$H_t f = s \lim_{\lambda \to \infty} e^{tA_\lambda} f \tag{2.23}$$

For $f \in \mathbf{B}_0$, let $\{f_n\}$ be a sequence from $\mathfrak{D}_A$ converging strongly to f. Then

$$H_t f = s \lim_{n \to \infty} H_t f_n$$

For any bounded-continuous f, we set

$$f_n = n \int_0^{1/n} H_s f\, ds \tag{2.24}$$

Then, $f_n \in \mathbf{B}_0$ for each n, and

$$(H_t f_n)(x) \xrightarrow[n \to \infty]{} (H_t f)(x) \tag{2.25}$$

for each $x \in S$ and each $t \ge 0$. Finally, we note that

$$f(u,x) = e^{iux}$$

is a bounded-continuous function in x for each real u. Therefore, the characteristic function

$$E_a e^{iuX_t} = F_a(u,t)$$

is uniquely determined by the generator A. This in turn implies that the transition function $P_a(E,t)$ is uniquely determined by A. What need to be shown are (2.23) and (2.25).

To prove (2.23), we note that

$$\frac{d}{dt} H_t f = A H_t f$$

and

$$\frac{d}{dt} e^{tA_\lambda} f = A_\lambda e^{tA_\lambda} f$$

Therefore,

$$\begin{aligned}
\|H_t f - e^{tA_\lambda} f\| &= \left\| \int_0^t (A H_s f - A_\lambda e^{sA_\lambda} f)\, ds \right\| \\
&\le \left\| \int_0^t (A - A_\lambda) H_s f\, ds \right\| + \left\| A_\lambda \int_0^t (H_s - e^{sA_\lambda}) f\, ds \right\| \\
&\le \int_0^t \|(A - A_\lambda) H_s f\|\, ds + \int_0^t \|A_\lambda (H_s - e^{sA_\lambda}) f\|\, ds
\end{aligned}$$

If we set $\varphi_t = \int_0^t \|H_s f - e^{sA_\lambda} f\|\, ds$ and make use of (2.22), we find

$$\dot{\varphi}_t \le 2\lambda \varphi_t + \int_0^t \|(A - A_\lambda) H_s f\|\, ds \tag{2.26}$$

or

$$\frac{d}{dt}(e^{-2\lambda t}\varphi_t) \le e^{-2\lambda t}\int_0^t \|(A - A_\lambda)H_sf\|\,ds$$

By direct integration, we get

$$e^{-2\lambda t}\varphi_t \le \int_0^t e^{-2\lambda\tau}\int_0^\tau \|(A - A_\lambda)H_sf\|\,ds\,d\tau = -\frac{1}{2\lambda}e^{-2\lambda t}$$

$$\int_0^t \|(A - A_\lambda)H_sf\|\,ds + \frac{1}{2\lambda}\int_0^t e^{-2\lambda s}\|(A - A_\lambda)H_sf\|\,ds \qquad (2.27)$$

Combining (2.26) and (2.27), we find

$$0 \le \|H_tf - e^{tA_\lambda}f\| = \dot{\varphi}_t \le \int_0^t e^{-2\lambda(t-s)}\|(A - A_\lambda)H_sf\|\,ds \xrightarrow[\lambda\to\infty]{} 0$$

which proves (2.23).

To prove (2.25), we write

$$\begin{aligned}(H_tf_n - H_tf)(a) &= E_af_n(X_t) - E_af(X_t)\\ &= \int_S P_a(dx,t)[f_n(x) - f(x)]\\ &= \int_S P_a(dx,t)\, n\int_0^{1/n}[E_xf(X_s) - f(x)]\,ds\end{aligned}$$

Because $\{X_t, 0 \le t < \infty\}$ is continuous in probability and f is bounded continuous,

$$|(H_tf_n)(a) - (H_tf)(a)| \xrightarrow[n\to\infty]{} 0$$

by dominated convergence. ▮

Example. As an example, we shall derive the generator for a standard Brownian motion. We recall that a Brownian motion has a transition-density function given by

$$p_a(x,t) = \frac{1}{\sqrt{2\pi t}}e^{-(1/2t)(x-a)^2} \qquad (2.28)$$

Let f be any function in **B** with a bounded-continuous second derivative f'', and let $\mathcal{C}^2$ denote the set of all such functions. Then,

$$\frac{1}{t}[E_af(X_t) - f(a)] = \frac{1}{t}\int_{-\infty}^{\infty}\frac{1}{\sqrt{2\pi t}}e^{-(1/2t)(x-a)^2}[f(x) - f(a)]\,dx$$

By using the Taylor's theorem at a, and by making a change in the variable of integration, we get

$$\frac{1}{t}[E_af(X_t) - f(a)] = \tfrac{1}{2}f''(a) + \frac{1}{2}\int_{-\infty}^{\infty}\frac{1}{\sqrt{2\pi}}[f''(\theta) - f''(a)]e^{-\frac{1}{2}z^2}\,dz$$

where θ lies between a and $a + \sqrt{t}\, z$. Because f'' is bounded continuous, we get

$$(Af)(a) = \tfrac{1}{2}f''(a) \tag{2.29}$$

for all $f \in \mathcal{C}^2$, and we have also shown that $\mathfrak{D}_A \supset \mathcal{C}^2$. What is more difficult is to determine $\mathbf{B}_0$ and $\mathfrak{D}_A$.

If we take the Laplace transform of (2.28), we get

$$\int_0^\infty e^{-\lambda t} p_a(x,t)\, dt = \frac{\exp(-\sqrt{2\lambda}\,|x - a|)}{\sqrt{2\lambda}} \qquad 0 < \lambda < \infty \tag{2.30}$$

Hence,

$$(R_\lambda f)(a) = \int_{-\infty}^{\infty} \frac{\exp(-\sqrt{2\lambda}\,|x - a|)}{\sqrt{2\lambda}} f(x)\, dx \tag{2.31}$$

For any bounded f, the right-hand side of (2.31) is a bounded-continuous function. If we denote the set of all bounded-continuous functions by $\mathcal{C}$, then we have

$$\mathfrak{D}_A = R_\lambda \mathbf{B}_0 \subset \mathcal{C}$$

Since $\mathbf{B}_0$ is the closure of $\mathfrak{D}_A$ under uniform convergence, we must also have

$$\mathbf{B}_0 \subset \mathcal{C}$$

On the other hand, for any bounded-continuous f

$$|E_a f(X_t) - f(a)| \xrightarrow[t \downarrow 0]{} 0$$

so that $\mathbf{B}_0 \supset \mathcal{C}$. Hence, $\mathbf{B}_0 = \mathcal{C}$. Now the only thing left to do is to find $\mathfrak{D}_A$.

For $f \in \mathcal{C} = \mathbf{B}_0$, let $g = R_\lambda f$. Then

$$g(a) = \frac{1}{\sqrt{2\lambda}} \left[\int_{-\infty}^{\infty} \exp(-\sqrt{2\lambda}\,|x - a|) f(x)\, dx\right]$$

$$g'(a) = -\int_a^\infty \exp[-\sqrt{2\lambda}\,(x - a)] f(x)\, dx - \int_{-\infty}^{a} \exp[+\sqrt{2\lambda}\,(x - a)] f(x)\, dx$$

$$g''(a) = -2f(a) + 2\lambda g(a)$$

Therefore, $g'' \in \mathcal{C}$, that is, $g \in \mathcal{C}^2$. We have shown that $\mathfrak{D}_A \subset \mathcal{C}^2$. But earlier we showed that $\mathfrak{D}_A \supset \mathcal{C}^2$. Hence, $\mathfrak{D}_A = \mathcal{C}^2$. To summarize, for a Brownian motion we have

$$\mathbf{B}_0 = \mathcal{C} \qquad \mathfrak{D}_A = \mathcal{C}^2 \tag{2.32}$$

and

$$Af = \tfrac{1}{2}f'' \tag{2.33}$$

Further, for every $t > 0$, $p_a(x,t)$ is $\mathcal{C}^2$ in a so that

$$\frac{\partial}{\partial t} p_a(x,t) = \frac{1}{2}\frac{\partial^2}{\partial a^2} p_a(x,t)$$

which is the familiar backward equation for Brownian motion.

3. STRONG MARKOV PROCESSES

Let $\{X_t,\ 0 \le t < \infty\}$ be a separable Markov process, continuous in probability, and with a stationary transition function $P_a(E,t)$. Let $\mathcal{A}_t$ denote the smallest σ algebra with respect to which $\{X_\tau,\ \tau \le t\}$ are all measurable. Then the Markov property can be stated as follows:

$$\mathcal{P}(X_{t+s} \in E|\mathcal{A}_t) = P_{X_t}(E,s) \qquad \text{a.s.} \tag{3.1}$$

for all $s \ge 0$. Often, we state (3.1) verbally as "future and present given the past and present depends only on the present." In (3.1) the present is a time t which is fixed. Roughly speaking, a strong Markov process satisfies (3.1) even when the present is a suitably restricted random time, i.e., it varies from sample function to sample function.

We assume from now on that all processes are sample right continuous with probability 1. Indeed, we shall shortly specialize to sample-continuous processes. With $\{\mathcal{A}_t,\ 0 \le t < \infty\}$ defined as before, a nonnegative random variable τ is said to be a **Markov time** if for every $t > 0$,

$$\{\omega\colon \tau(\omega) < t\} \in \mathcal{A}_t \tag{3.2}$$

This means that if we observe a sample function $X_s(\omega_0)$ on the interval $0 \le s \le t$, we can always determine where $\tau(\omega_0) < t$ or $\tau(\omega_0) \ge t$. However, we cannot always determine whether $\tau(\omega_0) = t$ or not. It is clear that deterministic times are always Markov times. Another important class of Markov times are the **first passage times** (for level a) defined by

$$\tau_a(\omega) = \min\,\{t\colon X_t(\omega) = a\} \tag{3.3}$$

To show that τ_a is a Markov time, we write

$$\begin{aligned}\{\omega\colon \tau_a(\omega) < t\} &= \bigcup_{n=1}^{\infty}\left\{\omega\colon \tau_a(\omega) \le t - \frac{1}{n}\right\}\\ &= \bigcup_{n=1}^{\infty}\left\{\omega\colon X_s(\omega) = a \text{ for some } s \text{ in } \left[0,\, t - \frac{1}{n}\right]\right\}\end{aligned}$$

Since for each n, the set $\{\omega: X_s(\omega) = a \text{ for some } s \text{ in } [0, t - 1/n]\}$ is in $\mathcal{A}_t$,

$$\{\omega: \tau_a(\omega) < t\} \in \mathcal{A}_t$$

Now, let S be the state space of the Markov process, and let a be in the interior of S. We define

$$\tau_{a+} = \lim_{b \downarrow a} \tau_b \tag{3.4}$$

$$\tau_{a-} = \lim_{b \uparrow a} \tau_b \tag{3.5}$$

and these are also Markov times. We should again note that if τ is a Markov time, the set $\{\omega: \tau(\omega) = t\}$ need not be in $\mathcal{A}_t$. For example, neither $\{\tau_{a+} = t\}$ nor $\{\tau_{a-} = t\}$ is necessarily in $\mathcal{A}_t$.

Let τ be a Markov time. We define the σ algebra $\mathcal{A}_{\tau+}$ as follows:

$$E \in \mathcal{A}_{\tau+} \quad \text{if and only if } E \cap \{\omega: \tau(\omega) < t\} \in \mathcal{A}_t$$

It is obvious that τ is $\mathcal{A}_{\tau+}$ measurable. If $\tau = t_0$ is a deterministic time, then

$$\mathcal{A}_{t_0+} = \bigcap_{n=1}^{\infty} \mathcal{A}_{t_0+\frac{1}{n}}$$

Thus, we see that if τ represents the present, then $\mathcal{A}_{\tau+}$ is a little bit more than the past and present.

Definition. $\{X_t, 0 \le t < \infty\}$ is said to be a **strong Markov** process if for every Markov time τ,

$$\mathcal{P}(X_{\tau+s} \in E | \mathcal{A}_{\tau+}) = P_{X_\tau}(E,s) \tag{3.6}$$

Every strong Markov process is Markov in the ordinary sense. This is because if (3.6) is satisfied, then

$$\begin{aligned} \mathcal{P}(X_{t+s} \in E | \mathcal{A}_t) &= E^{\mathcal{A}_t} \mathcal{P}(X_{t+s} \in E | \mathcal{A}_{t+}) \\ &= E^{\mathcal{A}_t} P_{X_t}(E,s) = P_{X_t}(E,s) \qquad \text{a.s.} \end{aligned}$$

which is just the ordinary Markov property.

For an example of a Markov process which is not a strong Markov process, consider the following:

$$X_t(\omega) = \max(0, t - T(\omega)) \qquad 0 \le t < \infty, \tag{3.7}$$

where $T(\omega)$ is a nonnegative random variable with

$$\mathcal{P}(T < t) = 1 - e^{-t} \tag{3.8}$$

This process is obviously Markov in the ordinary sense because:

1. Given $X_t = a > 0$, $X_{t+s} = a + s$ with probability 1

2. Given $X_t = 0$, X_s must be zero for $s \leq t$ so that it provides no further information

Now, the random variable T is a Markov time for this process because

$$\{\omega: T(\omega) \geq t\} = \{\omega: X_t(\omega) = 0\} \in \mathcal{A}_t \tag{3.9}$$

Given T, $X_{T+s} = s$ with probability 1. Therefore,

$$\mathcal{P}(X_{T+s} < x | \mathcal{A}_{T+}) = \begin{cases} 1 & x > s \\ 0 & x \leq s \end{cases} \tag{3.10}$$

On the other hand,

$$\begin{aligned} P_{X_T}(x,s) &= P_0(x,s) = \mathcal{P}(X_{t+s} < x | X_t = 0) \\ &= e^{-t} + \int_0^{\min(t,s)} e^{-(t-y)}\, dy \end{aligned} \tag{3.11}$$

Obviously, (3.10) and (3.11) are not the same, so $\{X_t, 0 \leq t < \infty\}$ cannot be a strong Markov process.

There are two extremely useful criteria for determining whether a Markov process is also a strong Markov process:

1. If (3.6) is satisfied for the following classes of Markov times, then the process is strongly Markov:

$$\begin{aligned} \tau &= \tau_a & a &\in S \\ \tau &= \tau_{a+} & a &\in \text{int } (S) \\ \tau &= \tau_{a-} & a &\in \text{int } (S) \end{aligned}$$

2. If for every $t \geq 0$, the operator H_t maps bounded-continuous functions into bounded-continuous functions, then the process is a strong Markov process.

Processes satisfying (2) are called **Feller processes.** For example, a Brownian motion is a Feller process, hence, it is a strong Markov process.

For strong Markov processes with continuous-sample functions, the generator A is a local operator, that is, Af at a depends only on f in a neighborhood of a. This fact can be deduced from the following proposition.

Proposition 3.1. Let $\{X_t, 0 \leq t < \infty\}$ be a strong Markov process. Let $\tau(\omega)$ be a Markov time.

(*a*) Let $f \in \mathbf{B}_0$ and define

$$u_\lambda(a) = (R_\lambda f)(a) = E_a \int_0^\infty e^{-\lambda t} f(X_t)\, dt \tag{3.12}$$

Then

$$u_\lambda(a) = E_a \int_0^\tau e^{-\lambda t} f(X_t)\, dt + E_a[e^{-\lambda \tau} u_\lambda(X_\tau)] \tag{3.13}$$

(*b*) Let $g \in \mathfrak{D}_A$, and let $E\tau < \infty$, then

$$E_a \int_0^\tau (Ag)(X_t)\,dt = E_a g(X_\tau) - g(a) \tag{3.14}$$

Remark: Equation (3.14) is generally known as **Dynkin's formula,** even though both are due to Dynkin [1965, pp. 132–133].

Proof: To prove (3.13), we only need to show

$$E_a \int_\tau^\infty e^{-\lambda t} f(X_t)\,dt = E_a e^{-\lambda\tau} u_\lambda(X_\tau)$$

which can be done as follows:

$$\begin{aligned} E_a \int_\tau^\infty e^{-\lambda t} f(X_t)\,dt &= E_a \int_0^\infty e^{-\lambda(t-\tau)} f(X_{t+\tau})\,dt \\ &= E_a \int_0^\infty e^{-\lambda(t+\tau)} E[f(X_{t+\tau})|\mathcal{A}_{\tau+}]\,dt \\ &= E_a \int_0^\infty e^{-\lambda(t+\tau)} E_{X_\tau}[f(X_t)]\,dt \\ &= E_a e^{-\lambda\tau} u_\lambda(X_\tau) \end{aligned}$$

To prove (3.14), set $f_\lambda = (\lambda - A)g$, so that

$$g = (\lambda - A)^{-1} f_\lambda = R_\lambda f_\lambda$$

From (3.13), we have

$$g(a) = E_a \int_0^\tau e^{-\lambda t}[(\lambda - A)g](X_t)\,dt + E_a e^{-\lambda\tau} g(X_\tau)$$

which becomes (3.14) as $\lambda \to 0$, provided that $E_a\tau < \infty$. ▮

Equation (3.14) reveals the local character of the generator A when the process is sample continuous. To see this, let $a \in \text{int}\,(S)$, and let

$$\tau = \min\,(\tau_{a+\epsilon}, \tau_{a-\epsilon}) \tag{3.15}$$

Then, starting from a at $t = 0$, $X_t \in (a - \epsilon,\ a + \epsilon)$, $0 \le t < \tau$. From (3.14) we have

$$(Ag)(a) = \lim_{\epsilon \downarrow 0} \frac{E_a g(X_\tau) - g(a)}{E_a\tau} \tag{3.16}$$

where τ is given by (3.16). It is easy to see that the right-hand side of (3.16) depends only on g in a neighborhood of a. It turns out that under some additional assumptions we can show that A is always a differential operator. This will be taken up in the next section.

4. CHARACTERISTIC OPERATORS

For the remainder of this chapter we shall restrict ourselves to processes $\{X_t, 0 \le t < \infty\}$ satisfying the following conditions:

Every point in the state space S is reachable from every point in int (S) with a nonzero probability, that is,

$$\mathcal{P}_a(\tau_b < \infty) > 0 \qquad \text{for every } a \in \text{int}\,(S) \text{ and every } b \in S \tag{4.1a}$$

$\{X_t,\ 0 \leq t < \infty\}$ is a strong Markov process with a stationary transition function. Starting from every point in S, every sample function is continuous with probability 1 (4.1b)

We observe that if (4.1a) is not satisfied, then S can be decomposed into sets, such that starting from a point in a set, X_t remains in that set for all t. In that case, the process can be decomposed into separate processes. Together, (4.1a) and (4.1b) imply that the process is a Feller process. We recall that this means the following: If f is bounded and continuous ($f \in C$), then $H_t f \in C$. In this case it is sufficient to consider the semigroup $\{H_t,\ 0 \leq t < \infty\}$ as acting on C rather than B. This is a great convenience, and we assume that this is done for the remainder of this chapter.

Let τ_a be the first passage time at the point a, defined as before by

$$\tau_a = \min\,(t\colon X_t = a) \tag{4.2}$$

Let (a,b) be an open interval such that its closure $[a,b]$ belongs to the state space S. Define the **first exit time** from (a,b),

$$\tau_{ab} = \min\,(t_a,t_b) \equiv t_a \wedge t_b \tag{4.3}$$

We have the following result on τ_{ab}.

Proposition 4.1. Under assumption (4.1) we have

$$\sup_{a \leq x \leq b} E_x \tau_{ab} < \infty \tag{4.4}$$

Proof: Under assumption (4.1), $\mathcal{P}_a(\tau_b < \infty) > 0$, so that for some $t < \infty$,

$$\mathcal{P}_a(\tau_b > t) = \alpha(t) < 1 \tag{4.5}$$

Now, for $a \leq x \leq b$,

$$\mathcal{P}_x(\tau_{ab} > t) \leq \mathcal{P}_x(\tau_b > t) \leq \mathcal{P}_a(\tau_b > t) = \alpha(t) < 1 \tag{4.6}$$

Next, by the Markov property, we have

$$\begin{aligned}\mathcal{P}_x(\tau_{ab} > t + s) &= \int_a^b \mathcal{P}_x(\tau_{ab} > s,\ X_s \subset dz)\mathcal{P}_z(\tau_{ab} > t)\\ &\leq \alpha(t) \int_a^b \mathcal{P}_x(\tau_{ab} > s,\ X_s \in dz)\\ &= \alpha(t)\mathcal{P}_x(\tau_{ab} > s)\end{aligned} \tag{4.7}$$

so that for some $t < \infty$,

$$\mathcal{P}_x(\tau_{ab} > nt) \leq \alpha^n(t) \tag{4.8}$$

Now, we write

$$
\begin{aligned}
E_x\tau_{ab} &= \int_0^\infty s\mathcal{P}_x(\tau_{ab} \in ds) \\
&= \sum_{n=0}^{\infty} \int_{nt}^{(n+1)t} s\mathcal{P}_x(\tau_{ab} \in ds) \\
&\le \sum_{n=0}^{\infty} (n+1)t[\mathcal{P}_x(\tau_{ab} > nt) - \mathcal{P}_x(\tau_{ab} > (n+1)t)] \\
&= t\sum_{n=0}^{\infty} \mathcal{P}_x(\tau_{ab} > nt) \le t\sum_{n=0}^{\infty} \alpha^n(t) = \frac{t}{1-\alpha(t)}
\end{aligned}
\tag{4.9}
$$

Since there exists some $t < \infty$ such that $\alpha(t) < 1$, we have proved the proposition. ∎

Now, let $[a,b] \subset S$ and $x \in (a,b)$. Dynkin's formula (3.14) yields

$$
E_x \int_0^{\tau_{ab}} (Ag)(X_t)\, dt = E_x g(X_{\tau_{ab}}) - g(x) \tag{4.10}
$$

Since we are considering $\{H_t, 0 \le t < \infty\}$ as acting on C, we can assume $g \in C$. It follows that $Ag \in C$, and by shrinking (a,b) down to x, we get

$$
(Ag)(x) = \lim_{(a,b)\downarrow\{x\}} \frac{E_x g(X_{\tau_{ab}}) - g(x)}{E_x\tau_{ab}} \tag{4.11}
$$

Equation (4.11) expresses the local nature of A in the interior of S, since $(Ag)(x)$ depends only on the values of g in a neighborhood of x. However, (4.11) does not completely specify A. For example, suppose that $S = [0,\infty)$ and for every $x \in (0,\infty)$,

$$
(Ag)(x) = \frac{1}{2}\frac{d^2g(x)}{dx^2} \tag{4.12}
$$

Both of the following two processes satisfy (4.12):

1. Absorbing Brownian motion

$$
\mathcal{P}_a(X_t = 0) = \frac{1}{\sqrt{2\pi}} \int_{-a/\sqrt{t}}^{a/\sqrt{t}} e^{-\frac{1}{2}z^2}\, dz
$$

$$
\mathcal{P}_a(0 < X_t < b) = \int_0^b \frac{1}{\sqrt{2\pi t}} [e^{-(1/2t)(z-a)^2} - e^{-(1/2t)(z+a)^2}]\, dz
$$

2. Reflecting Brownian motion

$$
\mathcal{P}_a(0 \le X_t < b) = \int_0^b \frac{1}{\sqrt{2\pi t}} [e^{-(1/2t)(z-a)^2} + e^{-(1/2t)(z+a)^2}]\, dz
$$

Obviously, these two processes have different generators.

We can now use (4.11) to define an extension to A as follows. Let J denote an interval which is the intersection of an open interval with S. We call such a J a **relatively open interval** (in S). Let $\mathfrak{D}_{\mathring{A}}$ denote the set of all functions such that

$$\lim_{J \downarrow \{x\}} \frac{E_x g(X_{\tau_J}) - g(x)}{E_x \tau_J} = (\mathring{A}g)(x) \tag{4.13}$$

exists at every $x \in S$. The difference between A and $\mathring{A}$ is that A is defined by a uniform limit, and $\mathring{A}$ by a pointwise limit. We adopt the convention that if $E_x\tau_J = \infty$ for every relatively open J containing x, then we set $(\mathring{A}g)(x) = 0$. Of course, this can only happen if x is a closed endpoint of S, because if $x \in \text{int}\,(S)$, then (4.4) applies. We shall denote by $\mathfrak{D}_{\mathring{A}}$ the set of all functions in C such that the limit in (4.13) exists at every $x \in S$.

Let p_{ab}^+ and $m_{ab}(x)$ be defined by

$$p_{ab}^+(x) = \mathcal{P}_x(\tau_b < \tau_a) \tag{4.14}$$

and

$$m_{ab} = E_x \tau_{ab} \tag{4.15}$$

For $x \in \text{int}\,(S)$, we can write (4.13) as

$$(\mathring{A}g)(x) = \lim_{\substack{b \downarrow x \\ a \uparrow x}} \frac{p_{ab}^+(x)g(b) + [1 - p_{ab}^+(x)]g(a) - g(x)}{m_{ab}(x)} \tag{4.16}$$

It is clear that knowing the two functions p_{ab}^+ and m_{ab} for every $[a,b] \subset S$ completely determines $(\mathring{A}g)(x)$ at every interior point of S. It will turn out that they also elucidate the behavior of $\mathring{A}g$ at any closed endpoints. The converse is also true. That is, knowing $\mathring{A}g$ in int (S) completely determines p_{ab}^+ and m_{ab} for every $[a,b] \subset S$, as the following proposition shows.

Proposition 4.2. For every $[a,b] \subset S$, the functions p_{ab}^+ and m_{ab} are, respectively, the unique continuous solutions to the equations

$$(\mathring{A}p_{ab}^+)(x) = 0 \qquad a < x < b \tag{4.17}$$
$$(\mathring{A}m_{ab})(x) = -1 \qquad a < x < b \tag{4.18}$$

subject to the boundary conditions

$$p_{ab}^+(b) = 1 = 1 - p_{ab}^+(a) \tag{4.19}$$
$$m_{ab}(a) = 0 = m_{ab}(b) \tag{4.20}$$

Proof: We can easily verify (4.17) as follows:

$$(\mathring{A}p_{ab}^{+})(x) = \lim_{\substack{d \downarrow x \\ c \uparrow x}} \frac{p_{cd}^{+}(x)p_{ab}^{+}(d) + [1 - p_{cd}^{+}(x)]p_{ab}^{+}(c) - p_{ab}^{+}(x)}{m_{cd}(x)} \tag{4.21}$$

If $x \in (c,d) \subset (a,b)$, then to exit from (a,b) starting at x the process must first exit from (c,d). Using the strong Markov property, we get

$$p_{ab}^{+}(x) = p_{cd}^{+}(x)p_{ab}^{+}(d) + [1 - p_{cd}^{+}(x)]p_{ab}^{+}(c) \tag{4.22}$$

whenever $x \in (c,d) \subset (a,b)$. Using (4.22) in (4.21) results in (4.17). Equation (4.18) can be verified in a similar way.

To prove continuity, let $y > x$, and set $c = a$, $d = y$ in (4.22). Then we get

$$p_{ab}^{+}(x) = p_{ay}^{+}(x)p_{ab}^{+}(y)$$

But $p_{ay}^{+}(x) \xrightarrow[x \uparrow y]{} 1$, so that $p_{ab}^{+}(x) \xrightarrow[x \uparrow y]{} p_{ab}^{+}(y)$. Similarly, we can show $p_{ab}^{+}(x) \xrightarrow[x \downarrow y]{} p_{ab}^{+}(y)$, so that p_{ab}^{+} is continuous. Continuity of m_{ab} can be proved in a similar way. Uniqueness is much more difficult to prove. It depends on the fact that $\mathring{A}$ satisfies a "minimum principle" [Dynkin, 1965, Chap. 1, p. 145]. ▮

If S is a closed interval, say $[0,1]$, let

$$u(x) = p_{01}^{+}(x) \tag{4.23}$$

Then, for $0 \leq a < b \leq 1$, we have from (4.22)

$$p_{ab}^{+}(x) = \frac{u(x) - u(a)}{u(b) - u(a)} \tag{4.24}$$

It turns out that (4.24) holds generally whether S is a closed interval or not. The function $u(x)$, unique only up to an affine transformation $u \to \alpha u + \beta$, is called the **scale function.** The scale function is always continuous and nondecreasing, but bounded only if S is closed. It is clear from (4.17) that

$$(\mathring{A}u)(x) = 0 \tag{4.25}$$

for all $x \in \text{int}\,(S)$. A process is said to be in its **natural scale** if $u(x)$ can be taken to be equal to x. For example, a Brownian motion is in its natural scale. If a process X_t has a scale function $u(x)$, then $u(X_t)$ is a process in its natural scale, so that it is no restriction to assume a process to be in its natural scale.

It also turns out that there exists a Borel measure μ, defined on the

Borel sets in int (S), such that

$$m_{ab}(x) = \int_{(a,b)} \frac{[u(x \wedge y) - u(a)][u(b) - u(x \vee y)]}{u(b) - u(a)} \mu(dy) \tag{4.26}$$

where $x \wedge y = \min(x,y)$ and $x \vee y = \max(x,y)$. In terms of u and μ, $\mathring{A}$ can be written as a generalized differential operator in int (S) as

$$A = \frac{d}{d\mu} \frac{d}{du} \tag{4.27}$$

Actually, we are primarily interested in diffusion processes for which both μ and u are absolutely continuous. In that case, the restriction of $\mathring{A}$ to twice-differentiable functions take on the form

$$(\mathring{A}g)(x) = \frac{1}{\mu'(x)} \frac{d}{dx} \left(\frac{1}{u'(x)} \frac{dg(x)}{dx} \right)$$

We won't give a precise definition to (4.27). Instead, we take up the subject of diffusion processes in the next section.

5. DIFFUSION PROCESSES

There is some disagreement in recent literature as to the definition of a diffusion process. Some authors call any process satisfying condition (4.1) a diffusion, while other restrict the name to a smaller class of processes. We shall adopt the more restrictive definition and define a **diffusion process** as any process satisfying (4.1) and for which the limits

$$\lim_{\substack{a \uparrow x \\ b \downarrow x}} \frac{E_x X_{\tau_{ab}} - x}{E_x \tau_{ab}} = m(x) \tag{5.1}$$

and

$$\lim_{\substack{a \uparrow x \\ b \downarrow x}} \frac{E_x (X_{\tau_{ab}} - x)^2}{E_x \tau_{ab}} = \sigma^2(x) \tag{5.2}$$

exist at every $x \in$ int (S). We shall always assume that m and σ^2 satisfy a Hölder condition (cf. 4.7) and

$$\sigma^2(x) > 0 \qquad x \in \text{int}(S) \tag{5.3}$$

Proposition 5.1. Let $\{X_t, 0 \le t < \infty\}$ be a diffusion process. Let $g \in C^2 \cap \mathfrak{D}_{\mathring{A}}$. Then for every $x \in$ int (S);

$$(\mathring{A}g)(x) = \tfrac{1}{2}\sigma^2(x) \frac{d^2 g(x)}{dx^2} + m(x) \frac{dg(x)}{dx} \tag{5.4}$$

Proof: By assumption g has a bounded-continuous second derivative so that by Taylor's theorem we can write

$$g(X_{\tau_{ab}}) = g(x) + g'(x)(X_{\tau_{ab}} - x) + \tfrac{1}{2}g''(\theta)(X_{\tau_{ab}} - x)^2 \qquad x,\ \theta \in (a,b)$$

Since $\sup_{\theta\in(a,b)} |g''(\theta) - g''(x)| \xrightarrow[(a,b)\downarrow\{x\}]{} 0$, (5.4) follows immediately from the definition of $\mathring{A}$. ∎

Because of Proposition 4.2, we know that if the equation

$$\tfrac{1}{2}\sigma^2(x)\frac{d^2g(x)}{dx^2} + m(x)\frac{dg(x)}{dx} = 0 \qquad a < x < b \tag{5.5}$$

has a continuous solution satisfying the boundary conditions $g(a) = 0 = 1 - g(b)$, then it must be $p_{ab}^+(x)$. Such a solution is easily constructed, and we find

$$p_{ab}^+(x) = \frac{\int_a^x \exp\left[-\int_a^y \frac{2m(z)}{\sigma^2(z)}\,dz\right]dy}{\int_a^b \exp\left[-\int_a^y \frac{2m(z)}{\sigma^2(z)}\,dz\right]dy} \tag{5.6}$$

It is clear that we can take the scale function to be

$$u(x) = \int_c^x \exp\left[-\int_c^y \frac{2m(z)}{\sigma^2(z)}\,dz\right]dy \tag{5.7}$$

where c is any point in int (S).

The function m_{ab} can be found by solving

$$\begin{aligned}\tfrac{1}{2}\sigma^2(x)\frac{d^2m_{ab}(x)}{dx^2} + m(x)\frac{dm_{ab}(x)}{dx} &= -1 \qquad a < x < b\\ m_{ab}(a) = 0 &= m_{ab}(b)\end{aligned} \tag{5.8}$$

To construct m_{ab}, define the Green's function

$$G_{ab}(x,y) = \frac{[u(x \wedge y) - u(a)][u(b) - u(x \vee y)]}{u(b) - u(a)} \qquad x,\ y \in (a,b) \tag{5.9}$$

Because $(\mathring{A}u)(x) = 0$, $x \in \text{int}\,(S)$,

$$\tfrac{1}{2}\sigma^2(x)\frac{\partial^2G_{ab}(x,y)}{\partial x^2} + m(x)\frac{\partial G_{ab}(x,y)}{\partial x} = 0$$

for $x \in (a,y)$ and $x \in (y,b)$. Furthermore, $G_{ab}(a,y) = G_{ab}(b,y) = 0$, and

$$\left.\frac{\partial G_{ab}(x,y)}{\partial x}\right|_{x=y^+} - \left.\frac{\partial G_{ab}(x,y)}{\partial x}\right|_{x=y^-} = -u'(y)$$

Therefore,

$$\tfrac{1}{2}\sigma^2(x)\,\frac{\partial^2 G_{ab}(x,y)}{\partial x^2} + m(x)\,\frac{\partial G_{ab}(x,y)}{\partial x} = -\tfrac{1}{2}\sigma^2(y)u'(y)\;\delta(x-y) \tag{5.10}$$

and

$$m_{ab}(x) = \int_a^b \frac{2}{\sigma^2(y)u'(y)}\,G_{ab}(x,y)\,dy \tag{5.11}$$

Given m and σ^2 in int (S), $p_{ab}{}^+$ and m_{ab} are determined, and thus $\mathring{A}$ is completely determined in int (S). However, we know that if S is closed at one or both endpoints, knowing $\mathring{A}$ in the interior of S may not be enough to determine the semigroup $\{H_t,\ 0 \le t < \infty\}$ (equivalently, the transition function) uniquely. To clarify the situation, we need to study the possible behavior at the boundaries. The first result in this direction is the following.

Proposition 5.2. Let a be the left endpoint of S. Then $a \in S$ if and only if $u(a) > -\infty$, and for some $c > a$,

$$\int_a^c \frac{u(y) - u(a)}{\sigma^2(y)u'(y)}\,dy < \infty \tag{5.12}$$

The right endpoint b of S belongs to S if and only if $u(b) < \infty$, and for some $c < b$,

$$\int_c^b \frac{u(b) - u(y)}{\sigma^2(y)u'(y)}\,dy < \infty \tag{5.13}$$

Proof: We shall only prove the first half, the proof for the second half being nearly identical.

First, suppose that $a \in S$. Let α be any point such that $[a,\alpha] \subset S$. Then from (4.4) we have

$$\sup_{a<x<\alpha} m_{a\alpha}(x) < \infty$$

From (5.11) we find that $\sup\limits_{a<x<\alpha} m_{a\alpha}(x)$ occurs at a point c satisfying

$$\int_a^c \frac{u(y) - u(a)}{\sigma^2(y)u'(y)}\,dy = \int_c^\alpha \frac{u(\alpha) - u(y)}{\sigma^2(y)u'(y)}\,dy$$

and $m_{a\alpha}(c)$ is given by

$$m_{ab}(c) = 2\int_a^c \frac{u(y) - u(a)}{\sigma^2(y)u'(y)}\,dy$$

Therefore, $a \in S$ implies (5.12).

Conversely, assume that (5.12) holds, but suppose that $a \notin S$. Then, for any point $c \in S$,

$$\begin{aligned} E_x\tau_c &= \lim_{z \downarrow a} m_{zc}(x) \qquad a < x < c \\ &= \lim_{z \downarrow a} \int_z^c G_{zc}(x,y) \frac{2}{\sigma^2(y)u'(y)}\, dy \\ &= 2\left[\frac{u(c) - u(x)}{u(c) - u(u)}\right] \int_a^x \frac{u(y) - u(a)}{\sigma^2(y)u'(y)}\, dy \\ &\qquad + 2\left[\frac{u(x) - u(a)}{u(c) - u(a)}\right] \int_x^c \frac{u(c) - u(y)}{\sigma^2(y)u'(y)}\, dy \end{aligned}$$

If (5.12) holds, the last expression goes to zero as $x \downarrow a$. But $E_x\tau_c$ must be nondecreasing as $x \downarrow a$, because it takes longer and longer to get from x to c. Hence, there is a contradiction and (5.12) implies that $a \in S$. ▌

Proposition 5.2 shows that the knowledge of m and σ^2 in the interior of S is sufficient to determine whether S is closed at one or the other of the endpoints. As an example, suppose that int $(S) = (0, \infty)$, and $m(x) = 0$, $\sigma^2(x) = 1$, $0 < x < \infty$. In this case, we can take $u(x) = x$. Since for every finite c,

$$\int_0^c \frac{u(y) - u(0)}{\sigma^2(y)u'(y)}\, dy = \int_0^c y\, dy < \infty$$

S must be closed at the left endpoint 0, that is, $S = [0, \infty)$. If S is closed at an endpoint, then $\mathring{A}$ at that point may not be determined by the knowledge of σ^2 and m in int (S). The situation is as follows: Suppose that a is a closed left endpoint of S and that for every $c > a$, $E_a\tau_c = \infty$. Then $(\mathring{A}g)(a) = 0$. However, if for some $c > a$, $E_a\tau_c < \infty$, then $(\mathring{A}g)(a)$ is not determined by m and σ^2 in int (S). We now state a criterion which separates these two cases.

Proposition 5.3. Let a be a closed left endpoint of S. If for some $c \in$ int (S),

$$\int_a^c \frac{1}{\sigma^2(y)u'(y)}\, dy < \infty \tag{5.14}$$

then for $x \in [a,c]$,

$$E_a\tau_x < \infty \tag{5.15}$$

On the other hand, if for every $c \in$ int (S),

$$\int_a^c \frac{1}{\sigma^2(y)u'(y)}\, dy = \infty \tag{5.16}$$

then $E_a\tau_c = \infty$ for every $c \in$ int (S).

Remarks:

(*a*) The point a is called a **regular boundary** if (5.14) is satisfied for some $c \in \text{int}\,(S)$, otherwise it is called an **exit boundary.**

(*b*) The criterion of Proposition 5.3 can be modified for a closed right endpoint in an obvious way, and we won't repeat it.

Let a be a regular left endpoint of S. Define

$$K_a = \lim_{c \downarrow a} \frac{E_a \tau_c}{c - a} \tag{5.17}$$

We note that K_a cannot be determined from m and σ^2. Indeed, each choice of K_a gives us a process with a different transition function. If b is a regular right endpoint of S, then we set

$$K_b = \lim_{c \downarrow b} \frac{E_b \tau_c}{b - c} \tag{5.18}$$

If a is a regular left endpoint, then

$$(\mathring{A}g)(a) = \frac{1}{K_a} g'(a^+) \tag{5.19}$$

and if b is a regular right endpoint, then

$$(\mathring{A}g)(b) = \frac{1}{K_b} g'(b^-) \tag{5.20}$$

To summarize the situation, we see that $\mathring{A}$ is completely summarized by the knowledge of m and σ^2 in int (S) and the values of K_a and K_b at regular boundaries, if there are any. The following proposition shows that the transition function is also uniquely determined.

Proposition 5.4. Let $g \in C$ (that is, g is bounded continuous). Let f_λ be defined by

$$f_\lambda(x) = \int_0^\infty e^{-\lambda t} E_x g(X_t)\, dt \tag{5.21}$$

Then for every $\lambda > 0$, $f_\lambda \in \mathfrak{D}_{\mathring{A}} \cap C^2$ and is the unique continuous solution to

$$(\mathring{A} f_\lambda)(x) = \lambda f_\lambda(x) - g(x) \qquad x \in S \tag{5.22}$$

Remark: We note that because $f_\lambda \in \mathfrak{D}_{\mathring{A}} \cap C^2$, (5.22) takes on a differential form in the interior of S, viz.,

$$\tfrac{1}{2}\sigma^2(x) \frac{d^2 f_\lambda(x)}{dx^2} + m(x) \frac{d f_\lambda(x)}{dx} = \lambda f_\lambda(x) - g(x) \qquad x \in \text{int}\,(S) \tag{5.23}$$

On a *closed boundary* of S, (5.22) takes on the form given in the following table:

Table 5.1

Nature of boundary	*Closed left endpoint a*	*Closed right endpoint b*
Exit	$\lambda f_\lambda(a^+) = g(a)$	$\lambda f_\lambda(b^-) = g(b)$
Regular	$\frac{1}{K_a} f'_\lambda(a^+) = \lambda f_\lambda(a^+) - g(a)$	$\frac{1}{K_b} f'_\lambda(b^-) = \lambda f_\lambda(b^-) - g(b)$

If $K_a = 0$, the boundary a is called a **reflecting boundary,** and we interpret the boundary condition to mean $f'_\lambda(a^+) = 0$. Similar comments apply if $K_b = 0$. On the other hand, if $K_a = \infty$ (or $K_b = \infty$), the boundary is called an **absorbing** boundary. We can now understand the two processes both satisfying (4.12). For the absorbing Brownian motion, the boundary at $x = 0$ is absorbing, and for the reflecting Brownian motion, the boundary at $x = 0$ is reflecting.

Let $F(x,u;\ t) = E_x e^{iuX_t}$ be the characteristic function. Because $g(x) = e^{iux}$ is bounded continuous, we can use (5.22) to obtain the Laplace transform of F. Hence, the characteristic function (hence, the transition function) can be uniquely determined by solving the differential equation (5.23) with the boundary conditions given in Table 5.1 for closed endpoints and by inverting a Laplace transform. Actually, the transition function can be found more directly.

As an example, let $S = [0, \infty)$ and set

$$\sigma^2(x) = 1 \qquad \mu(x) = 0 \qquad 0 < x < \infty \tag{5.24}$$

It is clear that $x = 0$ is a regular boundary. Thus, for a bounded-continuous g, $f_\lambda(x) = \int_0^\infty e^{-\lambda t} E_x g(X_t)\, dt$ is the unique bounded-continuous solution to

$$\tfrac{1}{2} f''_\lambda(x) - \lambda f_\lambda(x) = -g(x) \qquad 0 < x < \infty \tag{5.25}$$

subject to the boundary condition

$$\frac{1}{2K_0} f'_\lambda(0^+) = \lambda f_\lambda(0^+) - g(0) \tag{5.26}$$

We interpret (5.26) to mean $f'_\lambda(0^+) = 0$ if $K_0 = 0$, and $f_\lambda(0^+) = (1/\lambda)g(0)$ if $K_0 = \infty$. The solution for $f_\lambda(x)$ is rather easy. The best approach is to seek a solution of the form

$$f_\lambda(x) = \int_0^\infty F_\lambda(x,y) g(y)\, dy + v_\lambda(x) g(0) \tag{5.27}$$

where $F_\lambda(x,y)$ and $v_\lambda(x)$ are determined by the following differential equations and boundary conditions, together with the condition that both $F_\lambda(x,y)$ and $v_\lambda(x)$ are to be bounded in x:

$$\begin{aligned}
&\frac{1}{2}\frac{\partial^2}{\partial x^2} F_\lambda(x,y) - \lambda F_\lambda(x,y) = 0 \qquad \begin{matrix} 0 < y < x < \infty \\ 0 < x < y < \infty \end{matrix} \\
&F_\lambda(y^+,y) = F_\lambda(y^-,y) \\
&\frac{\partial}{\partial x} F_\lambda(x,y)\bigg|_{x=y^+} - \frac{\partial}{\partial x} F_\lambda(x,y)\bigg|_{x=y^-} = -2 \\
&\frac{1}{2K_0}\frac{\partial F_\lambda(x,y)}{\partial x}\bigg|_{x=0^+} = \lambda F_\lambda(0^+,y)
\end{aligned} \tag{5.28}$$

$$\begin{aligned}
&\frac{1}{2}\frac{d^2}{dx^2} v_\lambda(x) - \lambda v_\lambda(x) = 0 \qquad 0 < x < \infty \\
&\frac{1}{2K_0} v'_\lambda(0^+) = \lambda v_\lambda(0^+) - 1
\end{aligned} \tag{5.29}$$

It is obvious that the solutions must have the form

$$F_\lambda(x,y) = \begin{cases} A(y)\exp(-\sqrt{2\lambda}\,x) & x \ge y \\ B(y)\exp(-\sqrt{2\lambda}\,x) + C(y)\exp(\sqrt{2\lambda}\,x) & 0 \le x \le y \end{cases} \tag{5.30}$$

$$v_\lambda(x) = D\exp(-\sqrt{2\lambda}\,x) \qquad 0 \le x < \infty \tag{5.31}$$

where $A(y)$, $B(y)$, $C(y)$ and D are determined by the subsidiary conditions in (5.28) and (5.29).

Let $g(u,x) = e^{-ux}$. Then, $f_\lambda(u,x)$ is given by

$$\begin{aligned}
f_\lambda(u,x) &= \int_0^\infty e^{-\lambda t}(E_x e^{-uX_t})\,dt \\
&= \int_0^\infty e^{-\lambda t}\int_0^\infty e^{-ub} P_x(db,t)\,dt
\end{aligned} \tag{5.32}$$

which is seen to be the double Laplace transform of the transition function $P_x(E,t)$. It is easy to see from (5.28) and (5.29) that neither v_λ nor F_λ depends on u, and we can write

$$f_\lambda(u,x) = v_\lambda(x) + \int_0^\infty F_\lambda(x,y)e^{-uy}\,dy \tag{5.33}$$

It is clear now what the probabilistic interpretations of $F_\lambda(x,y)$ and $v_\lambda(x)$ are. If we write

$$P_x(E,t) = I_E(0)\mathcal{P}_x(X_t = 0) + \int_E p_x(b,t)\,db \tag{5.34}$$

where $I_E(x) = 0, 1$ according as $x \notin E$ or $x \in E$, then

$$v_\lambda(x) = \int_0^\infty e^{-\lambda t} \mathcal{P}_x(X_t = 0)\, dt \tag{5.35}$$

$$F_\lambda(x,y) = \int_0^\infty e^{-\lambda t} p_x(y,t)\, dt \tag{5.36}$$

It is easy to find v_λ and F_λ, but to invert the Laplace transforms to get $\mathcal{P}_x(X_t = 0)$ and $p_x(y,t)$ is tedious.

It is fairly clear how the above example generalizes. We shall sketch an outline of the situation below. First, the equation

$$\left[\tfrac{1}{2}\sigma^2(x) \frac{d^2}{dx^2} + \mu(x) \frac{d}{dx}\right] f_\lambda(x) = \lambda f_\lambda(x) \qquad x \in \operatorname{int}(S) \tag{5.37}$$

for $\lambda > 0$ has a pair of solutions $f_\lambda = \varphi_\lambda, \theta_\lambda$ with the following properties:

$$\left.\begin{array}{l} \varphi_\lambda(x) > 0 \qquad \theta_\lambda(x) > 0 \qquad x \in \operatorname{int}(S) \\ \varphi_\lambda(x) \text{ nondecreasing, } \theta_\lambda(x) \text{ nonincreasing} \\ \varphi_\lambda(x) \text{ bounded if and only if the right endpoint of } S \text{ is closed} \\ \theta_\lambda(x) \text{ bounded if and only if the left endpoint of } S \text{ is closed} \end{array}\right. \tag{5.38}$$

The two functions φ_λ and θ_λ are linearly independent, and the Wronskian Δ is given by

$$\Delta(x) = \theta_\lambda(x)\varphi_\lambda'(x) - \varphi_\lambda(x)\theta_\lambda'(x) = u'(x) \tag{5.39}$$

where $u(x)$ is the scale function. Therefore, every solution of (5.37) is a linear combination of φ_λ and θ_λ. We now seek a bounded solution of

$$\tfrac{1}{2}\sigma^2(x) \frac{d^2}{dx^2} f_\lambda(x) + \mu(x) \frac{df_\lambda(x)}{dx} = \lambda f_\lambda(x) - g(x) \qquad x \in \operatorname{int}(S) \tag{5.40}$$

subject to the boundary conditions at closed endpoints given in Table 5.1. Imitating the procedure in the example, we seek a solution in the form of

$$f_\lambda(x) = \int_a^b F_\lambda(x,y) g(y)\, dy + g(a) v_\lambda(x) + g(b) u_\lambda(x) \tag{5.41}$$

where a and b are endpoints of S. The functions F, u_λ, v_λ are of the form

$$F_\lambda(x,y) = \begin{cases} A(y)\varphi_\lambda(x) + B(y)\theta_\lambda(x) & x > y \\ C(y)\varphi_\lambda(x) + D(y)\theta_\lambda(x) & x < y \end{cases} \tag{5.42}$$

$$v_\lambda(x) = \alpha\varphi_\lambda(x) + \beta\theta_\lambda(x) \tag{5.43}$$

$$u_\lambda(x) = \gamma\varphi_\lambda(x) + \delta\theta_\lambda(x) \tag{5.44}$$

The unknowns in (5.42) to (5.44) are determined by requiring

$$F_\lambda(y^+,y) = F_\lambda(y^-,y) \tag{5.45}$$

$$\frac{\partial}{\partial x} F_\lambda(x,y) \Big|_{x=y^+} - \frac{\partial}{\partial x} F_\lambda(x,y) \Big|_{x=y^-} = -\frac{2}{\sigma^2(y)} \tag{5.46}$$

and by imposing the following boundary conditions:

$F_\lambda(x,y)$, $v_\lambda(x)$, $u_\lambda(x)$ stay bounded as x approaches an open endpoint (5.47*a*)

At a closed endpoint, $F_\lambda(x,y)$ (as a function of x) satisfies this boundary conditions of Table 5.1 for $g \equiv 0$ (5.47*b*)

If the left endpoint is closed, $v_\lambda(x)(u_\lambda(x))$ satisfies the boundary conditions of Table 5.1 corresponding to $g \equiv 1$ ($g \equiv 0$). If the right endpoint is closed, $v_\lambda(x)(u_\lambda(x))$ satisfies the boundary conditions of Table 5.1 corresponding to $g \equiv 0$ ($g \equiv 1$) (5.47*c*)

The probabilistic interpretation of u_λ, v_λ, and F_λ is given by

$$F_\lambda(x,y)\,dy = \int_0^\infty e^{-\lambda t}\mathcal{P}_x(X_t \in dy)\,dt \qquad a < x < y < b \tag{5.48a}$$

$$u_\lambda(x) = \int_0^\infty e^{-\lambda t}\mathcal{P}_x(X_t = b)\,dt \tag{5.48b}$$

$$v_\lambda(x) = \int_0^\infty e^{-\lambda t}\mathcal{P}_x(X_t = a)\,dt \tag{5.48c}$$

As to be expected, the simplest case results when both endpoints are open. In that case, both u_λ and v_λ are identically zero, and $F_\lambda(x,y)$ is given by

$$F_\lambda(x,y) = \begin{cases} \dfrac{2}{\sigma^2(y)u'(y)}\varphi_\lambda(y)\theta_\lambda(x) & b \geq x \geq y > a \\[2ex] \dfrac{2}{\sigma^2(y)u'(y)}\theta_\lambda(y)\varphi_\lambda(x) & a \leq x \leq y < b \end{cases} \tag{5.49}$$

From (5.49) and the fact that $F_\lambda(x,y)$ is the Laplace transform of the transition density $p_x(y,t)$, we can finally deduce that $p_x(y,t)$ must satisfy both the forward and the backward equations of diffusion. Why it is so hard to prove that $p_x(y,t)$ satisfy the diffusion equations (without assuming the differentiability conditions) is not well understood.

EXERCISES

1. Let $\{X_t,\ -\infty < t < \infty\}$ be a random telegraph process which we define as a stationary Markov process such that

   (*a*) $X_t = +1$ or -1 with $\mathcal{P}(X_t = 1) = \frac{1}{2} = \mathcal{P}(X_t = -1)$

   (*b*) $\mathcal{P}(X_t = X_s) = \frac{1}{2}(1 + e^{-|t-s|}) = 1 - \mathcal{P}(X_t = -X_s)$

   Find the generator A, and show that $H = e^{tA}$.

2. Let $\{X_t, 0 \leq t < \infty\}$ have a state space $S = [0, \infty)$, and let

$$\mathcal{P}(X_t < x | X_s = a)$$
$$= \int_0^x \frac{1}{\sqrt{2\pi(t-s)}} \left\{ \exp\left[-\frac{1}{2}\frac{(\xi - a)^2}{t-s} \right] - \exp\left[-\frac{1}{2}\frac{(\xi + a)^2}{t-s} \right] \right\} d\xi$$
$$+ \sqrt{\frac{2}{\pi}} \int_{a/\sqrt{t-s}}^{\infty} e^{-\frac{1}{2}z^2}\, dz \qquad \begin{matrix} t > s \\ x > 0 \\ a > 0 \end{matrix}$$

For $a = 0$ we have $\mathcal{P}(X_t = 0 | X_s = 0) = 1$. Find the generator A.

3. Let $\{X_t, t \geq 0\}$ be a Brownian motion starting from 0, that is, $X_0 = 0$ with probability 1. Let τ_c be defined by

$$\tau_c = \min_{t>0} \{t, X_t = c\} \qquad c > 0$$

(*a*) Introduce a function f defined as

$$f(x) = \begin{cases} 1 & x > c \\ 0 & x \leq c \end{cases}$$

and use (3.13) to find $Ee^{-\lambda \tau_c}$.

(*b*) Use the same method and prove the reflection principle of D. Andre,

$$\mathcal{P}(\tau_c \leq t) = 2\mathcal{P}(X_t \geq c)$$

4. Let $\{X_t, -\infty < t < \infty\}$ be a zero mean Gaussian process with $EX_tX_s = e^{-|t-s|}$, namely, X is an Ornstein-Uhlenbeck process. Use (3.13) to find

$$\mathcal{P}(X_t \geq 0, 0 \leq t \leq 1)$$

5. Let W_t and V_t be standard Brownian motion processes with $W_0 = V_0 = 0$. Let $X_t = W_t^2 + V_t^2$. Show that $X_t, t \geq 0$, is a diffusion process, and find the functions σ^2 and m.

6. For the process in Exercise 5 find a suitable scale function $u(x)$, and determine whether the state space S is closed at its left endpoint 0. If S is closed at 0, determine whether it is a regular boundary or an exit boundary. If it is a regular boundary, determine $(\mathring{A}g)(0)$.

7. Show that, except for notational changes, (5.22) can be viewed as a generalization of (2.12), that is, if $H_t g \in \mathfrak{D}_A$ for each t, then it follows from (2.12) that

$$Af_\lambda = \lambda f_\lambda - g$$

8. Let $X_t, t \geq 0$, be a diffusion process so that in int (S) $\mathring{A} = \frac{1}{2}\sigma^2(x)d^2/dx^2 + m(x)\, d/dx$. Let $h(x,t)$ be defined by

$$h(x,t) = E_x f(X_t) \exp\left[-\int_0^t k(X_s)\, ds \right]$$

Give a heuristic derivation of the equation

$$\frac{\partial}{\partial t} h(x,t) = \frac{1}{2}\sigma^2(x) \frac{\partial^2 h(x,t)}{\partial x^2} + m(x) \frac{\partial h(x,t)}{\partial x} - k(x)h(x,t) \qquad x \in \text{int}\,(S)$$

9. A precise and general statement of the results in Exercise 8 is known as Kac's theorem [Kac, 1951] stated as follows: Let k and f be in $\mathbf{B}_0$, and let k be nonnegative and f be continuous. Let $u_\lambda(x)$ be defined by

$$u_\lambda(x) = \int_0^\infty h(x,t)e^{-\lambda t}\,dt = E_x \int_0^\infty e^{-\lambda t} f(X_t) \exp\left[-\int_0^t k(X_s)\,ds\right] dt$$

Then u_λ is the unique solution of

$$\lambda u_\lambda - f = Au_\lambda - ku_\lambda$$

(which is roughly equivalent to $\partial h/\partial t = Ah - kh$).

Use Kac's theorem to prove that if τ is the amount of time that a Brownian motion is positive during $[0,1]$, then

$$\mathcal{P}(\tau \le t) = \frac{2}{\pi} \sin^{-1} \sqrt{t}$$

This is the celebrated arc-sine law of Lévy. [Hint: Set $k(x) = \beta((1 + \operatorname{sgn} x)/2)$ and $f(x) = 1$ in Kac's theorem.]

10. Let X_t, $t \ge 0$, be a standard Brownian motion. Use the result of Exercise 8 to find the distribution of the quadratic functional

$$\int_0^1 X_t^2\,dt$$

11. Suppose that int $(S) = (0, \infty)$. For the following pairs of σ^2 and m, determine:

(*a*) Whether S is closed at 0.

(*b*) If closed, is 0 a regular boundary?

$\sigma^2(x)$	$m(x)$
1	-1
1	$-x$
x	$-x$
x^2	$1 - \frac{3}{2}x$

6

Differentiability of Probability Measures and Its Applications

1. INTRODUCTION

Let Ω be a nonempty set and $\mathcal{A}$ be a σ algebra of subsets of Ω. The pair $(\Omega,\mathcal{A})$ is often referred to as a **measurable space** or a **preprobability space.** Now consider two nonnegative σ-additive set functions M_1 and M_2 defined on sets in $\mathcal{A}$ such that $M_1(\Omega) < \infty$ and $M_2(\Omega) < \infty$. We call M_1 and M_2 finite measures on $(\Omega,\mathcal{A})$. For a pair of finite measures defined on the same measurable space, we define the concepts of absolute continuity, equivalences and singularity as follows:

M_2 is said to be **absolutely continuous** with respect to M_1 if for every A in $\mathcal{A}$ such that $M_1(A) = 0$, we have $M_2(A) = 0$. In symbols we write $M_2 \ll M_1$.

M_1 and M_2 are said to be **equivalent** if they are mutually absolutely continuous. In symbols we write $M_1 \equiv M_2$.

M_1 and M_2 are said to be **singular** if there exists a set A such that $M_1(A) = 0$ and $M_2(\Omega - A) = 0$. We denote this by $M_1 \perp M_2$.

Two basic theorems related to these properties of measures are the

Lebesgue decomposition theorem and the Radon-Nikodym theorem. These can be combined and stated for the case of finite measures as follows.

Proposition 1.1. Let $(\Omega,\mathcal{A})$ be a measurable space, and let M_1 and M_2 be two finite measures defined on $(\Omega,\mathcal{A})$. Then there exists a nonnegative and $\mathcal{A}$-measurable function Λ and a finite measure μ such that for every $A \in \mathcal{A}$,

$$M_2(A) = \int_A \Lambda(\omega) M_1(d\omega) + \mu(A) \tag{1.1}$$

where Λ is integrable with respect to M_1, and μ is singular with respect to M_1.

Remark: It is clear that $M_2 \ll M_1$ if and only if $\mu \equiv 0$. In that case, we shall also often say that M_2 is **differentiable** with respect to M_1, and the function Λ will be called the **Radon-Nikodym derivative** of M_2 with respect to M_1 and denoted by

$$\Lambda = \frac{dM_2}{dM_1} \tag{1.2}$$

If $M_2 \ll M_1$ and $\mathcal{B}$ is a sub-σ algebra of $\mathcal{A}$, then the restriction of M_2 to $\mathcal{B}$ is always absolutely continuous with respect to the corresponding restriction of M_1. It is an easy exercise to verify this fact and the formula

$$\frac{dM_2{}^{\mathcal{B}}}{dM_1{}^{\mathcal{B}}} = E_1{}^{\mathcal{B}} \frac{dM_2}{dM_1} \qquad \text{(a.s. } M_1) \tag{1.3}$$

where $M_1{}^{\mathcal{B}}$ and $M_2{}^{\mathcal{B}}$ denote the restrictions of M_1 and M_2 to $\mathcal{B}$, respectively, and $E_1{}^{\mathcal{B}}$ denotes conditional expectation with respect to M_1 measure. Although our definition for conditional expectation given in Sec. 1.7 was for probability measures, extension to finite measures causes no difficulty.

We shall now restrict ourselves to probability measures which are generated by stochastic processes. By this we mean the following: Let $\{X_t, t \in T\}$ be a family of functions defined on a basic space Ω and taking values in R^n. Let $\mathcal{A}$ be the smallest σ algebra of subsets of Ω with respect to which X_t is measurable for each $t \in T$. Suppose that $\mathcal{P}_0$ and $\mathcal{P}$ are two probability measures on $(\Omega,\mathcal{A})$. We shall try to obtain conditions which guarantee that $\mathcal{P} \ll \mathcal{P}_0$ and to obtain an expression for $d\mathcal{P}/d\mathcal{P}_0\,(\omega)$ as a functional of $\{X_t(\omega),\ 0 \le t \le T\}$. The answers to these questions will play an important role in detection and filtering problems. There are basically two broad classes of problems that we shall consider. These are (1) Gaussian measures and (2) measures absolutely continuous with respect to the Wiener measure. The first of these two classes is closely related to the problem of detecting a Gaussian signal in Gaussian noise, and the second finds important applications in detection and filtering problems involving additive Gaussian white noise.

Let $\{X_t,\ t \in T\}$ be a family of R^m-valued functions defined on a basic space Ω, and let $\mathcal{A}$ denote the minimal σ algebra with respect to which X_t is measurable for every $t \in T$. We call the pair $(\Omega,\mathcal{A})$ the **sample space** of $\{X_t,\ t \in T\}$. Suppose that $\mathcal{P}_0$ and $\mathcal{P}$ are two probability measures defined on $(\Omega,\mathcal{A})$ such that with respect to each measure, $\{X_t,\ t \in T\}$ is a separable process continuous in probability. Let $\{T_n\}$ be an increasing sequence of partitions of T such that T_n becomes dense in T as $n \to \infty$. Let $\mathcal{A}_n$ denote the σ algebra generated by $\{X_t,\ t \in T_n\}$. It is clear that $\mathcal{A}_1 \subset \mathcal{A}_2 \subset \cdots$. Let $\mathcal{A}_\infty$ denote the smallest σ algebra containing $\bigcup_n \mathcal{A}_n$. It is clear that $\mathcal{A} \supset \mathcal{A}_\infty$. Now, suppose that $\mathcal{P} \ll \mathcal{P}_0$ (on $\mathcal{A}$). Let $\mathcal{A}'$ denote the completion of $\mathcal{A}_\infty$ with respect to $\mathcal{P}_0$. Because every dense set in T is a separant, $\mathcal{A}'$ must contain $\mathcal{A}$. It is clear that $\mathcal{P}$ and $\mathcal{P}_0$ can be uniquely extended to $\mathcal{A}'$, preserving absolute continuity. Thus, we can assume that $\mathcal{P}$ and $\mathcal{P}_0$ are probability measures defined on $\mathcal{A}'$ with $\mathcal{P} \ll \mathcal{P}_0$.

Now, for each n, let L_n be defined by

$$L_n = E_0^{\mathcal{A}_n} \frac{d\mathcal{P}}{d\mathcal{P}_0} \tag{1.4}$$

Because $\{\mathcal{A}_n\}$ is an increasing sequence, $\{L_n,\mathcal{A}_n\}$ is a martingale with respect to $\mathcal{P}_0$ measure, that is, when $p \geq n$,

$$E_0^{\mathcal{A}_n} L_p = L_n \qquad \text{a.s. } \mathcal{P}_0 \tag{1.5}$$

From (1.3) we also have

$$L_n = \frac{d\mathcal{P}^{\mathcal{A}_n}}{d\mathcal{P}_0^{\mathcal{A}_n}} \tag{1.6}$$

Let $\mathbf{P}_{T_n}$ and $\mathbf{P}_{T_n}{}^0$ denote the Borel probability measures generated by $\{X_t,\ t \in T_n\}$ under $\mathcal{P}$ and $\mathcal{P}_0$, respectively. Then, with $\mathcal{P}_0$-measure 1,

$$L_n(\omega) = \frac{d\mathbf{P}_{T_n}}{d\mathbf{P}_{T_n}{}^0} (X_t(\omega),\ t \in T_n) \tag{1.7}$$

Since for each t, X_t is an m vector, $d\mathbf{P}_{T_n}/d\mathbf{P}_{T_n}{}^0$ is a Borel function on R^{mn} if T_n contains n points. Now, if both $\mathbf{P}_{T_n}$ and $\mathbf{P}_{T_n}{}^0$ are absolutely continuous with respect to the Lebesgue measure, then (1.7) can be rewritten in terms of the density functions p_{T_n} and $p_{T_n}{}^0$ as

$$L_n(\omega) = \frac{p_{T_n}}{p_{T_n}{}^0} (X_t(\omega),\ t \in T_n) \tag{1.8}$$

Because $\{L_n, \mathcal{A}_n, n \geq 1\}$ is a nonnegative martingale with respect to $\mathcal{P}_0$, the martingale convergence theorem [Doob, 1953, p. 319] implies the existence of an $\mathcal{A}_\infty$-measurable function L such that

$$\lim_{n\to\infty} L_n = L \qquad \text{a.s. } \mathcal{P}_0 \tag{1.9}$$

and

$$E_0^{\mathcal{A}_n} L = L_n \qquad \text{a.s. } \mathcal{P}_0 \tag{1.10}$$

In particular, it follows from (1.10) that $E_0\Lambda = 1$. If we define a measure $\mathcal{P}'$ on $(\Omega, \mathcal{A}_\infty)$ by

$$\frac{d\mathcal{P}'}{d\mathcal{P}_0} = L$$

then $\mathcal{P}'$ is a probability measure. Now, take an arbitrary set $A \in \bigcup_n \mathcal{A}_n$. There exists some k sufficiently large so that $A \in \mathcal{A}_k$. Since

$$\begin{aligned}
\mathcal{P}(A) &= E_0 I_A \frac{d\mathcal{P}}{d\mathcal{P}_0} = E_0 I_A E_0^{\mathcal{A}_k} \frac{d\mathcal{P}}{d\mathcal{P}_0} \\
&= E_0 I_A \Lambda_k \\
&= E_0 I_A E_0^{\mathcal{A}_k} L \\
&= E_0 I_A L \\
&= \mathcal{P}'(A)
\end{aligned}$$

The two measures $\mathcal{P}$ and $\mathcal{P}'$ agree on $\bigcup_n \mathcal{A}_n$. Their extensions to $\mathcal{A}_\infty$ must also agree. Since $\mathcal{A}'$ is merely the completion of $\mathcal{A}_\infty$ with respect to $\mathcal{P}_0$, we finally arrive at the conclusion,

$$\frac{d\mathcal{P}}{d\mathcal{P}_0} = L \qquad \text{a.s. } \mathcal{P}_0 \tag{1.11}$$

Because $\mathcal{P} \ll \mathcal{P}_0$, (1.11) also holds with $\mathcal{P}$-measure 1.

We can summarize the foregoing by the following theorem.

Proposition 1.2. Let $(\Omega, \mathcal{A})$ be a measurable space, and let $\mathcal{P}$ and $\mathcal{P}_0$ be two probability measures defined on $(\Omega, \mathcal{A})$ such that $\mathcal{P} \ll \mathcal{P}_0$. Let $\{X_t, t \in T\}$ be a family of measurable (possibly, vector-valued) functions on $(\Omega, \mathcal{A})$, such that $\mathcal{A}$ is the completion with respect to $\mathcal{P}_0$ of the σ algebra generated $\{X_t, t \in T\}$. Suppose that $\{X_t, t \in T\}$ is separable and continuous in probability with respect to either measure. Then,

$$\frac{d\mathcal{P}}{d\mathcal{P}_0} = \lim_{n\to\infty} \frac{d\mathbf{P}_{T_n}}{d\mathbf{P}_{T_n}{}^0} (X_t, t \in T_n) \tag{1.12}$$

with $\mathcal{P}_0$-measure 1, where $\{T_n\}$ is an increasing $(T_{n+1} \supset T_n)$ sequence of partitions of T which become dense in T as $n \to \infty$, and $\mathbf{P}_{T_n}$ and $\mathbf{P}_{T_n}{}^0$ denote the Borel probability measures induced by $\{X_t, t \in T_n\}$ under $\mathcal{P}$ and $\mathcal{P}_0$, respectively.

For a specific example, consider the following. Suppose that under $\mathcal{P}_0$, $\{X_t, 0 \le t \le 1\}$ is a Brownian motion and under $\mathcal{P}$, $\{X_t - t, 0 \le t \le 1\}$ is a Brownian motion. Let $T_n = \{k/2^n, k = 0, 1, \ldots, 2^n\}$. Then,

$$\frac{d\mathbf{P}_{T_n}}{d\mathbf{P}_{T_n}{}^0}(x_0, x_1, \ldots, x_{2^n}) = \frac{p_{T_n}(x_0, x_1, \ldots, x_{2^n})}{p_{T_n}{}^0(x_0, x_1, \ldots, x_{2^n})}$$

$$= \frac{\prod_{\nu=0}^{2^n-1} (1/\sqrt{2\pi 2^{-n}})e^{-\frac{1}{2}(2^n)(x_{\nu+1}-x_\nu-2^{-n})^2}}{\prod_{\nu=0}^{2^n-1} (1/\sqrt{2\pi 2^{-n}})e^{-\frac{1}{2}(2^n)(x_{\nu+1}-x_\nu)^2}}$$

$$= \exp\left[+\sum_{\nu=0}^{2^n-1} (x_{\nu+1} - x_\nu)\right]e^{-\frac{1}{2}}$$

$$= e^{-\frac{1}{2}}e^{(x_1-x_0)}$$

Therefore, in this somewhat trivial example, we have with $\mathcal{P}_0$-measure 1,

$$\frac{d\mathcal{P}}{d\mathcal{P}_0}(\omega) = e^{-\frac{1}{2}}e^{X_1(\omega)-X_0(\omega)} = e^{-\frac{1}{2}}e^{X_1(\omega)}$$

As a check, we can verify that $E_0\, d\mathcal{P}/d\mathcal{P}_0 = 1$ by computing

$$E_0 e^{-\frac{1}{2}}e^{X_1} = e^{-\frac{1}{2}}\frac{1}{\sqrt{2\pi}}\int_{-\infty}^{\infty} e^x e^{-\frac{1}{2}x^2}\,dx = \frac{1}{\sqrt{2\pi}}\int_{-\infty}^{\infty} e^{-\frac{1}{2}(x-1)^2}\,dx = 1$$

2. GAUSSIAN MEASURES

Let $\{X_t, t \in T\}$ be a family of n-vector-valued functions defined on a basic space Ω. Let $\mathcal{A}$ be the smallest σ algebra with respect to which X_t is measurable for every $t \in T$. As before, we shall refer to such a pair $(\Omega,\mathcal{A})$ as the sample space of $\{X_t, t \in T\}$. A probability measure $\mathcal{P}$ defined on $(\Omega,\mathcal{A})$ is called a **Gaussian measure** if, with respect to $(\Omega,\mathcal{A},\mathcal{P})$, $\{X_t, t \in T\}$ is a Gaussian process. Suppose that $\mathcal{P}_0$ and $\mathcal{P}$ are two Gaussian measures defined on the same space $(\Omega,\mathcal{A})$. Let E_0 and E denote expectations with respect to $\mathcal{P}_0$ and $\mathcal{P}$. Since each probability measure is uniquely determined by the corresponding mean and covariance functions of $\{X_t, t \in T\}$, our aim is to investigate the mutual differentiability of $\mathcal{P}_0$ and $\mathcal{P}$ in terms of these functions.

The practical motivation of this class of problems is the detection problem which might be stated as follows. Suppose that we observe a waveform over some interval, say [0,1]. We assume that this waveform is either the sum of a signal plus a noise or noise alone. If we assume that both the noise and the signal can be considered as sample functions of Gaussian processes, then the observation can be considered to be a sample function of a stochastic process $\{X_t, 0 \leq t \leq 1\}$ which is Gaussian under either of the two hypotheses: (H_0) noise alone and (H_1) signal plus noise. Of course, the finite-dimensional distributions of $\{X_t, 0 \leq t \leq 1\}$ corresponding to the two hypotheses are different. Thus, the problem is one of deciding between two Gaussian measures $\mathcal{P}_0$ and $\mathcal{P}$ on the basis of observing a sample function $X_t(\omega)$, $0 \leq t \leq 1$. If $\mathcal{P} \ll \mathcal{P}_0$, then a test which is optimal in a very general sense is the likelihood ratio test which has the following form:

$$\frac{d\mathcal{P}}{d\mathcal{P}_0}(\omega) \begin{cases} > \rho & \text{decide in favor of } \mathcal{P} \\ < \rho & \text{decide against } \mathcal{P} \end{cases} \tag{2.1}$$

where ρ is a constant. Now it turns out that under very general conditions two Gaussian measures are either equivalent or singular. This means that if we write $\mathcal{P}$ in the form of (1.1) as

$$\mathcal{P}(A) = \int_A \Lambda(\omega)\mathcal{P}_0(d\omega) + \mu(A) \qquad A \in \mathcal{A} \tag{2.2}$$

then either $\mu \equiv 0$ or $\mu \equiv \mathcal{P}$. If $\mathcal{P}_0 \perp \mathcal{P}$, then by definition there is set A such that $\mathcal{P}_0(A) = 0$ and $\mathcal{P}(\Omega - A) = 0$. In this case, a perfect test can be constructed as follows. Given a sample function $X_t(\omega)$ determine whether $\omega \in A$ or $\omega \in \Omega - A$. If $\omega \in A$, decide in favor of $\mathcal{P}$, otherwise decide against $\mathcal{P}$.

We can now give a brief classification for the vast amount of literature on Gaussian measures published in the last 10 years. First, the basic dichotomy theorem stating that two Gaussian measures are either equivalent or singular is due independently to Feldman and Hajek. Each gave a different necessary and sufficient condition for equivalence, but neither condition is easy to verify in terms of the mean and covariance functions. A large number of papers have dealt with obtaining conditions which are either sufficient or necessary (but not both) for equivalence, and with finding necessary and sufficient conditions in special cases, e.g., when X_t is stationary or when $\mathcal{P}_0$ is the Wiener measure. Next, even when we know $\mathcal{P} \equiv \mathcal{P}_0$, to find $d\mathcal{P}/d\mathcal{P}_0\,(\omega)$ as an explicit functional on $\{X_t(\omega), 0 \leq t \leq 1\}$ is difficult. On the other hand, if $\mathcal{P} \perp \mathcal{P}_0$, then the problem is to find a separating set A and to express memberships in A in terms of a simple rule on $\{X_t(\omega), 0 \leq t \leq 1\}$. The singular case has been studied somewhat less than the equivalence case possibly for the reason that most people do not believe that perfect detection is possible in practice. If singularity arises, it is because the model is inadequate. In this section, we shall outline some of the main results on Gaussian measures starting with a statement of Kakutani's dichotomy theorem on product measures [Kakutani, 1948] restricted to the Gaussian case. Much of what we do closely follows the paper by Root [1962].

Proposition 2.1 (Kakutani). Let $X_1, X_2, \ldots,$ be a sequence of real-valued measurable functions defined on a fixed pair $(\Omega,\mathcal{A})$ such that $\mathcal{A}$ is the smallest σ algebra with respect to which every X_k is measurable. Let $\mathcal{P}_0$ and $\mathcal{P}$ be two probability measures on $(\Omega,\mathcal{A})$ such that under either measure, $X_1, X_2, \ldots,$ are independent and Gaussian random variables. Let $\mu_k, \mu_{0k}, \sigma_k^2, \sigma_{0k}^2$ be given by

$$\begin{aligned} \mu_k &= EX_k & \sigma_k^2 &= E(X_k - \mu_k)^2 \\ \mu_{0k} &= E_0X_k & \sigma_{0k}^2 &= E_0(X_k - \mu_{0k})^2 \end{aligned} \tag{2.3}$$

Then either $\mathcal{P}_0 \equiv \mathcal{P}$ or $\mathcal{P}_0 \perp \mathcal{P}$ according as

$$\sum_{k=1}^{\infty} \left[\frac{(\mu_k - \mu_{0k})^2}{\sigma_k^2 + \sigma_{0k}^2} + \ln\left(\frac{1}{2}\frac{\sigma_{0k}}{\sigma_k} + \frac{1}{2}\frac{\sigma_k}{\sigma_{0k}}\right)\right] \tag{2.4}$$

is less than ∞ or equal to ∞. Further, if we denote by $p_n(\mathbf{x}), p_n^0(\mathbf{x})$, $\mathbf{x} \subset R^n$, the probability density function of $X_1, \ldots, X_n$ corresponding to $\mathcal{P}$ and $\mathcal{P}_0$, respectively, then whenever $\mathcal{P} \equiv \mathcal{P}_0$, we have

$$\frac{d\mathcal{P}}{d\mathcal{P}_0}(\omega) = \lim_{n\to\infty} \frac{p_n(X_1(\omega), \ldots, X_n(\omega))}{p_n^0(X_1(\omega), \ldots, X_n(\omega))} \tag{2.5}$$

for all ω except on a set of measure zero ($\mathcal{P}_0$ or $\mathcal{P}$).

Proof: We shall only give an outline of the proof due to Kakutani. First, let ρ_n be given by

$$\begin{aligned} \rho_n &= \int_{R^n} \sqrt{p_n(\mathbf{x})p_n^0(\mathbf{x})}\, dx_1 \cdots dx_n \\ &= \exp\left\{-\sum_{k=1}^{n}\left[\frac{1}{4}\frac{(\mu_k - \mu_{0k})^2}{\sigma_k^2 + \sigma_{0k}^2} + \frac{1}{2}\ln\frac{1}{2}\left(\frac{\sigma_{0k}}{\sigma_k} + \frac{\sigma_k}{\sigma_{0k}}\right)\right]\right\} \end{aligned} \tag{2.6}$$

and define

$$\psi_n(\omega) = \left[\frac{p_n(X_1(\omega), \ldots, X_n(\omega))}{p_n^0(X_1(\omega), \ldots, X_n(\omega))}\right]^{\frac{1}{2}} \tag{2.7}$$

By direct computation we can show that if $\lim\limits_{n\to\infty} \rho_n = \rho > 0$, then

$$\sup_m E_0|\psi_{n+m} - \psi_n|^2 \xrightarrow[n\to\infty]{} 0$$

Hence there exists some ψ such that $E_0\psi^2 < \infty$ and

$$E_0|\psi_n - \psi|^2 \xrightarrow[n\to\infty]{} 0 \tag{2.8}$$

Now, let $\mathcal{B}$ denote the smallest Boolean algebra such that every X_k is measurable. Then every set in $\mathcal{B}$ involves only a finite number of X_k's so that for every set $A \in \mathcal{B}$, there exists some finite integer $n_0(A)$ such that for all $n > n_0(A)$,

$$EI_A = E_0\psi_n{}^2 I_A \xrightarrow[n\to\infty]{} E_0\psi^2 I_A$$

This means that if we define a probability measure $\tilde{\mathcal{P}}$ by

$$\tilde{\mathcal{P}}(A) = E_0\psi^2 I_A$$

then $\tilde{\mathcal{P}}$ and $\mathcal{P}$ agree on $\mathcal{B}$. But $\mathcal{B}$ generates $\mathcal{A}$, and the extension of any probability measure is unique. Hence

$$\mathcal{P}(A) = \int_A \psi^2\, d\mathcal{P}_0$$

which implies that $\mathcal{P} \ll \mathcal{P}_0$ and

$$\frac{d\mathcal{P}}{d\mathcal{P}_0} = \psi^2 \tag{2.9}$$

Reversing the roles of $\mathcal{P}$ and $\mathcal{P}_0$ proves $\mathcal{P} \equiv \mathcal{P}_0$.

Now we prove that $\rho_n \xrightarrow[n\to\infty]{} 0$ implies singularity. Let $\nu = \frac{1}{2}(\mathcal{P}_0 + \mathcal{P})$. Then ν is a probability measure with the property that $\mathcal{P}_0 \ll \nu$ and $\mathcal{P} \ll \nu$. Let E_ν denote expectation with respect to ν and let

$$\varphi_n(\omega) = \left\{\frac{p_n(X_1(\omega), \ldots, X_n(\omega))}{\frac{1}{2}[p_n{}^0(X_1(\omega), \ldots, X_n(\omega)) + p_n(X_1(\omega), \ldots, X_n(\omega))]}\right\}^{\frac{1}{2}}$$

$$\varphi_n{}^0(\omega) = \left\{\frac{p_n{}^0(\mathbf{X}(\omega))}{\frac{1}{2}[p_n{}^0(\mathbf{X}(\omega)) + p_n(\mathbf{X}(\omega))]}\right\}^{\frac{1}{2}}$$

Then, $E_\nu|\varphi_n - \sqrt{d\mathcal{P}/d\nu}|^2 \xrightarrow[n\to\infty]{} 0$ and $E_\nu|\varphi_n{}^0 - \sqrt{d\mathcal{P}_0/d\nu}|^2 \xrightarrow[n\to\infty]{} 0$. Hence, $E_\nu\varphi_n\varphi_n{}^0 = \rho_n \xrightarrow[n\to\infty]{} E_\nu \sqrt{d\mathcal{P}/d\nu\; d\mathcal{P}_0/d\nu}$. Therefore, $\rho_n \xrightarrow[n\to\infty]{} 0$ implies

$$\frac{d\mathcal{P}}{d\nu}\frac{d\mathcal{P}_0}{d\nu} = 0$$

everywhere except for a set of ν-measure zero. This implies singularity, because the set $\{\omega: d\mathcal{P}/d\nu\ (\omega) > 0\}$ has $\mathcal{P}$-measure 1, but $\mathcal{P}_0$-measure 0. ∎

Let $\{X_t, 0 \le t \le 1\}$ be a real-valued process defined on its sample space $(\Omega,\mathcal{A})$, that is, $\mathcal{A}$ is generated by $\{X_t, 0 \le t \le 1\}$. Let $\mathcal{P}_0$ and $\mathcal{P}$

be two Gaussian measures defined on $(\Omega,\mathcal{A})$ such that

$$E_0X_t = \mu_0(t) \qquad EX_t = \mu(t) \tag{2.10}$$

and

$$E_0[X_t - \mu_0(t)][X_s - \mu_0(s)] = E[X_t - \mu(t)][X_s - \mu(s)] = R(t,s) \tag{2.11}$$

In other words, the main assumption here is that the covariance function with respect to either probability measure is the same. It is clear that we can assume $\mu_0(t) \equiv 0$ with no loss of generality, because we can always consider $Y_t = X_t - \mu_0(t)$ instead of X_t. We shall also assume that $R(t,s)$ is continuous on $[0,1] \times [0,1]$. Now from Sec. 3.4 we know that the integral equation

$$\int_0^1 R(t,s)\varphi(s)\,ds = \lambda\varphi(t) \qquad 0 \le t \le 1 \tag{2.12}$$

yields a sequence of eigenvalues $\lambda_0, \lambda_1, \ldots,$ with corresponding orthonormalized (real) eigenfunctions φ_n. If the sequence $\{\lambda_n\}$ does not terminate, then $\lambda_n \xrightarrow[n\to\infty]{} 0$. The set of eigenfunctions spans a subspace S of the Hilbert space $L^2(0,1)$. Now, assume $\mu_0 \equiv 0$ and $\mu \in L^2(0,1)$. Define

$$Z_k = \int_0^1 \varphi_k(t)X_t\,dt \tag{2.13}$$

where the integral is a q.m. integral. Under either $\mathcal{P}_0$ or $\mathcal{P}$, $\{Z_k\}$ is a countable family of independent Gaussian random variables with

$$E_0Z_k = 0 \qquad EZ_k = \mu_k = \int_0^1 \mu(t)\varphi_k(t)\,dt$$
$$E_0Z_k^2 = E(Z_k - \mu_k)^2 = \lambda_k$$

Now let $\mathcal{A}'$ denote the smallest σ algebra with respect to which every Z_k is measurable. Then the restrictions of $\mathcal{P}$ and $\mathcal{P}_0$ to $\mathcal{A}'$ are either equivalent or singular according as

$$\sum_{k=0}^{\infty} \frac{\mu_k^2}{\lambda_k} < \infty \text{ or } = \infty \tag{2.14}$$

We assume that $\{X_t, 0 \le t \le 1\}$ is a separable process with respect to either $\mathcal{P}_0$ or $\mathcal{P}'$. Because R is continuous, this is no restriction. Three cases should be distinguished: (1) $\mu \in S$ and $\sum_{k=0}^{\infty} \mu_k^2/\lambda_k < \infty$, (2) $\mu \in S$ and $\sum_{k=0}^{\infty} \mu_k^2/\lambda_k = \infty$, and (3) $\mu \notin S$. We now show that $\mathcal{P} \equiv \mathcal{P}_0$ for the first case, and $\mathcal{P} \perp \mathcal{P}_0$ for the other two cases. First, suppose that $\mu \in S$.

Then, for every $t \in [0,1]$,

$$E\left[X_t - \sum_{k=0}^{\infty} Z_k\varphi_k(t)\right]^2 = 0 = E_0\left[X_t - \sum_{k=0}^{\infty} Z_k\varphi_k(t)\right]^2$$

This means that every X_t differs from an $\mathcal{A}'$-measurable function by a set of measure zero ($\mathcal{P}_0$ and $\mathcal{P}$). Because of separability, this implies that the completion of $\mathcal{A}'$ with respect to $\mathcal{P}_0$ and $\mathcal{P}$ both contain $\mathcal{A}$. Suppose

$$\sum_{k=0}^{\infty} \frac{\mu_k^2}{\lambda_k} < \infty$$

then $\mathcal{P}_0 \equiv \mathcal{P}$ on $\mathcal{A}'$, and the completion of $\mathcal{A}'$ with respect to $\mathcal{P}_0$ is the same as that with respect to $\mathcal{P}$. The extensions of $\mathcal{P}_0$ and $\mathcal{P}$ to the completion of $\mathcal{A}'$ are equivalent. Hence, $\mathcal{P}_0 \equiv \mathcal{P}$ on $\mathcal{A}$. Next suppose

$$\sum_{k=0}^{\infty} \frac{\mu_k^2}{\lambda_k} = \infty$$

Then, $\mathcal{P}_0 \perp \mathcal{P}$ on $\mathcal{A}'$ which is sufficient to imply that $\mathcal{P}_0 \perp \mathcal{P}$ on $\mathcal{A}$, because any set in $\mathcal{A}'$ which distinguishes $\mathcal{P}_0$ from $\mathcal{P}$ also belongs to $\mathcal{A}$. Finally, suppose $\mu \notin S$, then

$$\int_0^1 \mu^2(t)\, dt - \sum_{k=0}^{\infty} \mu_k^2 = K > 0$$

Now,

$$E_0 \int_0^1 \left[X_t - \sum_{k=0}^{\infty} Z_k\varphi_k(t)\right]^2 dt = 0$$

and

$$E \int_0^1 \left[X_t - \mu(t) - \sum_{k=0}^{\infty} (Z_k - \mu_k)\varphi_k(t)\right]^2 dt = 0$$

It follows that

$$\mathcal{P}_0\left(\int_0^1 \left[X_t - \sum_{k=0}^{\infty} Z_k\varphi_k(t)\right]^2 dt = 0\right)$$

$$= \mathcal{P}\left(\int_0^1 \left[X_t - \sum_{k=0}^{\infty} Z_k\varphi_k(t)\right]^2 dt = K > 0\right) = 1$$

hence $\mathcal{P} \perp \mathcal{P}_0$. Summarizing those results, we can state the following proposition [Grenander, 1950].

Proposition 2.2. Let $\{X_t, 0 \leq t \leq 1\}$ be a separable Gaussian process under either of the probability measures $\mathcal{P}_0$ and $\mathcal{P}$. Let $R(t,s)$ be the covariance function under either measure, and let $E_0X_t = 0$, $EX_t = \mu(t)$. Suppose that R is continuous on $[0,1] \times [0,1]$ and $\mu \in L^2(0,1)$. Let S be the subspace of $L^2(0,1)$ spanned by the eigenfunctions of R. Then $\mathcal{P}_0 \equiv \mathcal{P}$ if $\mu \in S$ and

$$\sum_{k=0}^{\infty} \frac{\mu_k{}^2}{\lambda_k} < \infty$$

otherwise $\mathcal{P}_0 \perp \mathcal{P}$.

We note that when $\mathcal{P}_0 \equiv \mathcal{P}$, the Radon-Nikodym derivative is given by

$$\frac{d\mathcal{P}}{d\mathcal{P}_0} = \exp\left(\sum_{k=0}^{\infty} \frac{\mu_k Z_k}{\lambda_k} - \frac{1}{2}\sum_{k=0}^{\infty} \frac{\mu_k{}^2}{\lambda_k}\right) \tag{2.15}$$

which appears to be simple, but is not easily computed since it involves the solution of an integral equation. This, unfortunately, is the case for most Radon-Nikodym derivatives. An important exception is the class of probability measures absolutely continuous with respect to the Wiener measure. Once again we have a strong motivation for using white noise as a model.

Proposition 2.2 is not as general as possible. In particular, the condition $\mu \in L^2(0,1)$ can be dropped, and the conclusion of the theorem would remain the same. Also, the condition for equivalence and the formula for the Radon-Nikodym derivative can both be restated more simply in terms of reproducing kernel Hilbert space [see Parzen, 1961]. We shall not do so here, but proceed to the case of unequal covariances instead.

The general dichotomy theorem on Gaussian measures is due independently to Feldman [1958] and Hajek [1958]. Their proofs and conditions for equivalence are quite different. We shall state the version due to Hajek. Let $\{X_t, t \in T\}$ be a Gaussian process with respect to either $\mathcal{P}_0$ or $\mathcal{P}$. Let T_n be an arbitrary finite subset of T and consider the finite collection of random variables $\{X_t, t \in T_n\}$. If R_0 and R are the covariance functions of $\{X_t, t \in T\}$ with respect to $\mathcal{P}_0$ and $\mathcal{P}$, then $\{R_0(t_i,t_j), t_i,t_j \in T_n\}\{R(t_i,t_j), t_i,t_j \in T_n\}$ are two nonnegative definite matrices. By a simultaneous diagonalization of these two matrices, we can find a set of random variables $Y_1, \ldots, Y_m$, $m \leq \dim(T_n)$, such that $\{Y_i\}$ are independent and Gaussian under either $\mathcal{P}_0$ or $\mathcal{P}$, and every X_t, $t \in T_n$ is a linear combination of the Y_i's. By an application of Proposition 2.1, the restrictions of $\mathcal{P}_0$ and $\mathcal{P}$ to $\mathcal{A}_{T_n}$, the σ algebra generated by $\{X_t, t \in T_n\}$, are either equivalent or singular. If these restrictions are singular for some T_n, then clearly $\mathcal{P}_0$ and $\mathcal{P}$ must be singular. Denote the restrictions of $\mathcal{P}_0$ and $\mathcal{P}$ to $\mathcal{A}_{T_n}$ by $\mathcal{P}_0^{T_n}$ and $\mathcal{P}^{T_n}$, and define

$$J_{T_n} = E \ln \frac{d\mathcal{P}^{T_n}}{d\mathcal{P}_0^{T_n}} - E_0 \ln \frac{d\mathcal{P}^{T_n}}{d\mathcal{P}_0^{T_n}} \tag{2.16}$$

We can now state Hajek's dichotomy theorem as follows.

Proposition 2.3. Let $\mathcal{P}_0$ and $\mathcal{P}$ be two Gaussian measures defined on the sample space of $\{X_t, t \in T\}$. Then either $\mathcal{P}_0 \equiv \mathcal{P}$ or $\mathcal{P}_0 \perp \mathcal{P}$. $\mathcal{P}_0 \equiv \mathcal{P}$ if and only if $\mathcal{P}_0^{T_n} \equiv \mathcal{P}^{T_n}$ for every finite $T_n \subset T$ and

$$J = \sup_{T_n \subset T} J_{T_n} < \infty \tag{2.17}$$

Remark: The quantity J appearing in (2.17) can also be shown to be given by

$$J = E \ln \frac{d\mathcal{P}}{d\mathcal{P}_0} - E_0 \ln \frac{d\mathcal{P}}{d\mathcal{P}_0} \tag{2.18}$$

whenever $\mathcal{P}_0 \equiv \mathcal{P}$.

We observe that Hajek's theorem (2.17) is easy to state, but difficult to verify. Feldman's condition is no better in this respect. In special cases much simpler conditions have been obtained. For example, suppose that $\{X_t, t \in T\}$ is a segment of a stationary Gaussian process with respect to either $\mathcal{P}_0$ or $\mathcal{P}$. Further, suppose that

$$E_0 X_t X_s = \int_{-\infty}^{\infty} \frac{A_0(\nu^2)}{B_0(\nu^2)} e^{i2\pi\nu(t-s)} \, d\nu \tag{2.19}$$

$$E X_t X_s = \int_{-\infty}^{\infty} \frac{A(\nu^2)}{B(\nu^2)} e^{i2\pi\nu(t-s)} \, d\nu \tag{2.20}$$

where A_0, B_0, A, B are polynomials. Then, it can be shown [see Feldman, 1960] that $\mathcal{P}_0 \equiv \mathcal{P}$ if and only if

$$\lim_{\nu \to \infty} \frac{A_0(\nu^2) B(\nu^2)}{A(\nu^2) B_0(\nu^2)} = 1 \tag{2.21}$$

The necessity of this condition can be inferred from a result known as Baxter's theorem which will be taken up in the next section. Another important special case is the class of Gaussian measures equivalent to the Wiener measure. For that case, all questions related to differentiability have been answered with great clarity by Shepp [1966]. More generally, measures absolutely continuous with respect to the Wiener measure constitute an important class of measures and will be examined in some detail in Sec. 4.

3. QUADRATIC VARIATION AND SUFFICIENT CONDITIONS FOR SINGULARITY

We begin by recalling from Proposition 2.3.4 that if $\{X_t, 0 \leq t \leq 1\}$ is a Brownian motion, then

$$\sum_{k=0}^{2^n-1} (X_{k+1/2^n} - X_{k/2^n})^2 \xrightarrow[n\to\infty]{\text{a.s.}} 1 \tag{3.1}$$

Therefore, if $\{X_t, 0 \leq t \leq 1\}$ is a Brownian motion with respect to $\mathcal{P}_0$ and a Gaussian process with respect to $\mathcal{P}$, then a necessary condition for $\mathcal{P} \equiv \mathcal{P}_0$ is

$$\sum_{k=0}^{2^n-1} (X_{k+1/2^n} - X_{k/2^n})^2 \xrightarrow[n\to\infty]{} 1 \tag{3.2}$$

with $\mathcal{P}$-measure 1. Obviously, negating this condition gives a sufficient condition for singularity.

A generalization of (3.1) is Baxter's theorem which can be stated as follows [Baxter, 1956].

Proposition 3.1. Let $\{X_t, t \in [0,1]\}$ be a Gaussian process with mean $\mu(t)$ and covariance function $R(t,s)$. Suppose that μ is of bounded variation, and $R(t,s)$ has a bounded second derivative $\partial^2 R(t,s)/\partial t\, \partial s$ everywhere on $[0,1] \times [0,1]$ except along the diagonal $t = s$. Let

$$\sigma^2(t) = \lim_{s\uparrow t^-} \frac{R(t,t) - R(s,t)}{t-s} - \lim_{s\downarrow t^+} \frac{R(t,t) - R(s,t)}{t-s} \tag{3.3}$$

Then,

$$\sum_{k=0}^{2^n-1} (X_{k+1/2^n} - X_{k/2^n})^2 \xrightarrow[n\to\infty]{\text{a.s.}} \int_0^1 \sigma^2(t)\, dt \tag{3.4}$$

It is clear how Proposition 3.1 can be used to give a sufficient condition for singularity [Slepian, 1958]. Suppose that $\{X_t, t \in [0,1]\}$ satisfies the hypothesis of Proposition 3.1 with respect to either $\mathcal{P}_0$ or $\mathcal{P}$, then $\mathcal{P}_0 \perp \mathcal{P}$, if on a set of t's with nonzero Lebesgue measure,

$$\lim_{s\uparrow t^-} \frac{R_0(t,t) - R_0(s,t)}{t-s} - \lim_{s\downarrow t^+} \frac{R_0(t,t) - R_0(s,t)}{t-s} \neq \lim_{s\uparrow t^-} \frac{R(t,t) - R(s,t)}{t-s} - \lim_{s\downarrow t^+} \frac{R(t,t) - R(s,t)}{t-s} \tag{3.5}$$

The necessity of condition (2.21) for the equivalence of a pair of stationary Gaussian measures can be deduced from Baxter's theorem by the following consideration:

$$R(t,s) = \int_{-\infty}^{\infty} \frac{A(\nu^2)}{B(\nu^2)} e^{i2\pi\nu(t-s)}\, d\nu \tag{3.6}$$

then

$$\sigma^2(t) = \lim_{\nu \to \infty} \frac{(2\pi\nu)^2 A(\nu^2)}{B(\nu^2)}$$

If A and B are rational, and the highest powers of ν^2 in A and B differ by $m + 1$, then X_t is m-times differentiable and

$$\sum_{k=0}^{2^n-1} [X^{(m)}_{k+1/2^n} - X^{(m)}_{k/2^n}]^2 = \lim_{\nu \to \infty} (2\pi\nu)^{2(m+1)} \frac{A(\nu^2)}{B(\nu^2)}$$

It follows that if $A_0(\nu^2)/B_0(\nu^2)$ and $A(\nu^2)/B(\nu^2)$ are the spectral-density functions corresponding to $\mathcal{P}_0$ and $\mathcal{P}$, then in order for $\mathcal{P}_0 \equiv \mathcal{P}$, it is necessary that

$$\lim_{\nu \to \infty} \frac{A_0(\nu^2)}{B_0(\nu^2)} \frac{B(\nu^2)}{A(\nu^2)} = 1$$

which is condition (2.21).

The basic idea of Baxter's theorem can be generalized in an important way to non-Gaussian processes. Suppose that we can write

$$X_t = X_0 + \int_0^t g_s\, ds + \int_0^t f_s\, dW_s \qquad 0 \le t \le 1 \tag{3.7}$$

where $\{g_t, 0 \le t \le 1\}$ and $\{f_t, 0 \le t \le 1\}$ are stochastic processes, $\{W_t, 0 \le t \le 1\}$ is a Brownian motion, and the last integral in (3.7) is a stochastic integral. Heuristically, if we write

$$(dX_t)^2 = (g_t\, dt + f_t\, dW_t)^2 = g_t^2(dt)^2 + (g_t f_t)\, dt\, dW_t + f_t^2(dW_t)^2$$

and note that $|dW_t| \sim \sqrt{dt}$, then we have

$$(dX_t)^2 = f_t^2\, dt + o(dt)$$

Therefore, we expect

$$\sum_{k=0}^{2^n-1} (X_{k+1/2^n} - X_{k/2^n})^2 \xrightarrow[n\to\infty]{\text{a.s.}} \int_0^1 f_t^2\, dt$$

The precise result can be stated as follows [see Wong and Zakai, 1965c].

Proposition 3.2. Let $\{W_t, \mathcal{A}_t, 0 \le t \le 1\}$ be a Brownian motion. Let $\{f_t, 0 \le t \le 1\}$ $\{g_t, 0 \le t \le 1\}$ be separable and measurable stochastic processes satisfying the following conditions:

(a) For each t, f_t is $\mathcal{A}_t$ measurable

(b) $E \int_0^1 (f_t^4 + g_t^4)\, dt < \infty$

Suppose that $\{X_t,\ 0 \le t \le 1\}$ is a process given by

$$X_t = X_0 + \int_0^t g_s\, ds + \int_0^t f_s\, dW_s \tag{3.8}$$

Then

$$\lim_{n\to\infty} \sum_{k=0}^{2^n-1} (X_{k+1/2^n} - X_{k/2^n})^2 = \int_0^1 f_t^2\, dt \qquad \text{a.s.} \tag{3.9}$$

Remark: The left-hand side of (3.9) is called the **quadratic variation** of the X process on [0,1]. It is clear that, more generally, the quadratic variation of the X process on $[0,t]$ is given by $\int_0^t f_s^2\, ds$. One interesting consequence is that for every t, $\int_0^t f_s^2\, ds$ is necessarily measurable with respect to the σ algebra generated by $\{X_s,\ 0 \le s \le t\}$.

Proposition 3.2 can be generalized to the case where X_t is the sum of a process of bounded variation and a martingale [see Fisk, 1966]. We shall not consider this generalization here. Proposition 3.2 can be used to supply a sufficient condition for singularity between two probability measures.

Proposition 3.3. Let $\mathcal{P}_0$ and $\mathcal{P}$ be two probability measures defined on the sample space of a process $\{X_t,\ 0 \le t \le 1\}$. Let $\mathcal{A}_t$ denote the σ algebra generated by $\{X_s,\ 0 \le s \le t\}$. Suppose that there exist processes $g_{0t}, f_{0t}, g_t, f_t, V_t, W_t, 0 \le t \le 1$ such that

(a) $\left.\begin{matrix}\{V_t, \mathcal{A}_t, 0 \le t \le 1\} \\ \{W_t, \mathcal{A}_t, 0 \le t \le 1\}\end{matrix}\right\}$ is a Brownian motion with respect to $\left\{\begin{matrix}\mathcal{P}_0 \\ \mathcal{P}\end{matrix}\right.$

(b) $\left.\begin{matrix}\{g_{0t}, f_{0t}, 0 \le t \le 1\} \\ \{g_t, f_t, 0 \le t \le 1\}\end{matrix}\right\}$ satisfy the hypotheses of Proposition 3.2 with respect to $\left\{\begin{matrix}\mathcal{P}_0 \\ \mathcal{P}\end{matrix}\right.$

(c) $X_t = \begin{cases} X_0 + \int_0^t g_{0s}\, ds + \int_0^t f_{0s}\, dV_s & \text{a.s. } \mathcal{P}_0 \\ X_0 + \int_0^t f_s\, ds + \int_0^t f_s\, dW_s & \text{a.s. } \mathcal{P} \end{cases}$

Then $\mathcal{P}_0 \perp \mathcal{P}$, if there exists a fixed $t \in [0,1]$ such that for all ω except a set of $\mathcal{P}_0 + \mathcal{P}$ measure zero,

$$\int_0^t f_{0s}^2(\omega)\, ds \ne \int_0^t f_s^2(\omega)\, ds$$

We observe that Proposition 3.3 includes the previous singularity condition for two Gaussian measures which was based on Baxter's theorem. Another useful special case is that of diffusion processes. Suppose that $\{X_t, 0 \leq t \leq 1\}$ is a sample-continuous Markov process with respect to either $\mathcal{P}_0$ or $\mathcal{P}$ with

$$\frac{1}{s} E_0^{\mathcal{a}_t}(X_{t+s} - X_t)^2 \xrightarrow[s \downarrow 0]{} \sigma_0^2(X_t,t)$$

$$\frac{1}{s} E^{\mathcal{a}_t}(X_{t+s} - X_t)^2 \xrightarrow[s \downarrow 0]{} \sigma^2(X_t,t)$$

Suppose that there exists a $t \subset [0,1]$ such that for all continuous functions $x(s)$, $0 \leq s \leq 1$,

$$\int_0^t \sigma^2(x(s), s)\, ds \neq \int_0^t \sigma_0^2(x(s), s)\, ds$$

Then, $\mathcal{P}_0 \perp \mathcal{P}$.

4. ABSOLUTE CONTINUITY WITH RESPECT TO WIENER MEASURE

A probability measure $\mathcal{P}$ defined on the sample space of a process $\{X_t, 0 \leq t \leq 1\}$ is called a **Wiener measure** if $\{X_t, 0 \leq t \leq 1\}$ is a separable Brownian motion with respect to $\mathcal{P}$. As usual, there is no loss in generality if we restrict ourselves to the standard Brownian motion ($EX_t^2 = t$). Let $\mathcal{P}_0$ be a Wiener measure defined on $(\Omega,\mathcal{a})$. In this section we shall study probability measures which are defined on $(\Omega,\mathcal{a})$ and absolutely continuous with respect to $\mathcal{P}_0$. There are two important results in this direction. First, we shall state a necessary and sufficient condition for a *Gaussian* measure to be absolutely continuous with respect to a Wiener measure. This condition is due to Shepp [1966]. It is clear that because of the dichotomy theorem this condition is also necessary and sufficient for equivalence. Secondly, we shall derive the Radon-Nikodym derivative of any probability measure that is absolutely continuous with respect to the Wiener measure.

Proposition 4.1. Let $\mathcal{P}$ be a Gaussian measure and $\mathcal{P}_0$ be a Wiener measure defined on the sample space of $\{X_t, 0 \leq t \leq 1\}$. Let $m(t) = EX_t$ and

$$R(t,s) = EX_tX_s - m(t)m(s) \tag{4.1}$$

Then $\mathcal{P} \equiv \mathcal{P}_0$ if and only if the following conditions are satisfied: (*a*) There exists a $K(t,s)$, t, $s \subset [0,1]$, such that

$$R(t,s) = \min(t,s) - \int_0^t \int_0^s K(u,v)\, du\, dv \tag{4.2}$$

$$\iint_0^1 K^2(t,s)\, dt\, ds < \infty \tag{4.3}$$

$$\iint_0^1 K(t,s)\varphi(t)\varphi(s)\, dt\, ds < \int_0^1 |\varphi(t)|^2\, dt$$

$$\text{for every nonzero } \varphi \in L^2[0,1] \tag{4.4}$$

(*b*) There exist $k \in L^2[0,1]$ such that

$$m(t) = \int_0^t k(s)\, ds \tag{4.5}$$

Remarks:

(*a*) The function K is given by $K(t,s) = -(\partial^2/\partial t\, \partial s)R(t,s)$ for almost all $(t,s) \in [0,1]^2$. If $(\partial/\partial t)R(t,s)$ is continuous everywhere except along the diagonal $t = s$, then (4.2) becomes, upon differentiation,

$$\left.\frac{\partial R(t,s)}{\partial t}\right|_{s=t^+} = 1 - \int_0^t K(t,v)\, dv$$

$$\left.\frac{\partial R(t,s)}{\partial t}\right|_{s=t^-} = 0 - \int_0^t K(t,v)\, dv$$

so that

$$\left.\frac{\partial R(t,s)}{\partial t}\right|_{s=t^+} - \left.\frac{\partial R(t,s)}{\partial t}\right|_{s=t^-} = 1 \tag{4.6}$$

which was stated in Proposition 3.1 as a necessary condition for absolute continuity with respect to the Wiener measure.

(*b*) Condition (4.4) is often expressed in the form

$$\sigma(K) \not\ni 1 \tag{4.7}$$

where $\sigma(K)$ is the spectrum of K defined by

$$\sigma(K) = \left\{\lambda: \int_0^1 K(t,s)\varphi(s)\, ds = \lambda\varphi(t),\ 0 \le t \le 1,\ \varphi \not\equiv 0,\ \varphi \in L^2\right\} \tag{4.8}$$

Conditions (4.4) and (4.7) can be shown to be equivalent.

Shepp has also obtained a formula for the Radon-Nikodym derivative of any Gaussian measure which is absolutely continuous with respect to the Wiener measure. We shall derive an equivalent formula in a more general context. The basic result here is due to Kunita and Watanabe [1967], and Duncan [1970]. First we give a heuristic derivation for this result. Let $\mathcal{P}_0$ be a Wiener measure defined on the sample space of a process $\{X_t, 0 \le t \le 1\}$. Let $\mathcal{P}$ be a probability measure defined on the same space and absolutely continuous with respect to $\mathcal{P}_0$. We shall derive a general expression for the Radon-Nikodym derivative $d\mathcal{P}/d\mathcal{P}_0$. Indeed, we shall derive an expression for $E^{\mathcal{A}_t}\, d\mathcal{P}/d\mathcal{P}_0$ where $\mathcal{A}_t$ denotes the smallest σ algebra with respect to which every X_s, $s \le t$, is measurable.

We begin by observing that $d\mathcal{P}/d\mathcal{P}_0 \ge 0$, and the set on which $d\mathcal{P}/d\mathcal{P}_0 = 0$ is of $\mathcal{P}$-measure zero, but possibly has positive $\mathcal{P}_0$ measure.

Let Λ denote the set

$$\Lambda = \left\{\omega: \frac{d\mathcal{P}}{d\mathcal{P}_0}(\omega) = 0\right\} \tag{4.9}$$

We shall now characterize $d\mathcal{P}/d\mathcal{P}_0$ on $\Omega - \Lambda$. Let $\{\mathcal{A}_t, 0 \leq t \leq 1\}$ be defined as before, and let

$$L_t = E_0^{\mathcal{A}_t} \frac{d\mathcal{P}}{d\mathcal{P}_0} \tag{4.10}$$

The process $\{L_t, \mathcal{A}_t, 0 \leq t \leq 1\}$ is a martingale with respect to $\mathcal{P}_0$. If we define

$$Z_t = \ln L_t \tag{4.11}$$

then $\{Z_t, \mathcal{A}_t, 0 \leq t \leq 1\}$ is a supermartingale ($\mathcal{P}_0$), that is,

$$E_0^{\mathcal{A}_t} Z_{t+s} \leq Z_t \tag{4.12}$$

We can now invoke Meyer's decomposition theorem for a class of supermartingales [Meyer, 1966, Chap. 6] and write

$$Z_t = Y_t - A_t \tag{4.13}$$

where $\{Y_t, \mathcal{A}_t, 0 \leq t \leq 1\}$ is again a martingale $\{\mathcal{P}_0\}$, and $\{A_t, 0 \leq t \leq 1\}$ is almost surely nondecreasing (cf. Sec. 4.6).

Now, because $\{\mathcal{A}_t, 0 \leq t \leq 1\}$ is generated by a Brownian motion ($\mathcal{P}_0$) $\{X_t, 0 \leq t \leq 1\}$, it follows from a result of Kunita and Watanabe [1967] that every second-order martingale with respect to $\{\mathcal{A}_t, 0 \leq t \leq 1\}$ is a stochastic integral with respect to the Brownian motion $\{X_t, 0 \leq t \leq 1\}$. Although $\{Y_t, 0 \leq t \leq 1\}$ in (4.13) need not be second order, it turns out that by using an argument involving stopping time, we can extend the result of Kunita and Watanabe and show that Y_t must be of the form

$$Y_t = \int_0^t \varphi_s \, dX_s \tag{4.14}$$

where $\int_0^1 \varphi_s^2 \, ds < \infty$ on the set $\Omega - \Lambda$. Now, using (4.11), (4.13), and (4.14), we get

$$L_t = e^{-A_t} e^{Y_t} = e^{-A_t} \exp\left(\int_0^t \varphi_s \, dX_s\right) \tag{4.15}$$

Applying the Ito differentiation formula to $\exp\left(\int_0^t \varphi_s \, dX_s\right)$, we get

$$dL_t = L_t(\varphi_t \, dX_t - dA_t + \tfrac{1}{2}\varphi_t^2 \, dt)$$

That is,

$$L_t - 1 - \int_0^t L_s \varphi_s \, dX_s = \frac{1}{2}\int_0^t L_s \varphi_s^2 \, ds - \int_0^t L_s \, dA_s \tag{4.16}$$

The left-hand side of (4.16) is a martingale and the right-hand side is of bounded variation and both are zero at $t = 0$. It follows that both sides of (4.16) are identically zero, so that $L_t\, dA_t/dt$ (as a Radon-Nikodym derivative) equals $\frac{1}{2}L_t\varphi_t^2$, and on the set $\{\omega\colon d\mathcal{P}/d\mathcal{P}_0\,(\omega) \neq 0\}$, we have

$$A_t = \frac{1}{2}\int_0^t \varphi_s^2\, ds \tag{4.17}$$

Now we can summarize these results in the following proposition.

Proposition 4.2. Let $\mathcal{P}_0$ be a Wiener measure defined on the sample space of $\{X_t,\ 0 \leq t \leq 1\}$. Let $\mathcal{P}$ be a probability measure absolutely continuous with respect to $\mathcal{P}_0$. Then on the set $\{\omega\colon d\mathcal{P}/d\mathcal{P}_0\,(\omega) \neq 0\}$, we can write

$$L_t = E^{\mathcal{a}_t}\frac{d\mathcal{P}}{d\mathcal{P}_0} = \exp\left[\int_0^t (\varphi_s\, dX_s - \tfrac{1}{2}\varphi_s^2\, ds)\right] \tag{4.18}$$

and

$$\frac{d\mathcal{P}}{d\mathcal{P}_0} = \exp\left(\int_0^1 \varphi_s\, dX_s - \frac{1}{2}\int_0^1 \varphi_s^2\, ds\right) \tag{4.19}$$

where $\{\varphi_t,\ 0 \leq t \leq 1\}$ satisfies:

(*a*) For each t, φ_t is $\mathcal{a}_t$ measurable

(*b*) $\int_0^1 \varphi_t^2\, dt < \infty$ a.s. $\mathcal{P}_0$

Equation (4.19) is an exceedingly general formula for the Radon-Nikodym derivative of any probability measure with respect to the Wiener measure. The expression is valid on the set $\{\omega\colon d\mathcal{P}/d\mathcal{P}_0\,(\omega) > 0\}$. Equivalently, we can say that (4.19) holds with $\mathcal{P}$-measure 1. If $\mathcal{P}$ is Gaussian, then the dichotomy theorem implies that (4.19) is also valid with $\mathcal{P}_0$-measure 1.

If $\mathcal{P}$ is Gaussian and zero mean, then φ_s in (4.19) is linear in $\{X_\tau,\ 0 \leq \tau \leq s\}$. By this we mean that for each s, there exists a sequence $\{t_{n\nu}\}$ in $[0,s]$ and a sequence of real numbers $\{a_{n\nu}\}$ such that

$$\varphi_s = \lim_{n\to\infty} \text{in q.m.} \sum_{\nu=1}^{n} a_{n\nu} X_{t_{n\nu}} \tag{4.20}$$

It follows that we can write

$$\varphi_s = \lim_{n\to\infty} \text{in q.m.} \int_0^s h_n(s,\tau)\, dX_s \tag{4.21}$$

where for each s, $\{h_n(s,\cdot)\}$ is a sequence of step functions and the integral on the right-hand side is a Wiener integral (a stochastic integral with a nonrandom integrand). Because every q.m. convergent sequence also converges *mutually* in q.m., we have

$$E\left|\int_0^s h_n(s,\tau)\,dX_s - \int_0^s h_m(s,\tau)\,dX_s\right|^2 = \int_0^s |h_n(s,\tau) - h_m(s,\tau)|^2\,ds \xrightarrow[m,n\to\infty]{} 0$$

From the completeness of $L^2(0,s)$, we conclude that φ_s must have the form

$$\varphi_s = \int_0^s h(s,\tau)\,dX_\tau \tag{4.22}$$

where $h(s,\cdot) \in L^2(0,s)$. Hence, in the Gaussian case, (4.19) is completely specified by the nonrandom function $\{h(s,\tau),\ 0 \le \tau \le s,\ 0 \le s \le 1\}$.[1]

The process $\{\varphi_t,\ 0 \le t \le 1\}$ occurring in (4.19) can be expressed in terms of the process $\{X_t,\ 0 \le t \le 1\}$ and the probability measure $\mathcal{P}$ in an interesting way. To do this involves a theorem due to Girsanov [1960]. We state Girsanov's theorem in a restricted form suitable for our use as follows.

Proposition 4.3. Let $\{X_t,\ 0 \le t \le 1\}$ be a (standard) Brownian motion defined on a probability space $\{\Omega,\mathcal{A},\mathcal{P}_0\}$. Let $\{\mathcal{A}_t,\ 0 \le t \le 1\}$ be a nondecreasing family of sub-σ algebras such that for each t, X_t is $\mathcal{A}_t$ measurable, and the aggregate $\{X_s - X_t,\ s \ge t\}$ is independent of $\mathcal{A}_t$. Let $\{\varphi_t,\ 0 \le t \le 1\}$ be a separable and measurable process satisfying the following conditions:

$$\begin{aligned}
&(a)\ \text{For each } t,\ \varphi_t \text{ is } \mathcal{A}_t \text{ measurable}\\
&(b)\ \int_0^1 \varphi_t^2\,dt < \infty \text{ with } \mathcal{P}_0\text{-measure } 1\\
&(c)\ E_0 \exp\left(\int_0^1 \varphi_s\,dX_s - \frac{1}{2}\int_0^1 \varphi_s^2\,ds\right) = 1
\end{aligned} \tag{4.23}$$

If we define a probability measure $\mathcal{P}$ by

$$\frac{d\mathcal{P}}{d\mathcal{P}_0} = \exp\left(\int_0^1 \varphi_s\,dX_s - \frac{1}{2}\int_0^1 \varphi_s^2\,ds\right) \tag{4.24}$$

then with respect to the measure $\mathcal{P}$, the process

$$\left\{X_t - \int_0^t \varphi_s\,ds,\ 0 \le t \le 1\right\}$$

is a Brownian motion.

[1] If $\mathcal{P}$ is Gaussian with a nonzero mean $EX_t = \int_0^t k(s)\,ds$, then it can be shown that (4.22) is modified to read $\varphi_s = \int_0^s h(s,\tau)\,dX_\tau + k(s)$.

Remarks: Girsanov also showed that if φ_t is bounded, then (4.23) is always satisfied. If $\varphi_t = f(X_t,t)$ and the function f satisfies a uniform Lipschitz condition in the first variable, then (4.23) is also satisfied automatically.

Let $\mathcal{P}_0$ and $\mathcal{P}$ be as in Proposition 4.2. That is, $\mathcal{P}_0$ is a Wiener measure defined on the sample space of $\{X_t, 0 \le t \le 1\}$ and $\mathcal{P} \ll \mathcal{P}_0$. If we combine Proposition 4.2 and 4.3 we conclude that *with respect to* $\mathcal{P}$, the process

$$Z_t = X_t - \int_0^t \varphi_s \, ds \tag{4.25}$$

must be a Brownian motion. We can now identify the term $\int_0^t \varphi_s \, ds = Y_t$ as the bounded-variation term in the unique decomposition of $\{X_t, 0 \le t \le 1\}$ into the sum of a martingale (Z_t in this case) and a process of bounded variation. This decomposition is with respect to $\mathcal{P}$ and the family $\{\mathcal{A}_t, 0 \le t \le 1\}$. Loosely speaking, we have

$$E^{\mathcal{A}_t} \, dX_t = \varphi_t \, dt \tag{4.26}$$

This completes the derivation of the Radon-Nikodym derivative $d\mathcal{P}/d\mathcal{P}_0$. Equation (4.26) suffices to yield an equation which uniquely determines the function h in (4.22) (see Exercise 4).

We can now apply these results to the detection problem involving additive Gaussian white noise [see, e.g., Duncan, 1968; Kailath, 1969]. First, we shall state the problem a little loosely as follows: Let $\{S_t, 0 \le t \le T\}$ be a stochastic process which we can identify as the signal. Let $\{\eta_t, 0 \le t \le T\}$ be a process that we identify as the observation. Given the observation, we have to decide between the following two hypotheses:

$$\begin{aligned} H_1&: \quad \eta_t = \text{Gaussian white noise} + S_t \\ H_0&: \quad \eta_t = \text{Gaussian white noise alone} \end{aligned} \tag{4.27}$$

We can avoid dealing with white noise and deal with Brownian motion instead, if we consider

$$X_t = \int_0^t \eta_s \, ds \qquad 0 \le t \le T \tag{4.28}$$

to be the observation. It is clear that T can be set equal to 1 with no loss of generality, and we shall do so. We now state the problem precisely as follows: Let $\{X_t, 0 \le t \le 1\}$ and $\{S_t, 0 \le t \le 1\}$ be two families of random variables defined on the same measurable space $(\Omega,\mathcal{A})$. Let $\mathcal{A}_{xt}$ be the sub-σ algebra generated by $\{X_s, 0 \le s \le t\}$, and we write $\mathcal{A}_x$ instead of $\mathcal{A}_{x1}$. Suppose that $\mathcal{P}$ and $\mathcal{P}_0$ are two probability measures such that

$$\begin{aligned} H_0&: \quad \text{Under } \mathcal{P}_0, \{X_t, 0 \le t \le 1\} \\ &\qquad\qquad \text{is a Brownian motion process} \\ H_1&: \quad \text{Under } \mathcal{P}, Y_t = X_t - \int_0^t S_\tau \, d\tau, 0 \le t \le 1, \\ &\qquad\qquad \text{is a Brownian motion} \end{aligned} \tag{4.29}$$

We want to decide between H_0 and H_1, given $\{X_t, 0 \le t \le 1\}$.

Statement (4.29) is a more precise rephrasing of (4.27). It can be shown that the Neyman-Pearson lemma can be generalized to this case so that for a fixed probability of error of one type, a test which minimizes the error of the second type is the **likelihood ratio test** given as follows: Denote the restrictions of $\mathcal{P}$ and $\mathcal{P}_0$ to $\mathcal{A}_x$ by $\mathcal{P}^x$ and $\mathcal{P}_0{}^x$ and assume $\mathcal{P}^x \ll \mathcal{P}_0{}^x$. Then the Radon-Nikodym derivative $d\mathcal{P}^x/d\mathcal{P}_0{}^x$ is a functional of the sample functions of $\{X_t, 0 \le t \le 1\}$. In other words, if we know $\{X_t(\omega), 0 < t < 1\}$, then we know $d\mathcal{P}^x/d\mathcal{P}_0{}^x\ (\omega)$. The likelihood ratio test is defined by

$$\frac{d\mathcal{P}^x}{d\mathcal{P}_0{}^x}(\omega)\begin{cases} >r & \text{accept } H_1 \\ \le r & \text{accept } H_0 \end{cases} \tag{4.30}$$

where the threshold r is chosen so that the fixed type-1 error is achieved. Therefore, implementation of the likelihood ratio test requires the knowledge of $d\mathcal{P}^x/d\mathcal{P}_0{}^x$ as a functional of the sample functions of $\{X_t, 0 \le t \le 1\}$.

From Proposition 4.2, we know that we can write

$$\frac{d\mathcal{P}^x}{d\mathcal{P}_0{}^x} = \exp\left(\int_0^1 \varphi_s\, dX_s - \frac{1}{2}\int_0^1 \varphi_s{}^2\, ds\right) \tag{4.31}$$

and from Proposition 4.3 we know that $\{\varphi_t, 0 \le t \le 1\}$ can be found by a decomposition of $\{X_t, 0 \le t \le 1\}$ with respect to $\mathcal{P}^x$ and $\{\mathcal{A}_{xt}, 0 \le t \le 1\}$. Loosely speaking, we can relate $\{\varphi_t, 0 \le t \le 1\}$ to $\{X_t, 0 \le t \le 1\}$ by the expression

$$E^{\mathcal{A}_{xt}}\, dX_t = \varphi_t\, dt \tag{4.32}$$

as was done in (4.26). If we accept (4.32), then the second statement of (4.29) gives us

$$E^{\mathcal{A}_{xt}}\, dX_t - (E^{\mathcal{A}_{xt}}S_t)\, dt = E^{\mathcal{A}_{xt}}\, dY_t \tag{4.33}$$

and $E^{\mathcal{A}_{xt}}\, dY_t = 0$, provided that we assume that for each t, $\{Y_s, s \ge t\}$ is independent of $\mathcal{A}_{xt}$. We note that a sufficient condition for this is the independence between $\{Y_s, s \ge t\}$ and $\{Y_\tau, S_\tau, 0 \le \tau \le t\}$ for each t. A comparison between (4.32) and (4.33) yields

$$\varphi_t = \hat{S}_t \equiv E^{\mathcal{A}_{xt}}S_t \tag{4.34}$$

Equation (4.31) now becomes

$$\frac{d\mathcal{P}^x}{d\mathcal{P}_0{}^x} = \exp\left(\int_0^1 \hat{S}_t\, dX_t - \frac{1}{2}\int_0^1 \hat{S}_t{}^2\, dt\right) \tag{4.35}$$

This line of argument is plausible, but not completely correct, because it rests on (4.33) which is a loose expression. Equation (4.35), however, is undoubtedly correct provided $\{\hat{S}_t, 0 \le t \le 1\}$ can be chosen to be a separable and measurable process. The interesting thing about (4.35) is that it relates the detection problem to the filtering problem, that is, to the problem of computing the conditional expectation

$$\hat{S}_t = E^{\mathcal{A}_{xt}} S_t \tag{4.36}$$

which is precisely the filtering problem of estimating S_t using data $\{X_s, 0 \le s \le t\}$.

5. RECURSIVE ESTIMATION

Equation (4.35) relates the detection problem to the filtering problem, but does not contribute to the solution of either. The goal of recursive filtering is the following. We want to find a process $\{U_t, 0 \le t \le 1\}$ taking values in a possibly infinitely dimensional vector space such that the following conditions are satisfied:

U_{t+dt} depends only on U_t and the new observation

$$dX_t = X_{t+dt} - X_t \tag{5.1a}$$

For each t, $E^{\mathcal{A}_{xt}} S_t = \hat{S}_t$ is completely determined by U_t (5.1b)

At this point it is convenient to generalize to the vector case and change the notation slightly. Suppose $\{\mathbf{X}_t, 0 \le t \le 1\}$ and $\{\mathbf{Z}_t, 0 \le t \le 1\}$ are two n-vector-valued processes defined on the same probability space $(\Omega, \mathcal{A}, \mathcal{P})$ such that

$$\mathbf{Y}_t = \mathbf{X}_t - \int_0^t \mathbf{Z}_s \, ds \tag{5.2}$$

is a vector process whose components are independent Brownian motions. Our objective is to obtain a recursive formula for

$$\hat{\mathbf{Z}}_t = E(\mathbf{Z}_t | \mathbf{X}_s, 0 \le s \le t) \tag{5.3}$$

We adopt a similar system of notations as before. Let $\mathcal{A}_{xt}$ and $\mathcal{A}_{zt}$ denote, respectively, the minimal σ algebra with respect to which $\{\mathbf{X}_s, 0 \le s \le t\}$ and $\{\mathbf{Z}_s, 0 \le s \le t\}$ are measurable. We write $\mathcal{A}_x$ and $\mathcal{A}_z$ rather than $\mathcal{A}_{x1}$ and $\mathcal{A}_{z1}$. If $\mathcal{A}$ and $\mathcal{B}$ are two σ algebras, $\mathcal{A} \vee \mathcal{B}$ denotes the smallest σ algebra containing both. If X is a random variable, then $\mathcal{A}(X)$ denotes the σ algebra generated by X. Finally, we denote

$$\mathcal{A}_t = \mathcal{A}_{xt} \vee \mathcal{A}_{zt}$$

and assume that $\mathcal{A}_1 = \mathcal{A}$ is the σ algebra of the probability triplet $(\Omega, \mathcal{A}, \mathcal{P})$.

We now proceed to derive a recursive formula for $\hat{\mathbf{Z}}_t$ under the following assumptions:

The vector process $\{\mathbf{Y}_t,\ 0 \le t \le 1\}$ as defined by (5.2) is independent of $\{\mathbf{Z}_t,\ 0 \le t \le 1\}$ under the probability measure $\mathcal{P}$ (5.4a)

$\int_0^1 \mathbf{Z}_t^T\mathbf{Z}_t\,dt = \int_0^1 \|\mathbf{Z}_t\|^2\,dt < \infty$ with probability 1 (T denotes transpose) (5.4b)

The process $\{\mathbf{Z}_t,\ 0 \le t \le 1\}$ is a Markov process (5.4c)

First, we introduce a new measure $\mathcal{P}_0$ on $(\Omega,\mathcal{A})$ by the formula

$$\frac{d\mathcal{P}_0}{d\mathcal{P}} = \exp\left(-\int_0^1 \mathbf{Z}_t^T\,d\mathbf{Y}_t - \frac{1}{2}\int_0^1 \mathbf{Z}_t^T\mathbf{Z}_t\,dt\right) \tag{5.5}$$

Proposition 5.1. $\mathcal{P}_0$ has the following properties:

$\mathcal{P}_0$ is a probability measure (5.6a)

Under $\mathcal{P}_0$, $\{\mathbf{X}_t,\ 0 \le t \le 1\}$ has components that are independent Brownian motions (5.6b)

Under $\mathcal{P}_0$ the processes $\{\mathbf{X}_t, 0 \le t \le 1\}$ and $\{\mathbf{Z}_t, 0 \le t \le 1\}$ are independent (5.6c)

The restriction of $\mathcal{P}_0$ to $\mathcal{A}_z$ is the same as the corresponding restriction of $\mathcal{P}$ (5.6d)

$\mathcal{P} \ll \mathcal{P}_0$ and (5.6e)

$$\frac{d\mathcal{P}}{d\mathcal{P}_0} = \exp\left(\int_0^1 \mathbf{Z}_t^T\,d\mathbf{X}_t - \frac{1}{2}\int_0^1 \mathbf{Z}_t^T\mathbf{Z}_t\,dt\right) \tag{5.7}$$

Proof: First, (5.6a) is proved by showing that

$$E\exp\left(-\int_0^1 \mathbf{Z}_t^T\,d\mathbf{Y}_t - \frac{1}{2}\int_0^1 \mathbf{Z}_t^T\mathbf{Z}_t\,dt\right) = 1$$

which can be done by showing that

$$E^{\mathcal{A}_z}\exp\left(-\int_0^1 \mathbf{Z}_t^T\,d\mathbf{Y}_t\right) = \exp\left(\frac{1}{2}\int_0^1 \mathbf{Z}_t^T\mathbf{Z}_t\,dt\right)$$

Because under $\mathcal{P}$, $\{\mathbf{Y}_t,\ 0 \le t \le 1\}$ is a Brownian motion independent of $\{\mathbf{Z}_t,\ 0 \le t \le 1\}$, this can be done easily. Next, by using (5.5) and (5.2) we can compute the characteristic functional

$$F(\mathbf{u},\mathbf{v}) = E_0\exp\left[i\int_0^1 \mathbf{u}^T(t)\,d\mathbf{X}_t + i\int_0^1 \mathbf{v}^T(t)\mathbf{Z}_t\,dt\right]$$

as follows,

$$F(\mathbf{u},\mathbf{v}) = E\left\{\exp\left(-\int_0^1 \mathbf{Z}_t^T\,d\mathbf{Y}_t - \frac{1}{2}\int_0^1 \mathbf{Z}_t^T\mathbf{Z}_t\,dt\right)\right.$$
$$\left.\exp\left[i\int_0^1 \mathbf{u}^T(t)(d\mathbf{Y}_t + \mathbf{Z}_t\,dt) + i\int_0^1 \mathbf{v}^T(t)\mathbf{Z}_t\,dt\right]\right\}$$

Because of (5.4), we can now write

$$F(\mathbf{u},\mathbf{v}) = E\left(\exp\left\{-\frac{1}{2}\int_0^1 \mathbf{Z}_t^T\mathbf{Z}_t\,dt + i\int_0^1 [\mathbf{u}(t) + \mathbf{v}(t)]^T\mathbf{Z}_t\,dt\right\}\right.$$
$$\left. E^{\alpha_z}\exp\left\{\int_0^1 [-\mathbf{Z}_t + i\mathbf{u}(t)]^T\,d\mathbf{Y}_t\right\}\right)$$

The term $E^{\alpha_z}\exp\left\{\int_0^1 [-\mathbf{Z}_t + i\mathbf{u}(t)]^T\,d\mathbf{Y}_t\right\}$ can be readily computed, because of (5.4*a*). In particular, given $\{\mathbf{Z}_t, 0 \le t \le 1\}$, the two integrals $I_1 = \int_0^1 \mathbf{Z}_t^T\,d\mathbf{Y}_t$ and $I_2 = \int_0^1 \mathbf{u}^T(t)\,d\mathbf{Y}_t$ are jointly Gaussian. If we denote

$$\mathbf{I} = \begin{bmatrix} I_1 \\ I_2 \end{bmatrix}$$

then

$$E^{\alpha_z}\mathbf{I} = 0$$

and

$$E^{\alpha_z}\mathbf{I}\mathbf{I}^T = \mathfrak{R} = \begin{bmatrix} \int_0^1 \mathbf{Z}_t^T\mathbf{Z}_t\,dt & \int_0^1 \mathbf{Z}_t^T\mathbf{u}(t)\,dt \\ \int_0^1 \mathbf{Z}_t^T\mathbf{u}(t)\,dt & \int_0^1 \mathbf{u}^T(t)\mathbf{u}(t)\,dt \end{bmatrix}$$

It now follows that

$$E^{\alpha_z}\exp\left\{\int_0^1 [-\mathbf{Z}_t + i\mathbf{u}(t)]^T\,dY_t\right\} = E^{\alpha_z}\exp\left(\begin{bmatrix} -1 \\ i \end{bmatrix}^{\mathbf{T}}\mathbf{I}\right)$$
$$= \frac{1}{2\pi}|\mathfrak{R}|^{-\frac{1}{2}}\int_{R^2} e^{-\frac{1}{2}\boldsymbol{\xi}^T\mathfrak{R}^{-1}\boldsymbol{\xi}}\exp\begin{bmatrix} -1 \\ i \end{bmatrix}^T\boldsymbol{\xi}\,d\boldsymbol{\xi} = \exp\left(\frac{1}{2}\begin{bmatrix} -1 \\ i \end{bmatrix}^T\mathfrak{R}\begin{bmatrix} -1 \\ i \end{bmatrix}\right)$$
$$= \exp\left[\frac{1}{2}\int_0^1 \mathbf{Z}_t^T\mathbf{Z}_t\,dt - \frac{1}{2}\int_0^1 \mathbf{u}^T(t)\mathbf{u}(t)\,dt - i\int_0^1 \mathbf{u}^T(t)\mathbf{Z}_t\,dt\right]$$

It follows that

$$F(\mathbf{u},\mathbf{v}) = E_0\exp\left[i\int_0^1 \mathbf{u}^T(t)\,d\mathbf{X}_t + i\int_0^1 \mathbf{v}^T(t)\mathbf{Z}_t\,dt\right]$$

and we find

$$F(\mathbf{u},\mathbf{v}) = \exp\left[-\frac{1}{2}\int_0^1 \mathbf{u}^T(t)\mathbf{u}(t)\,dt\right]E\exp\left[i\int_0^1 \mathbf{v}^T(t)\mathbf{Z}_t\,dt\right]$$

which proves (5.6*b*) to (5.6*d*), simultaneously. Finally, we note that $(d\mathcal{P}_0/d\mathcal{P})^{-1}$ is finite a.s. $\mathcal{P}_0$ and

$$E_0\left(\frac{d\mathcal{P}_0}{d\mathcal{P}}\right)^{-1} = 1$$

thus $\mathcal{P} \ll \mathcal{P}_0$ and $d\mathcal{P}/d\mathcal{P}_0 = (d\mathcal{P}_0/d\mathcal{P})^{-1}$, completing the proof of Proposition 5.1. ∎

Now, we find a U_t which has the property (5.1). Define

$$\Lambda_t = E_0{}^{\mathcal{A}_t} \frac{d\mathcal{P}}{d\mathcal{P}_0} = \exp\left(\int_0^t \mathbf{Z}_s{}^T \, d\mathbf{X}_s - \frac{1}{2}\int_0^t \mathbf{Z}_s{}^T \mathbf{Z}_s \, ds\right) \tag{5.8}$$

The conditional expectation

$$E_0(\Lambda_t | \mathcal{A}_{xt} \vee \mathcal{A}(\mathbf{Z}_t))$$

is a function of t, $\mathbf{Z}_t$, and $\{\mathbf{X}_s, 0 \leq s \leq t\}$, and nothing else. We abbreviate $\{\mathbf{X}_s, 0 \leq s \leq t\}$ by $\mathbf{X}_0{}^t$ and define the function $U_t(\mathbf{z},\mathbf{X}_0{}^t)$ by

$$U_t(\mathbf{Z}_t,\mathbf{X}_0{}^t) = E_0(\Lambda_t | \mathcal{A}_{xt} \vee \mathcal{A}(\mathbf{Z}_t)) \tag{5.9}$$

For a fixed t and $\mathbf{X}_0{}^t$, $U_t(\cdot,\mathbf{X}_0{}^t)$ can be considered an element of the function space of all functions mapping R^n into R and integrable with respect to the Borel measure on R^n defined by

$$P(d\mathbf{z},t) = \mathcal{P}(\mathbf{Z}_t \in d\mathbf{z}) = \mathcal{P}_0(\mathbf{Z}_t \in d\mathbf{z}) \tag{5.10}$$

Therefore, we can consider $U_t(\cdot,\mathbf{X}_0{}^t)$ to be an element of a countably infinite-dimensional linear space. Next, we shall show that U_t satisfies (5.1).

Proposition 5.2. Let $U_t(\mathbf{z},\mathbf{X}_0{}^t)$ be defined by (5.9), and let f be any Borel function such that $E|f(\mathbf{Z}_t)| < \infty$. Then,

$$E^{\mathcal{A}_{xt}} f(\mathbf{Z}_t) = \frac{\int_{R^n} f(\mathbf{z}) U_t(\mathbf{z},\mathbf{X}_0{}^t) P(d\mathbf{z},t)}{\int_{R^n} U_t(\mathbf{z},\mathbf{X}_0{}^t) P(d\mathbf{z},t)} \tag{5.11}$$

Remark: Equation (5.11) justifies the interpretation of $U_t(\cdot,\mathbf{X}_0{}^t)$ as an unnormalized conditional density of $\mathbf{Z}_t$ given $\mathbf{X}_0{}^t$.

Proof: Let A be any set in $\mathcal{A}_{xt}$. By the definition of conditional expectation, we have

$$\int_A E^{\mathcal{A}_{xt}} f(\mathbf{Z}_t) \, d\mathcal{P} = \int_A f(\mathbf{Z}_t) \, d\mathcal{P} \qquad A \in \mathcal{A}_{xt}$$

We can reexpress each side in terms of $\Lambda_t = E_0{}^{\mathcal{A}_t} \, d\mathcal{P}/d\mathcal{P}_0$ and an integral with respect to $\mathcal{P}_0$. First, we write

$$\begin{aligned}\int_A E^{\mathcal{A}_{xt}} f(\mathbf{Z}_t) \, d\mathcal{P} &= \int_A \Lambda_t E^{\mathcal{A}_{xt}} f(\mathbf{Z}_t) \, d\mathcal{P}_0 = E_0 I_A \Lambda_t E^{\mathcal{A}_{xt}} f(\mathbf{Z}_t) \\ &= E_0[E_0(I_A \Lambda_t E^{\mathcal{A}_{xt}} f(\mathbf{Z}_t) | \mathcal{A}_{xt})] = E_0[I_A E^{\mathcal{A}_{xt}} f(\mathbf{Z}_t) E_0{}^{\mathcal{A}_{xt}} \Lambda_t] \qquad A \in \mathcal{A}_{xt}\end{aligned}$$

Similarly, we have

$$\int_A f(\mathbf{Z}_t)\, d\mathcal{P} = E_0[I_A \Lambda_t f(\mathbf{Z}_t)] = E_0\{I_A E_0^{\mathcal{A}_{xt}}[\Lambda_t f(\mathbf{Z}_t)]\} \qquad A \in \mathcal{A}_{xt}$$

It follows that

$$E^{\mathcal{A}_{xt}} f(\mathbf{Z}_t) = \frac{E_0^{\mathcal{A}_{xt}}[\Lambda_t f(\mathbf{Z}_t)]}{E_0^{\mathcal{A}_{xt}} \Lambda_t}$$

which can be rewritten as (5.11). ▮

Equation (5.11) immediately yields

$$\hat{\mathbf{Z}}_t = \frac{\int_{R^n} \mathbf{z} U_t(\mathbf{z}, \mathbf{X}_0^t) P(d\mathbf{z}, t)}{\int_{R^n} U_t(\mathbf{z}, \mathbf{X}_0^t) P(d\mathbf{z}, t)} \tag{5.12}$$

Since $P(d\mathbf{z},t)$ is known a priori, we have satisfied (5.11), i.e., at each t, the knowledge of U_t completely determines $\hat{\mathbf{Z}}_t$.

To verify (5.1*a*), we invoke for the first time the Markov property of $\{\mathbf{Z}_t, 0 \le t \le 1\}$. Specifically, we shall derive the formula

$$U_t(\mathbf{z}, \mathbf{X}_0^t) = 1 + \int_0^t \left[\int_{R^n} \boldsymbol{\zeta} U_s(\boldsymbol{\zeta}, \mathbf{X}_0^s) P(d\boldsymbol{\zeta}, s|\mathbf{z}, t)\right]^T d\mathbf{X}_s \tag{5.13}$$

where

$$P(d\boldsymbol{\zeta}, s|\mathbf{z}, t) = \mathcal{P}(\mathbf{Z}_s \in d\boldsymbol{\zeta}|\mathbf{Z}_t = \mathbf{z}) \tag{5.14}$$

To derive (5.13), we begin with Λ_t as defined by (5.8) and apply Ito's differentiation rule to the term involving a stochastic integral. This yields

$$d\Lambda_t = \Lambda_t \mathbf{Z}_t^T\, d\mathbf{X}_t$$

or

$$\Lambda_t = 1 + \int_0^t \Lambda_s \mathbf{Z}_s^T\, d\mathbf{X}_s \tag{5.15}$$

Now,

$$U_t(\mathbf{z}, \mathbf{X}_0^t) = E_0(\Lambda_t | \mathbf{Z}_t = \mathbf{z}, \mathbf{X}_s, 0 \le s \le t)$$

Hence, from (5.15) we get

$$\begin{aligned} U_t(\mathbf{z}, \mathbf{X}_0^t) &= 1 + E_0\left(\int_0^t \Lambda_s \mathbf{Z}_s^T\, d\mathbf{X}_s | \mathbf{Z}_t = z, \mathbf{X}_s, 0 \le s \le t\right) \\ &= 1 + \int_0^t E_0(\Lambda_s \mathbf{Z}_s^T | \mathbf{Z}_t = \mathbf{z}, \mathbf{X}_0^t)\, d\mathbf{X}_s \end{aligned} \tag{5.16}$$

Now, under $\mathcal{P}_0$, $\{\mathbf{X}_t, 0 \le t \le 1\}$ is a vector Brownian motion process independent of $\{\mathbf{Z}_t, 0 \le t \le 1\}$, and

$$E_0(\Lambda_s \mathbf{Z}_s | \mathbf{Z}_t = \mathbf{z}, \mathbf{X}_0^t) = E_0(\Lambda_s \mathbf{Z}_s | \mathbf{Z}_t = \mathbf{z}, \mathbf{X}_0^s) \tag{5.17}$$

Because of the Markov property of $\{\mathbf{Z}_t, 0 \le t \le 1\}$ and the independence between $\{\mathbf{Z}_t, 0 \le t \le 1\}$ and $\{\mathbf{X}_t, 0 \le t \le 1\}$, for every $t > s$ $\mathbf{Z}_t$ is conditionally independent of $\{\mathbf{X}_s, \mathbf{Z}_s, 0 \le s \le t\}$ given $\mathbf{Z}_s$. This means that

$$\begin{aligned} E_0(\Lambda_s\mathbf{Z}_s|\mathbf{Z}_t = \mathbf{z},\mathbf{X}_0^s) &= E_0\{\mathbf{Z}_s E_0(\Lambda_s|\mathcal{A}(\mathbf{Z}_t) \vee \mathcal{A}(\mathbf{Z}_s) \vee \mathcal{A}_{xs})|\mathbf{Z}_t = \mathbf{z},\mathbf{X}_0^s\} \\ &= E_0\{\mathbf{Z}_s E_0(\Lambda_s|\mathcal{A}(\mathbf{Z}_s) \vee \mathcal{A}_{xs})|\mathbf{Z}_t = \mathbf{z},\mathbf{X}_0^s\} \\ &= E_0\{\mathbf{Z}_s U_s(\mathbf{Z}_s,\mathbf{X}_0^s)|\mathbf{Z}_t = \mathbf{z},\mathbf{X}_0^s\} \\ &= \int_{R^n} \boldsymbol{\zeta} U_s(\boldsymbol{\zeta},\mathbf{X}_0^s)\mathcal{P}_0(\mathbf{Z}_s \in d\boldsymbol{\zeta}|\mathbf{Z}_t = \mathbf{z},\mathbf{X}_0^s) \end{aligned} \tag{5.18}$$

Because under $\mathcal{P}_0$, $\{\mathbf{Z}_t, 0 \le t \le 1\}$ and $\{\mathbf{X}_t, 0 \le t \le 1\}$ are independent and because $\{\mathbf{Z}_t, 0 \le t \le 1\}$ has the same distribution under either $\mathcal{P}_0$ or $\mathcal{P}$, we have

$$\mathcal{P}_0(\mathbf{Z}_s \in d\boldsymbol{\zeta}|\mathbf{Z}_t = \mathbf{z},\mathbf{X}_0^s) = \mathcal{P}(\mathbf{Z}_s \in d\boldsymbol{\zeta}|\mathbf{Z}_t = \mathbf{z}) = P(d\boldsymbol{\zeta},s|\mathbf{z},t) \tag{5.19}$$

Combining (5.16) through (5.19) we find

$$U_t(\mathbf{z},\mathbf{X}_0^t) = 1 + \int_0^t \left[\int_{R^n} \boldsymbol{\zeta} U_s(\boldsymbol{\zeta},\mathbf{X}_0^s)P(d\boldsymbol{\zeta},s|\mathbf{z},t) \right]^T d\mathbf{X}_s$$

which is just (5.13).

To summarize, we find that under condition (5.4) if we define U_t by (5.9), then

$$\hat{\mathbf{Z}}_t = \frac{\int_{R^n} \mathbf{z} U_t(\mathbf{z},\mathbf{X}_0^t)P(d\mathbf{z},t)}{\int_{R^n} U_t(\mathbf{z},\mathbf{X}_0^t)P(d\mathbf{z},t)} \tag{5.20}$$

and

$$U_t(\mathbf{z},\mathbf{X}_0^t) = 1 + \int_0^t \left[\int_{R^n} \boldsymbol{\zeta} U_s(\boldsymbol{\zeta},\mathbf{X}_0^s)P(d\boldsymbol{\zeta},s|\mathbf{z},t) \right]^T d\mathbf{X}_s \tag{5.21}$$

In general, (5.21) represents the "equation of motion" for a system with an infinitely dimensional state space. There are some special cases for which (5.21) can be implemented by a system with a finite-dimensional state space. First, if $\mathbf{Z}_t$ assumes only finitely many values, say $\mathbf{z}_1$, $\mathbf{z}_2$, . . . , $\mathbf{z}_N$, then (5.21) becomes

$$U_t(\mathbf{z}_i,\mathbf{X}_0^t) = 1 + \sum_{j=1}^{N} \mathcal{P}(\mathbf{Z}_s = \mathbf{z}_j|\mathbf{Z}_t = \mathbf{z}_i)\mathbf{z}_j^T \int_0^t U_s(\mathbf{z}_j,\mathbf{X}_0^s)\, d\mathbf{X}_s \qquad i = 1, 2, \ldots, N \tag{5.22}$$

which is in the form of a vector stochastic integral equation [Wonham, 1964]. Another case for which (5.21) can be implemented by a system with a finite-dimensional state space is when $\{\mathbf{Z}_t, 0 \le t \le 1\}$ is not only Markov, but also Gaussian. We can then show that $\hat{\mathbf{Z}}_t$ itself satisfies a stochastic integral equation. This fact is not surprising since for the Gaussian case $\hat{\mathbf{Z}}_t$ is the same as the *linear* least-squares estimator. Since

$\{\mathbf{Z}_t, 0 \le t \le 1\}$ is also Markov, $\hat{\mathbf{Z}}_t$ satisfies the Kalman filtering equation studied in Sec. 3.9.

The "equation of evolution" (5.21) can be put in a more convenient form when the transition-density function of $\{\mathbf{Z}_t, 0 \le t \le 1\}$ satisfies a Fokker-Planck equation. Specifically, define $p(\mathbf{z},t|\boldsymbol{\zeta},s)$ by

$$\mathcal{P}(\mathbf{Z}_t \in d\mathbf{z}|\mathbf{Z}_s = \boldsymbol{\zeta}) = p(\mathbf{z},t|\boldsymbol{\zeta},s)\, d\mathbf{z} \tag{5.23}$$

and define $p(\mathbf{z},t)$ by

$$\mathcal{P}(\mathbf{Z}_t \in d\mathbf{z}) = p(\mathbf{z},t)\, d\mathbf{z} \tag{5.24}$$

Furthermore, set

$$V_t(\mathbf{z},\mathbf{X}_0^t) = p(\mathbf{z},t)U_t(\mathbf{z},\mathbf{X}_0^t) \tag{5.25}$$

Then we can rewrite (5.21) as

$$V_t(\mathbf{z},\mathbf{X}_0^t) = p(\mathbf{z},t) + \int_0^t \left[\int_{R^n} \boldsymbol{\zeta} p(\mathbf{z},t|\boldsymbol{\zeta},s) V_s(\boldsymbol{\zeta},\mathbf{X}_0^s)\, d\boldsymbol{\zeta}\right]^T d\mathbf{X}_s \tag{5.26}$$

Now, suppose that $p(\mathbf{z},t|\boldsymbol{\zeta},s)$ satisfies a Fokker-Planck equation

$$\frac{1}{2}\sum_{i,j} \frac{\partial^2}{\partial z_i\, \partial z_j} [\sigma_{ij}(\mathbf{z},t)p(\mathbf{z},t|\boldsymbol{\zeta},s)] - \sum_i \frac{\partial}{\partial z_i} [m_i(\mathbf{z},t)p(\mathbf{z},t|\boldsymbol{\zeta},s)] = \frac{\partial}{\partial t} p(\mathbf{z},t|\boldsymbol{\zeta},s) \tag{5.27}$$

with initial condition $p(\mathbf{z},s|\boldsymbol{\zeta},s) = \delta(\mathbf{z} - \boldsymbol{\zeta})$. We shall denote the differential operator on the left-hand side of (5.27) by $\mathcal{L}_{z,t}$. We note that since

$$p(\mathbf{z},t) = \int_{R^n} p(\mathbf{z},t|\boldsymbol{\zeta},s)p(\boldsymbol{\zeta},s)\, d\boldsymbol{\zeta}$$

we also have

$$\frac{\partial p(\mathbf{z},t)}{\partial t} = \mathcal{L}_{z,t}p(\mathbf{z},t) \tag{5.28}$$

Formally, from (5.26) we can write

$$V_{t+dt}(\mathbf{z},\mathbf{X}_0^{t+dt}) - V_t(\mathbf{z},\mathbf{X}_0^t) = dt \frac{\partial}{\partial t} p(\mathbf{z},t) + V_t(\mathbf{z},\mathbf{X}_0^t)\mathbf{z}^T\, d\mathbf{X}_t + dt \left\{\int_0^t \left[\frac{\partial}{\partial t} p(\mathbf{z},t|\boldsymbol{\zeta},s)\boldsymbol{\zeta} V_s(\boldsymbol{\zeta},\mathbf{X}_0^s)\, d\boldsymbol{\zeta}\right]^T d\mathbf{X}_s\right\} \tag{5.29}$$

Using (5.26), (5.27), and (5.28) in (5.29), we get

$$dV_t(\mathbf{z},\mathbf{X}_0^t) = V_t(\mathbf{z},\mathbf{X}_0^t)\mathbf{z}^T\, d\mathbf{X}_t + [\mathcal{L}_{z,t}V_t(\mathbf{z},\mathbf{X}_0^t)]\, dt \tag{5.30}$$

The precise meaning of (5.20) is

$$V_t(\mathbf{z},\mathbf{X}_0^t) = p(\mathbf{z},0) + \int_0^t V_s(\mathbf{z},\mathbf{X}_0^s)\mathbf{z}^T\, d\mathbf{X}_s + \int_0^t \mathcal{L}_{z,s}V_s(\mathbf{z},\mathbf{X}_0^s)\, ds \tag{5.31}$$

Although the derivation of (5.31) is formal, it can be fully justified under quite general conditions [see, e.g., Mortensen, 1966; Zakai, 1969].

Equation (5.31) has a twofold advantage over (5.21). First, the term involving the observation is now "diagonal" in the sense that the change in $V_t(\mathbf{z},\mathbf{X}_0^t)$ due to $d\mathbf{X}_t$ depends only on $V_t(\cdot,\mathbf{X}_0^t)$ at $\mathbf{z}$. Secondly, although the change due to dt is not diagonal, it is "local" in the sense that $\mathfrak{L}_{z,t}$ is a local operator. In practice, this means that to compute $V_{t+dt}(\mathbf{z},\mathbf{X}_0^{t+dt})$, we only need $d\mathbf{X}_t$ and the knowledge of $V_t(\cdot,\mathbf{X}_0^t)$ in a neighborhood of $\mathbf{z}$, resulting in considerable savings in computation.

Finally, we remark that it is trivial to generalize everything that we have done in this section to the case where

$$d\mathbf{X}_t = \mathbf{A}(t)\mathbf{Z}_t\,dt + d\mathbf{Y}_t \tag{5.32}$$

where $\mathbf{A}(t)$ is a matrix-valued nonrandom function. The advantage of (5.32) is that now $\mathbf{X}_t$ and $\mathbf{Z}_t$ need not be of the same dimension. This is a useful flexibility. Modification of our formulas to accommodate this case are easily obtained. For example, instead of (5.31), we would find

$$V_t(\mathbf{z},\mathbf{X}_0^t) = p(\mathbf{z},0) + \int_0^t \mathfrak{L}_{z,s}V_s(\mathbf{z},\mathbf{X}_0^s)\,ds + \int_0^t V_s(\mathbf{z},\mathbf{X}_0^s)[\mathbf{A}(s)\mathbf{z}]^T\,d\mathbf{X}_s \tag{5.33}$$

It is of interest to consider some examples. Unfortunately, the number of examples that permit complete analytic treatment is small.

Example 1. Let $Z_t = f(t)Z$ where f is a square-integrable deterministic function and Z is a single random variable with distribution function P_Z. For this case,

$$\Lambda_t = \exp\left[Z\int_0^t f(s)\,dX_s - \tfrac{1}{2}Z^2\int_0^t f^2(s)\,ds\right]$$

so that

$$U_t(z,X_0^t) = \exp\left[z\int_0^t f(s)\,dX_s - \tfrac{1}{2}z^2\int_0^t f^2(s)\,ds\right]$$

and

$$\hat{Z}_t = \frac{f(t)\int_{-\infty}^{\infty} zU_t(z,X_0^t)P_Z(dz)}{\int_{-\infty}^{\infty} U_t(z,X_0^t)P_Z(dz)}$$

Now, let $g(x,t)$ be defined by

$$g(x,t) = \int_{-\infty}^{\infty} \exp\left[zx - \tfrac{1}{2}z^2\int_0^t f^2(s)\,ds\right]P_Z(dz)$$

and let $h(x,t)$ be given by

$$h(x,t) = \frac{1}{g(x,t)}\frac{\partial g(x,t)}{\partial x}$$

Furthermore, set

$$\tilde{X}_t = \int_0^t f(s)\, dX_s$$

Then, we have

$$\hat{Z}_t = f(t)h(\tilde{X}_t,t)$$

and

$$d\tilde{X}_t = f(t)\, dX_t$$

Thus, for this example, $\hat{Z}_t$ can be implemented by a recursive equation with a one-dimensional state space.

Example 2 [Wonham, 1964]. Let $\{Z_t, -\infty < t < \infty\}$ be a random telegraph process defined as a two-state Markov process with

$$\begin{aligned}\mathcal{P}(Z_t = 1) &= \mathcal{P}(Z_t = -1) = \tfrac{1}{2}\\ \mathcal{P}(Z_t = Z_s) &= 1 - \mathcal{P}(Z_t = -Z_s)\\ &= \tfrac{1}{2}(1 + e^{-|t-s|})\end{aligned}$$

If we let U_t^+ and U_t^- denote $U_t(Z_t = 1, X_0{}^t)$ and $U_t(Z_t = -1, X_0{}^t)$, respectively, then (5.22) becomes

$$U_t^+ = 1 + \frac{1}{2}\int_0^t [1 + e^{-(t-s)}]U_s^+\, dX_s - \frac{1}{2}\int_0^t [1 - e^{-(t-s)}]U_s^-\, dX_s$$

$$U_t^- = 1 + \frac{1}{2}\int_0^t [1 - e^{-(t-s)}]U_s^+\, dX_s - \frac{1}{2}\int_0^t [1 + e^{-(t-s)}]U_s^-\, dX_s$$

and (5.20) can be written as

$$\hat{Z}_t = \frac{U_t^+ - U_t^-}{U_t^+ + U_t^-}$$

It is convenient to set $A_t = \frac{1}{2}(U_t^+ + U_t^-)$ and $B_t = \frac{1}{2}(U_t^+ - U_t^-)$ so that we have

$$A_t = 1 + \int_0^t B_s\, dX_s$$

$$B_t = \int_0^t e^{-(t-s)}A_s\, dX_s$$

In differential form, these can be written as

$$\begin{aligned}dA_t &= B_t\, dX_t\\ dB_t &= -B_t\, dt + A_t\, dX_t\end{aligned}$$

Because $\hat{Z}_t = B_t/A_t$ we can apply Ito's differentiation rule and get

$$d\hat{Z}_t = -\hat{Z}_t\, dt + \hat{Z}_t^3\, dt - 2\hat{Z}_t^2\, dt + (1 - \hat{Z}_t^2)\, dX_t$$

which is a recursive equation in $\hat{Z}_t$ directly.

Example 3. Let Z_t, $0 \leq t \leq 1$, be Gaussian as well as Markov. For this case the transition density $p(z,t|z_0,t_0)$ satisfies a Fokker-Planck equation of the form

$$\frac{\partial}{\partial t} p(z,t|z_0,t_0) = \tfrac{1}{2}\sigma^2(t) \frac{\partial^2 p}{\partial z^2} - \frac{\partial}{\partial z} [\beta(t) z p]$$

Equation (5.31) becomes

$$dV_t(z,X_0^t) = zV_t(z,X_0^t)\, dX_t + \left\{ \frac{\partial}{\partial z} \left[\frac{\sigma^2(t)}{2} \frac{\partial V_t(z,X_0^t)}{\partial z} - \beta(t) V_t(z,X_0^t) \right] \right\} dt$$

We know that Z_t, given X_0^t, is Gaussian. Since the density of Z_t, given X_0^t, is

$$p(z,t|X_0^t) = \frac{V_t(z,X_0^t)}{\int_{-\infty}^{\infty} V_t(z,X_0^t)\, dz}$$

$V_t(z,X_0^t)$ must have the form

$$V_t(z,X_0^t) = \frac{A_t}{\sqrt{2\pi \Sigma_t^2}} \exp \left[-\frac{1}{2\Sigma_t^2} (z - \hat{Z}_t)^2 \right]$$

Thus, we find

$$dV_t(z,X_0^t) = zV_t(z,X_0^t)\, dXt + V_t(z,X_0^t) \left[-\beta(t) + \beta(t) \frac{z - \hat{Z}_t}{\Sigma_t^2} \right] dt + V_t(z,X_0^t) \frac{\sigma^2(t)}{2} \left[\frac{(z - \hat{Z}_t)^2}{\Sigma_t^4} - \frac{1}{\Sigma_t^2} \right] dt$$

Now, Ito's differentiation rule yields

$$\begin{aligned} d \ln V_t(z,X_0^t) &= d \ln A_t - \tfrac{1}{2} d \ln \Sigma_t^2 - \tfrac{1}{2} d \frac{(z - \hat{Z}_t)^2}{\Sigma_t^2} \\ &= \frac{1}{V_t} dV_t - \frac{1}{V_t^2} z^2 V_t^2\, dt \\ &= z\, dX_t + \beta(t) \left(\frac{z - \hat{Z}_t}{\Sigma_t^2} - 1 \right) dt \\ &\quad + \frac{\sigma^2(t)}{2} \left(\frac{z - \hat{Z}_t}{\Sigma_t^4} - \frac{1}{\Sigma_t^2} \right) dt - z^2\, dt \end{aligned}$$

We can now equate coefficients of z and z^2 and get

$$d \frac{\hat{Z}_t}{\Sigma_t^2} = dX_t + \frac{\beta(t)}{\Sigma_t^2} dt - \frac{\sigma^2(t) \hat{Z}_t}{\Sigma_t^4} dt$$

$$-\tfrac{1}{2} d \frac{1}{\Sigma_t^2} = \frac{\sigma^2(t)}{2} \frac{1}{\Sigma_t^4} dt - dt$$

The first of these equations is a linear equation in $\hat{Z}_t$ and is precisely the same equation for the Kalman-Bucy filter for this case. The second of these equations does not involve the observation dX_t and can be solved off-line for Σ_t^2. Obviously, it must be the same as the Riccati equation (3.9.39) in Kalman-Bucy filtering theory, except for different notations. Finally, we note that the quantity A_t in $V_t(z,X_0^t)$ is already known, because

$$\begin{aligned} A_t &= \int_{-\infty}^{\infty} V_t(z,X_0^t)\, dz \\ &= E_0(\Lambda_t|\mathcal{A}_{xt}) \end{aligned}$$

which from our discussions in Sec. 4 [for example, (4.35)] is given by

$$A_t = \exp\left(\int_0^t \hat{Z}_s\, dX_s - \frac{1}{2}\int_0^t \hat{Z}_s^2\, ds\right)$$

EXERCISES

1. Verify (1.3).

2. Let $\mathcal{P}$ and $\mathcal{P}_0$ be two *Gaussian* measures on the sample space of $\{X_t,\ 0 \le t \le 1\}$ such that

$$\begin{aligned} E_0X_t &= 0 = EX_t \\ E_0X_tX_s &= \min(t,s) + \tfrac{1}{2} \\ EX_tX_s &= \tfrac{1}{2}e^{-|t-s|} \end{aligned}$$

(*a*) Show that under $\mathcal{P}$, $W_t = X_t - X_0 + \int_0^t X_s\, ds$ is a standard Brownian motion.

(*b*) Using (1.12) and the fact that the results of (*a*) implies

$$p(x, t + \delta|x',t) \cong \frac{1}{\sqrt{2\pi\delta}} \exp\left[-\frac{1}{2\delta}(x - x' + \delta x')^2\right]$$

for small δ, show that

$$\frac{d\mathcal{P}}{d\mathcal{P}_0} = \exp\left(-\int_0^1 X_t\, dX_t - \frac{1}{2}\int_0^1 X_t^2\, dt\right)$$

3. Let $X_t = e^{W_t}$ where W_t is a Brownian motion. Compute the quadratic variation of the X process on [0,1] in terms of $\{X_s,\ 0 \le s \le 1\}$.

4. (*a*) Using (4.26) and the fact that $EX_s(E^{\mathcal{A}_t}\, dX_t - dX_t) = 0,\ 0 < s < t$, show that the function h in (4.22) satisfies

$$h(t,s) - \int_0^t K(s,\tau)h(t,\tau)\, d\tau = -K(t,s) \qquad s < t$$

where K is the function defined in Proposition 4.1 and assumed to be continuous.

(*b*) If $K(s,\tau) = f(s)f(\tau)$ with

$$\int_0^t f^2(s)\, ds < \infty$$

find the function h.

5. Let $\mathcal{P}$ and $\mathcal{P}_0$ be two probability measures defined on the sample space of $\{X_t, 0 \leq t \leq 1\}$ such that

$$\begin{aligned} X_t &= e^t e^{W_t} && \text{under } \mathcal{P} \\ X_t &= e^{-t} e^{V_t} && \text{under } \mathcal{P}_0 \end{aligned}$$

where W_t, V_t are standard Brownian motion processes under $\mathcal{P}$ and $\mathcal{P}_0$, respectively.

(*a*) Is the quadratic variation the same under $\mathcal{P}$ and $\mathcal{P}_0$?

(*b*) Is $\mathcal{P}$ absolutely continuous with respect to $\mathcal{P}_0$? Conversely?

(*c*) If $\mathcal{P} \ll \mathcal{P}_0$, find $d\mathcal{P}/d\mathcal{P}_0$.

6. Prove the second equality in (5.8), viz.,

$$E_0^{\mathcal{A}_t} \exp\left(\int_0^1 \mathbf{Z}_s^T \, d\mathbf{X}_s - \frac{1}{2}\int_0^1 \mathbf{Z}_s^T \mathbf{Z}_s \, ds\right) = \exp\left(\int_0^t \mathbf{Z}_s^T \, d\mathbf{X}_s - \frac{1}{2}\int_0^t \mathbf{Z}_s^T \mathbf{Z}_s \, ds\right)$$

7. Starting from (5.15), prove that

$$\int_{R^n} U_t(\mathbf{z}, \mathbf{X}_0^t) P(d\mathbf{z}, t) = E_0^{\mathcal{A}_{xt}} \Lambda_t = \exp\left(\int_0^t \hat{\mathbf{Z}}_s^T \, d\mathbf{X}_s - \frac{1}{2}\int_0^t \hat{\mathbf{Z}}_s^T \hat{\mathbf{Z}}_s \, ds\right)$$

which provides a direct proof of (4.35) in the case where the hypothesis H_1 in (4.29) is modified to read that the Y process is a Brownian motion independent of the S process.

8. Suppose that $\{Z_t, t \geq 0\}$ satisfies

$$dZ_t = m(t) Z_t \, dt + \sigma(t) \, dV_t \qquad Z_0 = 0$$

where V_t is a Brownian motion, then $\{Z_t, t \geq 0\}$ is Gauss-Markov. It follows from Example 3 in Sec. 5 that $\hat{Z}_t$ satisfies an equation

$$\begin{aligned} d\hat{Z}_t &= \alpha(t) \hat{Z}_t \, dt + \beta(t) \, dX_t \\ &= \alpha(t) \hat{Z}_t \, dt + \beta(t) Z_t \, dt + \beta(t) \, dW_t \end{aligned}$$

This means that $\begin{bmatrix} Z_t \\ \hat{Z}_t \end{bmatrix}$ is a Gaussian-Markov process.

(*a*) The joint density function $p(z, \hat{z}; t)$ satisfies a Fokker-Planck equation. Find this equation.

(*b*) From the requirement $E(Z_t | \hat{Z}_t) = \hat{Z}_t$, obtain a condition on $\alpha(\cdot)$ and $\beta(\cdot)$ in terms of $m(\cdot)$ and $\sigma(\cdot)$.

(*c*) Obtain a second condition on $\alpha(\cdot)$ and $\beta(\cdot)$ by minimizing $E(Z_t - \hat{Z}_t)^2$ over all $\alpha(\cdot)$ and $\beta(\cdot)$ which satisfy the requirement in (*b*).

9. Suppose that $m(\cdot)$ and $\sigma(\cdot)$ in Exercise 8 are given by

$$m(\cdot) = -1 \qquad \sigma(\cdot) = 1$$

Find the asymptotic error $\lim_{t \to \infty} E(Z_t - \hat{Z}_t)^2$

10. If Z_t is a random telegraph process as in Example 2, find the limiting distribution of $\hat{Z}_t$ as $t \to \infty$.

7
Random Fields

1. INTRODUCTION

We have used the term stochastic process to denote a collection of random variables indexed by a single real parameter. In other words, the parameter space is a subset of the real line and usually an interval. In most applications, this parameter is interpreted as time. There are many applications where it is more appropriate to consider collections of random variables indexed by points in a more general parameter space. For example, in problems involving propagation of electromagnetic waves through random media, the natural parameter space is a subset of R^4, representing space and time. A similar example is the velocity field in turbulence theory. The term random field is often used to denote a collection of random variables with a parameter space which is a subset of R^n. There are other possible parameter spaces. For example, the parameter space can be taken to be a function space of some kind. Such is the case with **generalized processes.** Alternatively, we can also take the parameter space to be a collection of subsets of R^n. Such, for example, is the situation for **random measures,** which have already been made use of in connection with second-order stochastic integrals. Generally speaking, the kind of assumptions that we make concerning mutual dependence of the collection of random variables reflects something of the parameter space. For example, if the parameter space is an interval, we usually assume continuity in probability. If the parameter space is a linear topological space, we usually assume that the collection of random variables as a function of the parameter is both linear and continuous in probability. If the parameter space is a σ algebra, then we usually assume that the collection of random variable is σ additive, and so on.

Compared to the one-parameter case, relatively little is known concerning processes with a more general parameter space. Of course, a great deal of the results concerning stochastic processes with a one-dimensional parameter space do not depend on the fact that the parameter is one dimensional. These results are easily generalized to more general collections of random variables. Such generalizations require little elaboration. In this chapter, we shall focus our attention on problems of the following two kinds: (1) problems which arise only when the parameter space is more complex than one-dimensional, and (2) important properties of one-dimensional processes which are not easily extended, because they depend on the parameter space being one dimensional. As an example of category (1), we have the rich interplay between the probabilistic properties of a random field and the geometry of its parameter space. Although this interplay already appears in the one-dimensional case in the form of stationarity, the geometry of the real line is obviously both degenerate and rather trivial by comparison with the geometry of higher dimensions. As an example of category (2), consider Markov processes. The definition of a Markov process makes explicit use of the well orderedness of the real line. It is difficult to see how it can be generalized to a multidimensional parameter space. The way that it is done is one of the most interesting problems that we shall discuss in this chapter.

To avoid confusion, we shall adopt the following terminology: A collection of random variables defined on a common probability space will be called a **stochastic process** or a **random field** accordingly as its parameter space is one-dimensional or multidimensional. We should note that while this terminology is widely used, it is by no means universal. For example, a random field is often called a stochastic process with a several-dimensional time.

2. HOMOGENEOUS RANDOM FIELDS

The simplest generalization is to take the parameter space to be an n-dimensional interval, that is, $A = \{(z_1, z_2, \ldots, z_n): z_i \in I_i\}$ where each I_i is an interval, closed, open, or semiclosed. For a random field $\{X(\omega,z),\ z \in A\}$ most of our discussion in Chaps. 2 and 3 carry over with no difficulty, e.g., continuity questions, separability, Karhunen-Loève expansion. Results on wide-sense stationary process can also be generalized to random fields with parameter space R^n. Here, the generalization is interesting and not entirely trivial.

Let $\{X_z,\ z \in R^n\}$ be a family of second-order random variables defined on a probability space $(\Omega,\mathcal{A},\mathcal{P})$. We assume that X_z is continuous in probability, i.e., for every $\epsilon > 0$,

$$\mathcal{P}(|X_z - X_{z'}| \geq \epsilon) \xrightarrow[\|z=z'\|\to 0]{} 0 \tag{2.1}$$

where $\|z\|$ denotes the Euclidean norm

$$\|z\| = \left(\sum_{i=1}^{n} z_i^2\right)^{\frac{1}{2}} \tag{2.2}$$

As in Chap. 3, we allow X_z to be complex valued. The most straightforward generalizations to wide-sense stationary processes are **homogeneous random fields** defined as follows: We say that $\{X_z,\ z \in R^n\}$ is homogeneous if $EX_z = \mu$ does not depend on z and

$$EX_z\overline{X}_{z'} = EX_{z+z_0}\overline{X}_{z'+z_0} \tag{2.3}$$

for all z_0, z, z' in R^n. Setting $z_0 = -z'$ in (2.3), we see that the covariance function

$$E(X_z - EX_z)\overline{(X_{z'} - EX_{z'})} = R(z - z') \tag{2.4}$$

depends only on $z - z'$. Of course, $R(z - z')$ is also nonnegative definite; i.e., for any finite number of points $z_1, z_2, \ldots, z_N$ in R^n, and any collection of complex constants $a_1, a_2, \ldots, a_N$, we have

$$\sum_{i,j=1}^{N} a_i\bar{a}_jR(z_i - z_j) \geq 0 \tag{2.5}$$

Bochner's theorem (Proposition 3.5.1) can now be generalized as follows.

Proposition 2.1. A function $R(z)$, $z \in R^n$, is the covariance function of a homogeneous q.m. continuous random field if and only if it is of the form

$$R(z) = \int_{R^n} e^{i2\pi(\nu,z)}F(d\nu) \tag{2.6}$$

Where F is a finite Borel measure on R^n, and (ν,z) denotes the inner product

$$(\nu,z) = \sum_{i=1}^{n} \nu_i z_i \tag{2.7}$$

Proposition 2.1 can be proved in exactly the same way as Proposition 3.5.1, and we won't repeat it here. Similarly, we can obtain a spectral representation for $\{X_z, z \in R^n\}$ of the form

$$X_z = \int_{R^n} e^{i2\pi(\nu,z)} \hat{X}(d\nu) \tag{2.8}$$

where $\hat{X}(\cdot)$ is a random set function defined on Borel sets of R^n such that

$$E\hat{X}(A)\hat{X}(B) = F(A \cap B) \tag{2.9}$$

F being defined by (2.6). An integral of the form

$$\int_{R^n} f(\nu)\hat{X}(d\nu)$$

can be defined for any $f \in L^2(F)$ as the q.m. limit of a sequence of random variables resulting from approximating f by a sequence $\{f_k\}$ where each f_k is a linear combination of indicator functions of Borel sets and

$$\int_{R^n} |f(\nu) - f_k(\nu)|F(d\nu) \xrightarrow[k\to\infty]{} 0$$

The details are nearly identical to those given in Sec. 3.6.

Translations: $z \to z + z_0$ are transformations of R^n onto R^n which leave the Euclidean distance $\|z - z'\|$ between any pair of points unchanged. However, these are not the only transformations of R^n onto R^n which have this property. Transformations t: $R^n \to R^n$ of the form

$$\begin{aligned} t(z) &= (t_1(z), \ . \ . \ . \ , t_n(z)) \\ t_i(z) &= \sum_{j=1}^{1} a_{ij}z_1 \end{aligned} \tag{2.10}$$

where $\mathbf{A} = (a_{ij})$ is an orthogonal matrix, also preserve Euclidean distances. Such a transformation is called a **rotation,** and is often further designated as being **proper** or **improper** according as the determinant of $\mathbf{A}$ is $+1$ or -1. It follows that any succession of rotations and translations also preserves Euclidean distances. It also turns out that every transformation t of R^n onto R^n preserving Euclidean distances can be represented as a translation followed by a rotation. Such distance preserving transformations are called **rigid-body motions.** The collection of all rigid-body motions forms a group G which is called the **Euclidean group.**

Now consider a second-order random field $\{X_z, z \in R^n\}$ such that $EX_z = \mu$ is a constant which we assume to be zero and

$$EX_z\bar{X}_{z'} = EX_{t(z)}\bar{X}_{t(z')} \tag{2.11}$$

for all $z, z' \in R^n$ and all rigid-body motions t. Suppose that we move z'

to the origin **0** by translation, then z is moved to $z - z'$. If we follow this by a rotation which takes $z - z'$ into $(\|z - z'\|, 0, 0, \ldots, 0)$, the origin **0** is left invariant. Therefore, from (2.11) we have

$$\begin{aligned} EX_z\bar{X}_{z'} &= EX_{(\|z-z'\|,0,\ldots,0)}\bar{X}_{(0,0,\ldots,0)} \\ &= R(\|z - z'\|) \end{aligned} \tag{2.12}$$

which is a function of a single, real positive variable $\|z - z'\|$. A random field satisfying (2.11) will be called a **homogeneous and isotropic random field.** It is obviously homogeneous.

For an isotropic and homogeneous random field, it is advantageous to adopt a polar-coordinate system. We define the $(n - 1)$-dimensional unit sphere S^{n-1} as the set of all points z in R^n such that $\|z\| = 1$. An arbitrary point z in R^n can be uniquely identified by its distance from the origin **0** which is $\|z\|$ and the point on S^{n-1} intersected by a line connecting z and **0** which is

$$\frac{z}{\|z\|} = \left(\frac{z_1}{\|z\|}, \frac{z_2}{\|z\|}, \ldots, \frac{z_n}{\|z\|}\right)$$

Therefore any $z \in R^n$ is uniquely identified by $\|z\|$ and any $n - 1$ coordinates identifying a point of S^{n-1}. We choose a coordinate system for S^{n-1}, recursively, as follows: S^1 is the unit circle in R^2, and every point on S^1 has the form

$$z = (\cos\theta_1, \sin\theta_1) \qquad 0 \le \theta_1 < 2\pi \tag{2.13}$$

Hence, S^1 has a single coordinate θ_1. For S^n, consider the point $z = (1, 0, \ldots, 0) \in R^{n+1}$ and call it the **north pole.** The set of all points on S^n with $z_1 = \cos\theta_n$ forms an $(n - 1)$-dimensional sphere with radius $\sin\theta_n$. If a coordinate system $(\theta_1, \theta_2, \ldots, \theta_{n-1})$ has already been chosen for S^{n-1}, then this automatically produces a coordinate system $(\theta_1, \theta_2, \ldots, \theta_n)$ for S^n. Starting from (2.13), we can thus generate a coordinate system $(\theta_1, \ldots, \theta_n)$ for S^n with the property that θ_n = constant represents a sphere of dimension $n - 1$. We have also generated a coordinate system for R^n where a point z is represented by $(r, \theta_1, \ldots, \theta_{n-1})$ with $r = \|z\|$, and $(\theta_1, \ldots, \theta_{n-1})$ is the intersection between S^{n-1} and a line connecting z and the origin **0**. The origin **0** has a degenerate representation in this system given by $r = 0$ with $\theta_1, \ldots, \theta_{n-1}$ undefined.

Starting from (2.6), we can now find the form of Bochner's theorem for an isotropic and homogeneous random field. Since an isotropic and homogeneous random field is necessarily homogeneous, we have

$$R(\|z\|) = \int_{R^n} e^{i2\pi(\nu,z)}F(d\nu) \tag{2.14}$$

For any *rotation* t,

$$R(\|t(z)\|) = R(\|z\|)$$
$$= \int_{R^n} e^{i2\pi(\nu,t(z))} F(d\nu)$$
$$= \int_{R^n} e^{i2\pi(\nu,z)} F(d\nu) \tag{2.15}$$

Since $(\nu,t(z)) = (t(t^{-1}(\nu)),t(z)) = (t^{-1}(\nu),z)$, we have, upon a change in the variable of integration,

$$\int_{R^n} e^{i2\pi(\nu,t(z))} F(d\nu) = \int_{t(R^n)} e^{i2\pi(\lambda,z)} F(t(d\lambda))$$
$$= \int_{R^n} e^{i2\pi(\nu,z)} F(t(d\nu)) \tag{2.16}$$

It follows from (2.15) and (2.16) that for every $z \in R^n$ and every rotation t,

$$\int_{R^n} e^{i2\pi(\nu,z)} F(d\nu) = \int_{R^n} e^{i2\pi(\nu,z)} F(t(d\nu)) \tag{2.17}$$

Therefore, the Borel measure F must be isotropic, that is,

$$F(A) = F(t(A)) \tag{2.18}$$

for every Borel set A and every rotation t. We can now take z to be directed along the north pole (that is, $z_1 = \|z\|$, $z_2 = 0$, $z_3 = 0, \ldots$), and if we make use of the isotropy of F, then (2.14) becomes

$$R(\|z\|) = \int_0^\infty \int_{S^{n-1}} e^{i2\pi\lambda\|z\| \cos \varphi_{n-1}} \, d\Omega_{\varphi_1,\ldots,\varphi_{n-1}} F_0(d\lambda) \tag{2.19}$$

where $d\Omega$ is the differential surface element of S^{n-1}.

The set of all points on S^{n-1} with a fixed φ_{n-1} (that is, at a fixed distance from the north pole) is a sphere of dimension $n - 2$ with radius $\sin \varphi_{n-1}$, the surface area of which must be $\sin^{n-2} \varphi_{n-1}$ area (S^{n-2}). Since the integrand in (2.19) depends only on φ_{n-1}, we have

$$R(\|z\|) = K \int_0^\infty \int_0^\pi e^{i2\pi\lambda\|z\| \cos \varphi_{n-1}} \sin^{n-2} \varphi_{n-1} \, d\varphi_{n-1} F_0(d\lambda) \tag{2.20}$$

The constant K is just the total area of S^{n-2} and can be absorbed into F_0. The inside integral in (2.20) can be evaluated to be

$$\int_0^\pi e^{i2\pi\lambda\|z\| \cos \varphi} \sin^{n-2} \varphi \, d\varphi = \frac{J_{(n-2)/2}(2\pi\lambda\|z\|)}{(2\pi\lambda\|z\|)^{(n-2)/2}} \tag{2.21}$$

where J is the Bessel function. Hence, the isotropic version of Bochner's theorem is given as follows [see, e.g., Yaglom, 1962, pp. 81–86].

Proposition 2.2. A function $R(r)$, $0 \le r < \infty$, is the covariance function of an isotropic and homogeneous q.m. continuous random field if and only if

$$R(r) = \int_0^\infty \frac{J_{(n-2)/2}(\lambda r)}{(\lambda r)^{(n-2)/2}} F_0(d\lambda) \tag{2.22}$$

where F_0 is a finite Borel measure on $[0, \infty)$.
We shall call F_0 the **spectral measure** for the random field. We note that the constant 2π in (2.21) is absorbed into λ in (2.22) to result in a simpler formula.

For an isotropic and homogeneous random field, the spectral representation (2.8) assumes a special form which involves a countable family of random set functions defined on the Borel sets of $[0, \infty)$ in place of the n-dimensional set function $\hat{X}$. This spectral-representation formula is expressible in terms of spherical harmonics which will be introduced in the next section. In the process, we shall also introduce isotropic random fields which are not necessarily homogeneous.

3. SPHERICAL HARMONICS AND ISOTROPIC RANDOM FIELDS

We have already defined a unit n-dimensional sphere S^n as the set of all points in R^{n+1} at unit distance from the origin. Starting from (2.13), we can find a natural coordinate system for S^n so that an arbitrary point $\boldsymbol{\theta} \in S^n$ has a representation $\boldsymbol{\theta} = (\theta_1, \theta_2, \ldots, \theta_n)$ in which θ_n is the angle of $\boldsymbol{\theta}$ from the north pole. The north pole is thus defined by having $\theta_n = 0$, and for the north pole $\theta_1, \theta_2, \ldots, \theta_{n-1}$ are not specified.

The sphere S^n is obviously invariant under any rotation (proper or improper) of R^{n+1}, and the set of all rigid-body motions of R^{n+1} which leave S^n invariant is just the set of rotations. Suppose that we define a metric d on S^n by

$$d(\boldsymbol{\theta},\boldsymbol{\theta}') = \psi(\boldsymbol{\theta},\boldsymbol{\theta}') \tag{3.1}$$

where $\psi(\boldsymbol{\theta},\boldsymbol{\theta}')$ is just the angle between the two straight lines connecting the origin in R^{n+1} to $\boldsymbol{\theta}$ and $\boldsymbol{\theta}'$. It is easy to verify that (3.1) defines a metric. Now, suppose that we consider the set of all transformations $t\colon S^n \to S^n$ such that

$$d(\boldsymbol{\theta},\boldsymbol{\theta}') = d(t(\boldsymbol{\theta}),t(\boldsymbol{\theta}')) \tag{3.2}$$

It turns out that this set is just the set of rotations in R^{n+1}. This set is denoted by $G(S^n)$ and is easily verified to be a group. If we denote by $G(R^{n+1})$, the group of rigid-body motions in R^{n+1}, and by $T(R^{n+1})$, the group of all translations in R^{n+1}, then we have

$$G(R^{n+1}) = G(S^n) \times T(R^{n+1}) \tag{3.3}$$

We also note that $T(R^{n+1})$ is isomorphic to R^{n+1}, since to each $t \in R^{n+1}$

$$t(z) = z + t \tag{3.4}$$

This observation and (3.3) allow us to identify R^{n+1} as $G(R^{n+1})/G(S^n)$ where the latter denotes the collection of equivalence classes of rigid motions, each equivalence class being made up of the set of all motions which take the origin to the same point.

Now, consider the Laplacian operator on R^{n+1} which we denote by $\Delta(R^{n+1})$. In Cartesian coordinates, we have

$$\Delta(R^{n+1}) = \frac{\partial^2}{\partial z_1^2} + \frac{\partial^2}{\partial z_2^2} + \cdots + \frac{\partial^2}{\partial z_{n+1}^2} \tag{3.5}$$

As domain of $\Delta(R^{n+1})$, we take the set of all complex-valued functions on R^{n+1} with bounded-continuous second partials with respect to each z_i, and we denote this set by $C^2(R^{n+1})$. For each rigid-body motion $g \in G(R^{n+1})$, we can define a mapping T_g: $C^2(R^{n+1})$ onto $C^2(R^{n+1})$ by

$$T_g f(z) = f(g(z)) \tag{3.6}$$

It is both well known and easily verified that for each $g \in G(R^{n+1})$, T_g commutes with $\Delta(R^{n+1})$. That is,

$$T_g(\Delta f) = \Delta(T_g f) \qquad g \in G(R^{n+1}) \tag{3.7}$$

The easiest proof of this fact is probably by representing each function in $C^2(R^{n+1})$ in terms of its Fourier integral.

In terms of polar coordinates, whereby each point in R^{n+1} is represented by a pair $(r,\boldsymbol{\theta}) \in [0, \infty) \times S^n$, the Laplacian can be rewritten as

$$\Delta(R^{n+1}) = \frac{1}{r^n}\frac{\partial}{\partial r}\left(r^n \frac{\partial}{\partial r}\right) + \frac{1}{r^2}\Delta(S^n) \tag{3.8}$$

where $\Delta(S^n)$, the Laplacian of the n sphere, can be recursively defined as follows:

$$\begin{aligned} \Delta(S^1) &= \frac{\partial^2}{\partial \theta_1^2} \\ \Delta(S^n) &= \frac{1}{\sin^{n-1}\theta_n}\frac{\partial}{\partial \theta_n}\left(\sin^{n-1}\theta_n \frac{\partial}{\partial \theta_n}\right) + \frac{1}{\sin^2\theta_n}\Delta(S^{n-1}) \end{aligned} \tag{3.9}$$

From (3.8) it is easy to see that $\Delta(S^n)$ commutes with each T_g, $g \in G(S^n)$, since r is left invariant by any $g \in G(S^n)$, and T_g commutes with $\Delta(R^{n+1})$. It is now convenient to take as the domain of $\Delta(S^n)$ the space $C^2(S^n)$ of functions in $C^2(R^{n+1})$ which do not depend on the radial distance r. Now, consider the eigenvalues and eigenfunctions of $\Delta(S^n)$. An **eigenvalue** λ of $\Delta(S^n)$ is any complex number such that the equation

$$\Delta(S^n)f = \lambda f \qquad f \in C^2(S^n) \tag{3.10}$$

has a solution f which is not identically zero. For an eigenvalue λ, any not-identically-zero f satisfying (3.10) is called an **eigenfunction** corresponding to (3.10). Because $\Delta(S^n)$ is self-adjoint, the eigenvalues must be real. To find the eigenvalues, first we seek those eigenvalues such that (3.10) has a solution f which depends only on θ_n and not on $(\theta_1, \theta_2, \ldots, \theta_{n-1})$. For such a function we have

$$\frac{1}{\sin^{n-1}\theta_n}\frac{d}{d\theta_n}\left(\sin^{n-1}\theta_n\frac{df(\theta_n)}{d\theta_n}\right) = \lambda f(\theta_n) \tag{3.11}$$

If we set $\cos\theta_n = x$ and $f(\theta_n) = h(\cos\theta_n)$, then (3.11) becomes

$$(1-x^2)\frac{d^2h(x)}{dx^2} - nx\frac{dh(x)}{dx} - \lambda h(x) = 0 \tag{3.12}$$

which is an equation of familiar form and is sometimes called the **Gegenbauer equation.** By seeking a power series solution, it is easy to discover that a bounded h exists if and only if λ is of the form

$$\lambda_m = -m(m+n-1) \qquad m = 0, 1, 2, \ldots \tag{3.13}$$

and for $n \geq 2$, the corresponding eigenfunctions are the Gegenbauer polynomials defined by

$$\begin{aligned} f_n(\theta_n) &= h_m(\cos\theta_n) = C_m^{(n-1)/2}(\cos\theta_n) \\ &= K_n\int_0^\pi (\cos\theta_n + i\sin\theta_n\cos\varphi)^m \sin^{n-2}\varphi\, d\varphi \end{aligned} \tag{3.14}$$

where the constant K_n is such that $f_m(0) = 1$.

We have thus found a subset of eigenvalues for (3.10), and for each $\lambda_m = -m(m+n-1)$ we have found an eigenfunction of the form

$$f_m(\boldsymbol{\theta}) = C_m^{(n-1)/2}(\cos\theta_n) \qquad \boldsymbol{\theta} \in S^n \tag{3.15}$$

For each $g \in G(S^n)$ we can define a linear operator $T_g: C^2(S^n) \to C^2(S^n)$ by

$$(T_g f)(\boldsymbol{\theta}) = f(g(\boldsymbol{\theta})) \tag{3.16}$$

Since each T_g commutes with $\Delta(S^n)$, for each g and each f_m given by (3.15), the function $T_g f_m$ is again an eigenfunction corresponding to the same eigenvalue $\lambda_m = -m(m+n-1)$. Now consider the Hilbert space $L^2(S^n)$ generated by complex-valued functions on S^n which are square integrable with respect to the uniform measure $d0$ $\left(\int_{S^n} d0 = 1\right)$. For each m, the set of functions $\{T_g f_m,\ g \in G(S^n)\}$ spans a finite-dimensional subspace $\mathcal{H}_m$ of $L^2(S^n)$ with dimension d_m. For $n \geq 2$, we have

$$1 = d_0 < d_1 < d_2 < \cdots \tag{3.17}$$

For $m \neq m'$, $\mathcal{H}_m$ and $\mathcal{H}_{m'}$ are orthogonal. We now choose a real orthonormal basis $\{h_{ml}^{(n)}, l = 1, \ldots, d_m\}$ for $\mathcal{H}_m$, where we always take $h_{m1}^{(n)}$ to be proportional to f_m as defined by (3.15). It turns out that the set of functions

$$\{h_{ml}^{(n)}, l = 1, \ldots, d_m, m = 0, 1, \ldots\} \tag{3.18}$$

is a basis for $L^2(S^n)$. Since $L^2(S^n)$ contains $C^2(S^n)$, this means that (3.13) exhausts all eigenvalues for (3.10), and for each eigenvalue λ_m every eigenfunction is a linear combination of $\{h_{ml}^{(n)}, l = 1, 2, \ldots, d_m\}$. We shall call the functions $\{h_{ml}^{(n)}\}$ **spherical harmonics** [Erdélyi, 1953, Chap. 11].

Let F be a function defined on $S^n \times S^n$ such that for every $g \in G(S^n)$ and every $(\theta,\theta') \in S^n \times S^n$,

$$F(g(\boldsymbol{\theta}),g(\boldsymbol{\theta}')) = F(\boldsymbol{\theta},\boldsymbol{\theta}') \tag{3.19}$$

Let $\mathbf{0}$ denote the north pole as before. Then for every rotation τ which leaves $\mathbf{0}$ fixed, we have

$$F(\boldsymbol{\theta},\mathbf{0}) = F(\tau(\boldsymbol{\theta}),\mathbf{0}) \tag{3.20}$$

Therefore, $F(\boldsymbol{\theta},\mathbf{0})$ can depend only on θ_n, that is, the last component of $\boldsymbol{\theta}$. Suppose that $F(\cdot,\mathbf{0}) \in \mathcal{H}_m$. That is, suppose that

$$\Delta(S^n)F(\cdot,\mathbf{0}) = -m(m + n - 1)F(\cdot,\mathbf{0}) \tag{3.21}$$

Then it follows that $F(\boldsymbol{\theta},\mathbf{0})$ as a function θ_n must satisfy (3.11) corresponding to $\lambda = -m(m + n - 1)$. This means that we must have

$$F(\boldsymbol{\theta},\mathbf{0}) = Kf_m(\boldsymbol{\theta}) = KC_m^{(n-1)/2}(\cos \theta_n) \tag{3.22}$$

where K is a constant. Now, for any fixed pair $(\boldsymbol{\theta},\boldsymbol{\theta}')$ there always exists a motion g which simultaneously takes $\boldsymbol{\theta}'$ into $\mathbf{0}$ and $\boldsymbol{\theta}$ into $(0, 0, \ldots, \psi(\boldsymbol{\theta},\boldsymbol{\theta}'))$ where $\psi(\boldsymbol{\theta},\boldsymbol{\theta}')$ denotes the angle between $\boldsymbol{\theta}$ and $\boldsymbol{\theta}'$. This means that

$$F(\boldsymbol{\theta},\boldsymbol{\theta}') = KC_m^{(n-1)/2}(\cos \psi(\boldsymbol{\theta},\boldsymbol{\theta}')) \tag{3.23}$$

What we have shown is that if F satisfies (3.19) and (3.21) then it must be of the form (3.23).

Let $\{h_{ml}^{(n)}\}$ denote the spherical harmonics that we defined earlier. Consider the bilinear sum

$$F(\boldsymbol{\theta},\boldsymbol{\theta}') = \sum_{l=1}^{d_m} h_{ml}^{(n)}(\boldsymbol{\theta})h_{ml}^{(n)}(\boldsymbol{\theta}') \tag{3.24}$$

We shall show that the F in (3.24) satisfies both (3.19) and (3.21). For an arbitrary $g \in C(S^n)$, $\mathcal{H}_m$ is invariant under T_g so that we can write

$$(T_g h_{ml}^{(n)})(\boldsymbol{\theta}) = h_{ml}^{(n)}(g(\boldsymbol{\theta})) = \sum_{l'=1}^{d_m} \alpha_{ll'}(g)h_{ml'}^{(n)}(\boldsymbol{\theta}) \tag{3.25}$$

Because

$$\int_{S^n} h_{mk}^{(n)}(g(\boldsymbol{\theta}))h_{ml}^{(n)}(g(\boldsymbol{\theta}))\,dO = \int_{S^n} h_{mk}^{(n)}(\boldsymbol{\theta})h_{ml}^{(n)}(\boldsymbol{\theta})\,dO = \delta_{kl} \tag{3.26}$$

we find that for each $g \in C(S^n)$,

$$\sum_{l'=1}^{d_m} \alpha_{ll'}(g)\alpha_{kl'}(g) = \delta_{kl} \tag{3.27}$$

This means that $\mathbf{A}(g) = [\alpha_{ij}(g)]$ is an orthogonal matrix so that we also have

$$\sum_{l'=1}^{d_m} \alpha_{l'l}(g)\alpha_{l'k}(g) = \delta_{kl} \tag{3.28}$$

It follows that

$$\begin{aligned}\sum_{l=1}^{d_m} h_{ml}^{(n)}(g(\boldsymbol{\theta}))h_{ml}^{(n)}(g(\boldsymbol{\theta}')) &= \sum_{l'=1}^{d_m}\sum_{k'=1}^{d_m} h_{ml'}^{(n)}(\boldsymbol{\theta})h_{mk'}^{(n)}(\boldsymbol{\theta}) \sum_{l=1}^{d_m} \alpha_{ll'}(g)\alpha_{lk'}(g) \\ &= \sum_{l=1}^{d_m} h_{ml}^{(n)}(\boldsymbol{\theta})h_{ml}^{(n)}(\boldsymbol{\theta}')\end{aligned} \tag{3.29}$$

Hence, the function F defined by (3.24) satisfies (3.19). It is obvious that it also satisfies (3.21) so that

$$\sum_{l=1}^{d_m} h_{ml}^{(n)}(\boldsymbol{\theta})h_{ml}^{(n)}(\boldsymbol{\theta}') = KC_m^{(n-1)/2}(\cos\psi(\boldsymbol{\theta},\boldsymbol{\theta}')) \tag{3.30}$$

The constant K can be evaluated as follows:

$$\begin{aligned}\int_{S^n} \sum_{l=1}^{d_m} h_{ml}^{(n)}(\boldsymbol{\theta})h_{ml}^{(n)}(\theta)\,dO &= d_m \\ &= KC_m^{(n-1)/2}(1) = K\end{aligned} \tag{3.31}$$

Therefore,

$$C_m^{(n-1)/2}(\cos\psi(\boldsymbol{\theta},\boldsymbol{\theta}')) = \frac{1}{d_m}\sum_{l=1}^{d_m} h_{ml}^{(n)}(\boldsymbol{\theta})h_{ml}^{(n)}(\boldsymbol{\theta}') \tag{3.32}$$

We can now apply the results on spherical harmonics to the problem of representing isotropic random fields. A second-order random field $\{X_z,\ g \in R^n\}$ is said to be isotropic if for every rotation ρ

$$EX_{\rho(z)} = EX_z \qquad z \in R^n \tag{3.33}$$

and

$$EX_{\rho(z)}\bar{X}_{\rho(z')} = EX_z\bar{X}_{z'} \qquad z,\ z' \in R^n \tag{3.34}$$

If we adopt a polar-coordinate system $z = (r,\boldsymbol{\theta})$, $r \in [0,\infty)$, $\boldsymbol{\theta} \in S^{n-1}$, then

(3.33) and (3.34) imply that

$$EX(r,\boldsymbol{\theta}) = \mu(r) \tag{3.35}$$

and

$$EX(r,\boldsymbol{\theta})\bar{X}(r',\boldsymbol{\theta}) = R(r, r', \cos \psi(\boldsymbol{\theta},\boldsymbol{\theta}')) \tag{3.36}$$

where $\psi(\boldsymbol{\theta},\boldsymbol{\theta}')$ is the angle between $\boldsymbol{\theta}$ and $\boldsymbol{\theta}'$. Suppose that we now seek a representation of the form

$$X(r,\boldsymbol{\theta}) = \sum_{m=0}^{\infty} \sum_{l=1}^{d_m} X_{ml}(r) h_{ml}^{(n-1)}(\boldsymbol{\theta}) \tag{3.37}$$

where $h_{ml}^{(n-1)}$ are the spherical harmonics. The orthonormality of the spherical harmonics yields

$$X_{ml}(r) = \int_{S^{n-1}} X(r,\boldsymbol{\theta}) h_{ml}^{(n-1)}(\boldsymbol{\theta})\, d0 \tag{3.38}$$

so that

$$\begin{aligned} EX_{ml}(r)\bar{X}_{m'l'}(r') &= \int_{S^{n-1}\times S^{n-1}} EX(r,\boldsymbol{\theta})\bar{X}(r',\boldsymbol{\theta}') h_{ml}^{(n-1)}(\boldsymbol{\theta}) \bar{h}_{m'l'}^{(n-1)}(\boldsymbol{\theta}')\, d0\, d0' \\ &= \int_{S^{n-1}\times S^{n-1}} R[r, r', \cos \psi(\boldsymbol{\theta},\boldsymbol{\theta}')] h_{ml}^{(n-1)}(\boldsymbol{\theta}) \bar{h}_{m'l'}^{(n-1)}(\boldsymbol{\theta}')\, d0\, d0' \end{aligned} \tag{3.39}$$

If $\{X_z, z \in R^n\}$ is q.m. continuous, then $R(r, r', \cos \psi)$ can be expanded in terms of $C_m^{(n-1)/2}(\cos \psi)$ as

$$R(r, r', \cos \psi) = \sum_{m=0}^{\infty} d_m R_m(r,r') C_m^{(n-1)/2}(\cos \psi) \tag{3.40}$$

The bilinear form (3.32) and (3.40) can now be used in (3.39) to yield

$$EX_{ml}(r)X_{m'l'}(r') = \delta_{mm'}\delta_{ll'}R_m(r,r') \tag{3.41}$$

This means that $\{X_{ml}(r)\}$ is a countable family of orthogonal one-dimensional stochastic processes.

Suppose that $\{X_z, z \in R^n\}$ is not only isotropic, but also homogeneous; then we know from (2.8) that we can write

$$X_z = \int_{R^n} e^{i2\pi(\nu,z)} \hat{X}(d\nu) \tag{3.42}$$

where $\hat{X}$ is a random set function defined on the Borel sets of R^n. Now, adopt a polar-coordinate system for both ν and z in (3.42) so that

$$2\pi\nu = (\lambda,\boldsymbol{\phi}) \qquad \begin{aligned} &0 \le \lambda < \infty \\ &\boldsymbol{\phi} \in S^{n-1} \end{aligned}$$

$$z = (r,\boldsymbol{\theta}) \qquad \begin{aligned} &0 \le r < \infty \\ &\boldsymbol{\theta} \in S^{n-1} \end{aligned}$$

It is obvious that

$$\Delta(R^n)e^{i2\pi(\nu,\cdot)} = -\lambda^2 e^{i2\pi(\nu,\cdot)}$$

It follows that we must be able to write

$$e^{i\lambda r\cos\psi(\boldsymbol{\theta},\mathbf{0})} = e^{i\lambda r\cos\theta_{n-1}} = \sum_{m=0}^{\infty} C_m{}^{(n-2)/2}(\cos\theta_{n-1})f_m(\lambda r) \tag{3.43}$$

where f_m satisfies

$$\frac{1}{r^{n-1}}\frac{d}{dr}\left(r^{n-1}\frac{d}{dr}f_m(\lambda r)\right) - \frac{m(m+n-2)}{r^2}f_m(\lambda r) = -\lambda^2 f_m(\lambda r) \tag{3.44}$$

$$f_m(\lambda r) = K_m\frac{J_{(n-2)/2+m}(\lambda r)}{(\lambda r)^{(n-2)/2}} \tag{3.45}$$

Therefore, from (3.43) and (3.32), we have

$$\begin{aligned} e^{i\lambda r\cos\psi(\theta,\varphi)} &= \sum_{m=0}^{\infty} K_m\frac{J_{(n-2)/2+m}(\lambda r)}{(\lambda r)^{(n-2)/2}}C_m{}^{(n-2)/2}(\cos\psi(\boldsymbol{\theta},\boldsymbol{\phi}))\\ &= \sum_{m=0}^{\infty}\sum_{l=1}^{d_m}\frac{K_m}{d_m}\frac{J_{(n-2)/2+m}(\lambda r)}{(\lambda r)^{(n-2)/2}}h_{ml}{}^{(n-1)}(\boldsymbol{\theta})h_{ml}{}^{(n-1)}(\boldsymbol{\phi}) \end{aligned} \tag{3.46}$$

We can now use (3.46) in (3.42) and get

$$X(r,\boldsymbol{\theta}) = \sum_{m=0}^{\infty}\sum_{l=1}^{d_m} h_{ml}{}^{(n-1)}(\boldsymbol{\theta})\int_0^{\infty}\frac{J_{(n-2)/2+m}(\lambda r)}{(\lambda r)^{(n-2)/2}}\hat{X}_{ml}(d\lambda) \tag{3.47}$$

where $\{\hat{X}_{ml}\}$ is a family of random set functions defined on the Borel sets of $[0,\infty)$ by the formula

$$\hat{X}_{ml}(A) = \frac{K_m}{d_m}\int_{A\times S^{n-1}} h_{ml}{}^{(n-1)}(\boldsymbol{\phi})\hat{X}\left(\frac{d\lambda}{2\pi}\,d\boldsymbol{\phi}\right) \tag{3.48}$$

It is easy to verify that

$$E\hat{X}_{ml}(A)\bar{\hat{X}}_{m'l'}(B) = \frac{K_m{}^2}{d_m{}^2}\delta_{mm'}\delta_{ll'}\int_{A\cap B}F_0(d\lambda) \tag{3.49}$$

where F_0 is the Borel measures appearing in (2.22).

A comparison of (3.37) and (3.47) shows that for an isotropic and homogeneous random field, we have

$$X_{ml}(r) = \int_0^{\infty}\frac{J_{(n-2)/2+m}(\lambda r)}{(\lambda r)^{(n-2)/2}}\hat{X}_{ml}(d\lambda) \tag{3.50}$$

with

$$R_m(r,r') = \frac{K_m{}^2}{d_m{}^2}\int_0^{\infty}\frac{J_{(n-2)/2+m}(\lambda r)}{(\lambda r)^{(n-2)/2}}\frac{J_{(n-2)/2+m}(\lambda r')}{(\lambda r')^{(n-2)/2}}F_0(d\lambda) \tag{3.51}$$

Thus we see that an isotropic random field can be decomposed into a countable number of mutually uncorrelated stochastic processes with a one-dimensional parameter. If the random field is also homogeneous, then the correlation functions of these component processes can all be expressed in terms of a single Borel measure, as is done in (3.51). Results very similar to these can be obtained for random fields with parameter spaces which are certain differentiable manifolds with a Riemannian metric, when these random fields are homogeneous with respect to the motions which preserve this metric. More generally, representation results for random fields defined on homogeneous spaces can be obtained [Yaglom, 1961; Gangoli, 1968; Wong, 1969].

4. MARKOVIAN RANDOM FIELDS

One characterization of Markov processes is that the future and the past should be independent given the present (see 2.5.17). Lévy [1956] generalized the concept of Markov property to random fields with parameter space R^n by identifying the present with any smooth, closed $(n - 1)$ surface separating the parameter space into a bounded part (past) and an unbounded part (future). We should immediately note that this correspondence does not quite reduce to the usual identification when the parameter space is R^1, since a "closed" surface in R^1 corresponds to a pair of points rather than to a single point. If we had identified the present with any $(n - 1)$ surface separating the parameter space into two parts, and not just with closed surfaces, the definition for a Markov random field would correspond more closely to that of Markov processes. However, this would lead to other difficulties.

Let ∂D be a smooth (infinitely differentiable), closed surface separating R^n into a bounded part D^- and an unbounded part D^+. A random field $\{X_z, z \in R^n\}$ is said to be Markov if for any such ∂D, X_z and $X_{z'}$, $z \in D^-$, $z' \in D^+$, are independent, given $\{X_z, z \in \partial D\}$. Now suppose that $\{X_z, z \in R^n\}$ is a q.m. continuous zero-mean Gaussian isotropic and homogeneous random field. From (2.22), we know that the probability law of such a random field is completely specified by a finite Borel measure F_0 on $[0, \infty)$. A question of obvious interest is what must F_0 be so that the random field $\{X_z, z \in R^n\}$ is Markov? We shall answer this question in this section. The answer, unfortunately, is not very satisfactory in that the only such Markov fields turn out to be degenerate ones.

Suppose that $\{X_z, z \in R^n\}$ is a q.m. continuous Gaussian isotropic random field. From Sec. 3, we know that if we rewrite

$$X(r,\boldsymbol{\theta}) = \sum_{m=0}^{\infty} \sum_{l=1}^{d_m} X_{ml}(r) h_{ml}^{(n-1)}(\boldsymbol{\theta}) \tag{4.1}$$

then $\{X_{ml}(r)\}$ are mutually uncorrelated processes. Since $\{X_z, z \in R^n\}$ is Gaussian, $\{X_{ml}(\cdot)\}$ are in fact independent Gaussian processes with parameter space $[0, \infty)$. Now, suppose that $\{X_z, z \in R^n\}$ is Markov. Then each $X_{ml}(r), 0 \leq r < \infty$, is a Gaussian Markov process from the following reasoning: Because each $X_{ml}(\cdot)$ is Gaussian, we only need to prove that for each m and l,

$$E[X_{ml}(r)|X_{ml}(\rho),\ \rho \leq r_0 < r] = E[X_{ml}(r)|X_{ml}(r_0)] \tag{4.2}$$

Because $\{X_{ml}(\cdot)\}$ are mutually independent, we have

$$\begin{aligned} E[X_{ml}(r)|X_{ml}(\rho),\ \rho \leq r_0 < r] \\ &= E[X_{ml}(r)|X_{m'l'}(\rho), \text{ for all } m', l' \text{ and } \rho \leq r_0 < r] \\ &= E[X_{ml}(r)|X(\rho,\boldsymbol{\theta}),\ \rho \leq r_0 < r,\ \boldsymbol{\theta} \in S^{n-1}] \end{aligned}$$

Since $X_{ml}(r)$ is given by

$$X_{ml}(r) = \int_{S^{n-1}} h_{ml}^{(n-1)}(\boldsymbol{\theta}) X(r,\boldsymbol{\theta})\, d0$$

we can apply the Markov property of $\{X_z, z \in R^n\}$ with $\partial D =$ sphere with radius r_0, and find

$$\begin{aligned} E[X_{ml}(r)|X(\rho,\boldsymbol{\theta}),\ \rho \leq r_0 < r,\ \boldsymbol{\theta} \in S^{n-1}] &= E[X_{ml}(r)|X(r_0,\boldsymbol{\theta}),\ \boldsymbol{\theta} \in S^{n-1}] \\ &= E[X_{ml}(r)|X_{m'l'}(r_0), \text{ all } m', l'] \\ &= E[X_{ml}(r)|X_{ml}(r_0)] \end{aligned}$$

which is just (4.2). We can summarize this result as follows.

Proposition 4.1. Let $\{X_z, z \in R^n\}$ be a q.m. continuous Gaussian isotropic Markov random field. Define

$$X_{ml}(r) = \int_{S^{n-1}} X(r,\boldsymbol{\theta}) h_{ml}^{(n-1)}(\boldsymbol{\theta})\, d0 \qquad 0 \leq r < \infty \tag{4.3}$$

Then $\{X_{ml}(r),\ 0 \leq r < \infty,\ m = 0, 1, 2, \ldots ; l \leq d_m\}$ is a family of independent Gaussian Markov processes.

From (3.41) we know that the correlation function of $X_{ml}(\cdot)$ only depends on m, and from (2.5.14) we know that it must have the form

$$EX_{ml}(r)X_{ml}(r') = R_m(r,r') = f_m(\max(r,r'))g_m(\min(r,r')) \tag{4.4}$$

If $\{X_z, z \in R^n\}$ is also homogeneous, then we know further that R_m must be of the form [cf. (3.51)]

$$\begin{aligned} R_m(r,r') &= f_m(\max(r,r'))g_m(\min(r,r')) \\ &= A_m \int_0^\infty \frac{J_{(n-2)/2+m}(\lambda r)}{(\lambda r)^{(n-2)/2}} J_{(n-2)/2+m}(\lambda r') F_0(d\lambda) \end{aligned} \tag{4.5}$$

where A_m are positive constants. Let Δ_m denote the operator

$$(\Delta_m f)(r) = \frac{1}{r^{n-1}} \frac{d}{dr}\left(r^{n-1} \frac{df(r)}{dr}\right) - \frac{m(m+n-2)}{r^2} f(r) \tag{4.6}$$

For $r > r'$ consider

$$[\Delta_m R_m(\cdot,r')](r) = g_m(r')(\Delta_m f_m)(r) \tag{4.7}$$

and

$$[\Delta_m R_m(r,\cdot)](r') = f_m(r)(\Delta_m g_m)(r') \tag{4.8}$$

Because of (4.5), (3.44), and (3.45), we have

$$\begin{aligned}[\Delta_m R_m(\cdot,r')](r) &= A_m \int_0^\infty (-\lambda^2) \frac{J_{(n-2)/2+m}(\lambda r)}{(\lambda r)^{(n-2)/2}} \frac{J_{(n-2)/2+m}(\lambda r')}{(\lambda r')^{(n-2)/2}} F_0(d\lambda) \\ &= [\Delta_m R_m(r,\cdot)](r')\end{aligned} \tag{4.9}$$

so that for $r > r'$,

$$\frac{1}{f_m(r)} (\Delta_m f_m)(r) = \frac{1}{g(r')} (\Delta_m g_m)(r') \tag{4.10}$$

Since the two sides of (4.10) are functions of different variables, they must be equal to a constant, i.e.,

$$(\Delta_m f_m)(r) = \nu_m f_m(r) \qquad r > 0 \tag{4.11}$$

$$(\Delta_m g_m)(r) = \nu_m g_m(r) \qquad r > 0 \tag{4.12}$$

Now, consider

$$\begin{aligned} R(r) &= EX(r,\boldsymbol{\theta})X(0,\cdot) = EX_z X_0 \\ &= \sum_{m,l} R_{ml}(r,0) h_{ml}^{(n-1)}(\boldsymbol{\theta}) h_{ml}^{(n-1)}(\boldsymbol{\hat{\phi}}) \end{aligned}$$

It is obvious that we must have

$$R_{ml}(r,0) = 0 \qquad m \geq 1$$

because $h_{01}^{(n-1)}(\boldsymbol{\theta}) = 1$ and all other $h_{ml}^{(n-1)}$ are orthogonal to it. Therefore,

$$R(r) = R_{01}(r,0) = f_0(r)g_0(0) \tag{4.13}$$

It follows from (4.12) that $R(\cdot)$ must satisfy

$$\frac{1}{r^{n-1}} \frac{d}{dr}\left(r^{n-1} \frac{d}{dr} R(r)\right) = \nu_0 R(r) \qquad r > 0 \tag{4.14}$$

The only solution of (4.14) such that $R(0) < \infty$ and $R(\|z - z'\|)$ is nonnegative definite is given by

$$R(r) = \frac{AJ_{(n-2)/2}(\lambda_0 r)}{(\lambda_0 r)^{(n-2)/2}} \tag{4.15}$$

where λ_0 and A are nonnegative constants.

If $\lambda_0 = 0$ in (4.15) then the random field is just a single Gaussian random variable. It is Markov but only trivially so. If $\lambda_0 > 0$ then

$$X_{ml}(r) = X_{ml} \frac{J_{(n-2)/2+m}(\lambda_0 r)}{(\lambda_0 r)^{(n-2)/2}} \tag{4.16}$$

where $\{X_{ml}\}$ is a collection of independent Gaussian random variables. Since $J_{(n-2)/2+m}(\lambda_0 r)$ has zeros on $0 \leq r < \infty$, the processes $X_{ml}(r)$, $0 \leq r < \infty$ are not Markov.[1] By virtue of Proposition 4.1 the corresponding random field is not Markov.

Equation (4.14) has a second solution of the form

$$R(r) = A \frac{K_{(n-2)/2}(\lambda_0 r)}{(\lambda_0 r)^{(n-2)/2}} \tag{4.17}$$

for which $R(\|z - z'\|)$ is positive definite, but $R(0) = \infty$. Equation (4.17) is counterpart of the exponential-correlation function $R(\tau) = e^{-\alpha|\tau|}$ of a stationary Gaussian Markov process (Ornstein-Uhlenbeck process), and the correspondence becomes more evident when we write

$$\frac{K_{(n-2)/2}(\lambda_0 r)}{(\lambda_0 r)^{(n-2)/2}} = \int_0^\infty \frac{J_{(n-2)/2}(\lambda r)}{(\lambda r)^{(n-2)/2}} \frac{1}{(\lambda^2 + \lambda_0^2)} \lambda^{n-1}\, d\lambda \tag{4.18}$$

which suggests that the spectral density is of the form

$$\varphi(\lambda) = \frac{1}{\lambda^2 + \lambda_0^2} \tag{4.19}$$

Of course, (4.18) corresponds to an unbounded spectral measure

$$F_0(d\lambda) = \frac{\lambda^{n-1}\, d\lambda}{(\lambda^2 + \lambda_0^2)} \tag{4.20}$$

and there is no q.m. continuous random field with a correlation function given by (4.17). It is possible to define a generalized random field with a correlation function given by (4.17). It turns out that this generalized random field, if Gaussian, is Markovian in a well-defined sense. To extend the definition of a Markov random field to generalized random fields requires some care, but it can be done [Wong, 1969].

[1] This point was clarified for me by Professor Frank Spitzer.

Lévy [1948] defined a **Brownian motion** with parameter space R^n as a Gaussian random field $\{X_z, z \in R^n\}$ with zero mean and covariance function

$$EX_zX_{z'} = \tfrac{1}{2}(\|z\| + \|z'\| - \|z - z'\|) \tag{4.21}$$

For $n = 1$, it reduces to the ordinary Brownian motion on $(-\infty, \infty)$ as defined by (2.3.17). For odd-dimensional parameter spaces ($n = 2p + 1$), Lévy [1956] conjectured that $\{X_z, z \in R^{2p+1}\}$ is Markovian of order $p + 1$ in the following sense. A random field $\{X_z, z \in R^n\}$ is said to be Markov of order $\leq p + 1$ if, for any smooth, closed $(n - 1)$ surface ∂D, every approximation $\tilde{X}_z$ to X_z in a neighborhood of ∂D which has the property

$$\lim_{\delta \downarrow 0} \frac{1}{\delta^p} |\tilde{X}_z - X_z| = 0 \qquad \delta = \text{distance } (z, \partial D) \tag{4.22}$$

also has the property that given $\tilde{X}$, X_z and $X_{z'}$ are independent whenever $z \in D^-$ and $z' \in D^+$. If X_z is Markov of order $\leq p + 1$ but not $\leq p$, then it is said to be Markov of order $p + 1$. This conjecture in a somewhat different formulation was proved by McKean [1963]. McKean showed that for a Brownian motion on R^{2p+1}, given X and its "normal derivatives" $\partial^k X$, $k = 1, 2, \ldots, p$, on ∂D, X_z, $z \in D^-$ is independent of $X_{z'}$, $z \in D^+$. However, a Brownian motion is not even once differentiable, so that the normal derivatives need to be defined, which McKean has done. Brownian motions with an even-dimensional parameter space have no Markov property at all.

One way in which Markovian random fields (of some order) arise naturally is through differential equations driven by white noise, very much in the same way that diffusion processes are generated by white noise. It is not difficult to define white noise, Gaussian or not, with a multidimensional parameter. However, stochastic partial-differential equations as extensions of Ito equations have not been studied, except for linear equations of the form

$$\Delta X_z = kX_z + \eta_z$$

where η_z is a white noise. Several examples of this type have been given by Whittle [1963].

References

1. Baxter, G. (1956): A strong theorem for Gaussian processes, *Proc. Am. Math. Soc.*, **7**:522–528.

1*a*. Birkoff, G. and S. MacLane (1953): "A Survey of Modern Algebra," Macmillan, New York.

2. Breiman, L. (1968): "Probability," Addison-Wesley, Reading, Mass.

3. Bucy, R. S. and P. D. Joseph (1968): "Filtering for Stochastic Processes with Applications to Guidance," Interscience Publishers, Wiley, New York.

4. Cramér, H. (1966): On stochastic processes whose trajectories have no discontinuities of the second kind, *Ann. di Matematica (iv)*, **71**:85–92.

5. Davenport, W. B., Jr. and W. L. Root (1958): "An Introduction to the Theory of Random Signals and Noise," McGraw-Hill, New York.

6. Doob, J. L. (1953): "Stochastic Processes," Wiley, New York.

7. Duncan, T. E. (1968): Evaluation of likelihood functions, *Information and Control*, **13**:62–74.

8. Duncan, T. E. (1970): On the absolute continuity of measures, *Annals Math. Stat.*, **41**:30–38.

9. Dynkin, E. B. (1965): "Markov Processes" (2 vols.), Academic, New York, Springer-Verlag, Berlin.

10. Erdelyi, A. (1953): "Higher Transcendental Functions," vol. II, Bateman Manuscript Proj., McGraw-Hill, New York.

11. Feldman, J. (1958): Equivalence and perpendicularity of Gaussian processes, *Pacific J. Math.*, **8**:699–708.

12. Feldman, J. (1960): Some classes of equivalent Gaussian processes on an interval, *Pacific J. Math.*, **10**:1211–1220.

13. Fisk, D. L. (1963): Quasi-martingales and stochastic integrals, *Michigan State University, Dept. of Stat. Tech. Rept. No.* 1.

14. Fisk, D. L. (1965): Quasi-martingales, *Trans. Am. Math. Soc.*, **120**:369–389.

15. Fisk, D. L. (1966): Sample quadratic variation of sample continuous second-order martingales, *Z. Wahrscheinlichkeitstheorie verw. Geb.*, **6**:273–278.

16. Gangoli, R. (1967): Abstract harmonic analysis and Lévy's Brownian motion of several parameters, *Proc. 5th Berkeley Symp. Math. Stat. and Prob.*, **II-1**:13–30.

17. Girsanov, I. V. (1960): On transforming a certain class of stochastic processes by absolutely continuous substitution of measures, *Theory of Prob. and Appl.*, **5**:285–301.

18. Grenander, Ulf (1950): Stochastic processes and statistical inference, *Arkiv für Mathematik*, **1**:195–277.

19. Hajek, J. (1968): On a property of normal distribution of any stochastic process, *Cy. Math. J.*, **8**:610–617. (Also selected translations in *Math. Stat. Prob.*, **1**:245–253.)

20. Halmos, P. R. (1950): "Measure Theory," Van Nostrand, Princeton, N.J.

21. Ito, K. (1944): Stochastic integrals, *Proc. Imp. Acad. Tokyo*, **20**:519–524.

22. Ito, K. (1951*a*): On stochastic differential equations, *Mem. Amer. Math. Soc.*, **4**.

23. Ito, K. (1951*b*): On a formula concerning stochastic differentials, *Nagoya Math. J.*, **3**:55–65.

24. Jazwinski, A. H. (1970): "Stochastic Processes and Filtering Theory," Academic, New York.

25. Kac, M. (1951): On some connections between probability theory and differential and integral equations, *Proc. 2nd Berkeley Symp. on Math. Stat. and Prob.*, 189–215.

26. Kailath, T. (1969): A general likelihood-ratio formula for random signals in Gaussian noise, *IEEE Trans. Inf. Th.*, **IT-5**:350–361.

27. Kakutani, S. (1948): On equivalence of infinite product measures, *Ann. Math.*, **47**:214–224.

28. Kalman, R. E. and R. S. Bucy (1961): New results in linear filtering and prediction theory, *Trans. Am. Soc. Mech. Engn. Series D, J. Basic Eng.*, **83**:95–108.

29. Karhunen, K. (1947): Über linear Methoden in der Wahrscheinlichkeits-rechnung, *Ann. Acad. Sci. Fenn.*, **37**.

30. Kolmogorov, A. N. (1931): Über die analytische Methoden in der Wahrscheinlichkeitsrechnung, *Math. Ann.*, **104**:415–458.

31. Kunita, H. and S. Watanabe (1967): On square integrable martingales, *Nagoya Math. J.*, **30**:209–245.

32. Lévy, P. (1956): A special problem of Brownian motion, and a general theory of Gaussian random function, *Proc. 3rd Berkeley Symp. Math. Stat. and Prob.*, **2**:133–175.

33. Loève, M. (1963): "Probability Theory," 3d ed., Van Nostrand, Princeton, N.J.

34. McKean, H. P., Jr. (1960): The Bessel motion and a singular integral equation, *Mem. Coll. Sci. Univ. Kyota, Series A*, **33**:317–322.

35. McKean, H. P., Jr. (1963): Brownian motion with a several dimensional time, *Theory of Prob. and Appl.*, **8**:335–365.

36. McKean, H. P., Jr. (1969): "Stochastic Integrals," Academic, New York.

37. McShane, E. J. (1969): Toward a stochastic calculus, II, *Proc. National Academy of Sciences*, **63**:1084–1087.

38. McShane, E. J. (1970): On the use of stochastic differential equations in models of random processes, *Proc. 6th Berkeley Symp. Math. Stat. and Prob.*, to be published.

39. Meyer, P. A. (1966): "Probability and Potentials," Blaisdell, Waltham, Mass.

40. Mortensen, R. E. (1966): "Optimal control of continuous time stochastic systems," doctoral dissertation, Dept. of Electrical Engineering, University of California, Berkeley.

41. Neveu, J. (1965): "Mathematical Foundations of the Calculus of Probability," Amiel Feinstein (trans.), Holden-Day, Inc., San Francisco.

42. Paley, R. E. A. C. and N. Wiener (1934): "Fourier Transforms in the Complex Domain," *Amer. Math. Soc. Coll. Pub., Am. Math. Soc.*, **19.**

43. Parzen, E. (1961): An approach to time series analysis, *Ann. Math. Statist.*, **32**:951–989.

44. Prokhorov, Yu. V. (1956): Convergence of random processes and limit theorems in probability theory, *Theory of Prob. and Appl.*, **1**:157–214.

45. Riesz, F. and B. Sz.-Nagy (1955): "Functional Analysis," Ungar, New York.

46. Root, W. L. (1962): Singular Gaussian measures in detection theory, *Proc. Symp. Time Series Analysis*, Brown University, 1962, Wiley, New York, 1963, pp. 292–316.

47. Rudin, Walter (1966): "Real and Complex Analysis," McGraw-Hill, New York.

48. Shepp, L. A. (1966): Radon-Nikodym derivatives of Gaussian processes, *Ann. Math. Stat.*, **37**:321–354.

49. Skorokhod, A. V. (1965): "Studies in the Theory of Random Processes" (trans. from Russian), Addison-Wesley, Reading, Mass.

50. Slepian, D. (1958): Some comments on the detection of Gaussian signals in Gaussian noise, *IRE Trans. Inf. Th.*, **4**:65–68.

51. Stratonovich, R. L. (1966): A new form of representation of stochastic integrals and equations, *SIAM J. Control*, **4**:362–371.

52. Taylor, A. E. (1961): "Introduction to Functional Analysis," Wiley, New York.

53. Thomasian, A. J. (1969): "The Structure of Probability Theory with Applications," McGraw-Hill, New York.

54. Whittle, P. (1963): Stochastic processes in several dimensions, *Bull. Inst. Int. Statist.*, **40**:974–994.

55. Wiener, N. (1949): "Extrapolation, Interpolation, and Smoothing of Stationary Time Series," Wiley, New York.

56. Wiener, N. and P. Masani (1958): The prediction theory of multivariate stochastic processes—II, the linear predictor, *Acta Mathematica*, **99**:93–137.

57. Wong, E. (1964): The construction of a class of stationary Markoff processes, *Proc. Symp. in Appl. Math., Am. Math. Soc.*, **16**:264–276.

58. Wong, E. (1969): Homogeneous Gauss-Markov random fields, *Ann. Math. Stat.*, **40**:1625–1634.

59. Wong, E. and J. B. Thomas (1961): On the multidimensional prediction and filtering problem and the factorization of spectral matrices, *J. Franklin Institute*, **272**:87–99.

60. Wong, E. and M. Zakai (1965*a*): On the relationship between ordinary and stochastic differential equations, *Int. J. Engng. Sci.*, **3**:213–229.

61. Wong, E. and M. Zakai (1965*b*): On the convergence of ordinary integrals to stochastic integrals, *Ann. Math. Stat.*, **36**:1560–1564.

62. Wong, E. and M. Zakai (1965*c*): The oscillation of stochastic integrals, *Z. Wahrscheinlichkeitstheorie verw. Geb.*, **4**:103–112.

63. Wong, E. and M. Zakai (1966): On the relationship between ordinary and stochastic differential equations and applications to stochastic problems in control theory, *Proc. 3rd IFAC Congress*, paper 3B.

64. Wong, E. and M. Zakai (1969): Riemann-Stieltjes approximations of stochastic integrals, *Z. Wahrscheinlichkeitstheorie verw. Geb.*, **12**:87–97.

65. Wonham, W. M. (1964): Some applications of stochastic differential equations to optimal nonlinear filtering, *Tech. Rept.* 64-3, Feb. 1964, *RIAS*, Baltimore.

66. Wonham, W. M. (1970): Random Differential Equations in Control Theory, in "Probabilistic Methods in Applied Math.," vol. 2, Academic, New York.

67. Yaglom, A. M. (1961): Second-order homogeneous random fields, *Proc. 4th Berkeley Symp. Math. Stat. and Prob.*, **2**:593–620.

68. Yaglom, A. M. (1962): "An Introduction to the Theory of Stationary Random Functions," R. A. Silverman (trans.), Prentice-Hall, Englewood Cliffs, N.J.

69. Youla, D. C. (1961): On the factorization of rational matrices, *IRE Trans. Inf. Th.*, **IT-7**:172–189.

70. Zakai, M. (1969): On the optimal filtering of diffusion processes, *Z. Wahrscheinlichkeitstheorie verw. Geb.*, **11**:230–243.

Solutions to Exercises

CHAPTER 1

1. (*a*) We need only note that $[0,a) \cap [0, a + 1) = [a, a + 1) \notin C_1$.

(*b*) Let $A = \bigcup_{i=1}^{m} [a_i,b_i)$ and $B = \bigcup_{i=m+1}^{m+n} [a_i,b_i)$

Then

$$A \cup B = \bigcup_{i=1}^{m+n} [a_i,b_i) \qquad \text{and} \qquad A \cap B = \bigcup_{i=1}^{m} \bigcup_{j=m+1}^{m+n} [a_i,b_i) \cap [a_j,b_j)$$

But $[a_i,b_i) \cap [a_j,b_j)$ is either empty or of the form $[\min (a_i,b_j), \max (b_i,b_j))$. Hence, $A \cap B$ is again a finite union of intervals of the form $[a,b)$.

(*c*) Let C be any Boolean algebra containing C_1. Because $[a,b) = [0,b) \cap [0,a)$, C must contain all sets of the form $[a,b)$ and, hence, all finite unions of such sets. Hence, every Boolean algebra containing C_1 must also contain C_2. Because C_2 is a Boolean algebra, it must also be the smallest.

(*d*) $$[a,b] = \bigcap_{n=1}^{\infty} \left[a, b + \frac{1}{n}\right)$$

$$(a,b) = \bigcup_{n=1}^{\infty} \left[a + \frac{1}{n}, b\right)$$

$$(a,b] = \bigcap_{n=1}^{\infty} \left(a, b + \frac{1}{n}\right)$$

2. (*a*) See Exercise 1.

(*b*) If $\mathcal{P}$ is σ additive, then it is also sequentially continuous so that from Solution 1.1*d*, we have

$$
\begin{aligned}
\mathcal{P}([a,b]) &= \mathcal{P}\left(\lim_{n\to\infty}\left[a, b + \frac{1}{n}\right)\right) \\
&= \lim_{n\to\infty} \mathcal{P}\left(\left[a, b + \frac{1}{n}\right)\right) \\
&= \lim_{n\to\infty} P\left(b + \frac{1}{n}\right) - P(a) \\
&= P(b^+) - P(a) \\
\mathcal{P}((a,b)) &= \lim_{n\to\infty}\left\{P(b) - P\left(a + \frac{1}{n}\right)\right\} \\
&= P(b) - P(a^+) \\
\mathcal{P}((a,b]) &= P(b^+) - P(a^+)
\end{aligned}
$$

3. (*a*) Set $f^{-1}(A) = \{\mathbf{x}, f(\mathbf{x}) \in A\}$. If $A \in \mathcal{R}$, then $f^{-1}(A) \in \mathcal{R}^n$, because f is a Borel function. Hence,

$$
\begin{aligned}
P(f^{-1}(A)) &= \mathcal{P}(\{\omega: \mathbf{X}(\omega) \in f^{-1}(A)\} \\
&= \mathcal{P}(\{\omega: f(\mathbf{X}(\omega)) \in A\})
\end{aligned}
$$

(*b*) We can assume that f is nonnegative, because otherwise we can write $f = f^+ - f^-$ where f^+ and f^- are both nonnegative. If, in addition, f is also simple, i.e., there exist disjoint Borel sets $A_1, \ldots, A_m$ such that

$$\bigcup_{k=1}^{m} A_m = R^n$$

and

$$f(\mathbf{x}) = f_k \qquad \mathbf{x} \in A_k, \quad k = 1, \ldots, m,$$

then, by definition, we have

$$\int_{R^n} f(\mathbf{x})P(d\mathbf{x}) = \sum_{k=1}^{m} f_k P(A_k)$$

If we set $X^{-1}(A) = \{\omega: \mathbf{X}(\omega) \in A\}$, then we can write

$$
\begin{aligned}
\int_{R^n} f(\mathbf{x})P(d\mathbf{x}) &= \sum_{k=1}^{m} \int_{X^{-1}(A_k)} f(\mathbf{X}(\omega))\mathcal{P}(d\omega) \\
&= \int_{\Omega} f(\mathbf{X}(\omega))\mathcal{P}(d\omega)
\end{aligned}
$$

If f is not a simple function, then by definition

$$\int_{R^n} f(\mathbf{x})P(d\mathbf{x}) = \lim_{m\to\infty} \int_{R^n} f_m(\mathbf{x})P(d\mathbf{x})$$

when $\{f_m\}$ is a nondecreasing sequence of simple functions converging pointwise to f. We now have

$$\begin{aligned}\int_{R^n} f(\mathbf{x})P(d\mathbf{x}) &= \lim_{m\to\infty} \int_{R^n} f_m(\mathbf{x})P(d\mathbf{x}) \\ &= \lim_{m\to\infty} \int_\Omega f_m(\mathbf{X}(\omega))\mathcal{P}(d\omega) \\ &= \int_\Omega f(\mathbf{X}(\omega))\mathcal{P}(d\omega)\end{aligned}$$

where the last equality follows from monotone convergence.

6. Let $X_1 = Y\cos\theta$, $X_2 = Y\sin\theta\cos\Phi$, and $X_3 = Y\sin\theta\sin\Phi$. If we denote the joint density function of Y, θ, and Φ by p, then

$$p(y,\theta,\varphi) = \begin{vmatrix} \cos\theta & -y\sin\theta & 0 \\ \sin\theta\cos\varphi & y\cos\theta\cos\varphi & -y\sin\theta\sin\varphi \\ \sin\theta\sin\varphi & y\cos\theta\sin\varphi & y\sin\theta\cos\varphi \end{vmatrix}$$

$$p_X(y\cos\theta, y\sin\theta\cos\varphi, y\sin\theta\sin\varphi) = \frac{y^2\sin\theta}{(2\pi)^{\frac{3}{2}}} e^{-\frac{1}{2}y^2}$$

Therefore,

$$\begin{aligned}p_Y(y) &= \int_0^{2\pi} d\varphi \int_0^\pi d\theta \frac{1}{(2\pi)^{\frac{3}{2}}} y^2 \sin\theta\, e^{-\frac{1}{2}y^2} \\ &= \sqrt{\frac{2}{\pi}}\, y^2 e^{-\frac{1}{2}y^2} \qquad y \geq 0\end{aligned}$$

7. Since $Y_k = \sum_{j=1}^k X_j$, we have $X_1 = Y_1$ and

$$Y_k - Y_{k-1} = X_k \qquad k = 2, \ldots, n$$

Therefore,

$$p_Y(y_1, \ldots, y_n) = \begin{vmatrix} 1 & 0 & 0 & \cdots & 0 \\ -1 & 1 & 0 & \cdots & \cdot \\ 0 & -1 & 1 & \cdots & \cdot \\ \cdots & \cdots & \cdots & \cdots & \cdots \\ 0 & \cdots & \cdots & -1 & 1 \end{vmatrix} p_X(y_1, y_2 - y_1, \ldots, y_n - y_{n-1})$$

$$= \frac{1}{(2\pi)^{\pi/2}} \exp\left[-\frac{1}{2}\sum_{k=1}^n (y_k - y_{k-1})^2\right]$$

where $y_0 = 0$.

8. $$E\frac{|X|}{1+|X|} = \int_{|X|\geq\epsilon} \frac{|X|}{1+|X|}\, d\mathcal{P} + \int_{|X|<\epsilon} \frac{|X|}{1+|X|}\, d\mathcal{P}$$

This implies that

$$E\frac{|X|}{1+|X|} \geq \frac{\epsilon}{1+\epsilon}\mathcal{P}(|X| \geq \epsilon)$$

and

$$E\frac{|X|}{1+|X|} \leq \mathcal{P}(|X| \geq \epsilon) + \frac{\epsilon}{1+\epsilon}\mathcal{P}(|X| < \epsilon)$$

14. (*a*) Because Y is independent of X_2, we have

$$E(Y|X_2) = EY = EX_1 + \alpha EX_2 = 0$$

On the other hand

$$X_1 = Y_1 - \alpha X_2$$

so that

$$E(X_1|X_2) = E(Y_1|X_2) - \alpha X_2 = -\alpha X_2$$

15. Let $X_1 = Y\cos\Phi$. The joint density function of Y and Φ is given by

$$\begin{aligned} p(y,\varphi) &= \begin{vmatrix} \cos\varphi & -y\sin\varphi \\ \sin\varphi & y\cos\varphi \end{vmatrix} \frac{1}{2\pi} e^{-\frac{1}{2}(y^2\cos^2\varphi + y^2\sin^2\varphi)} \\ &= \frac{1}{2\pi} y e^{-\frac{1}{2}y^2} \end{aligned}$$

In other words, Y and Φ are independent, with Φ uniformly distributed on $[0,2\pi)$. Therefore,

$$\begin{aligned} E(X_1|Y) &= E(Y\cos\Phi|Y) = YE\cos\Phi \\ &= Y\frac{1}{2\pi}\int_0^{2\pi}\cos\varphi\,d\varphi = 0 \end{aligned}$$

16. By definition $\mathcal{A}_X$ contains every set of the form $\{\omega: X_i(\omega) \in A\}$, $A \in \mathcal{R}^1$, $i = 1, \ldots, n$. It follows that if $A_1, \ldots, A_n$ are one-dimensional Borel sets then

$$X^{-1}\left(\prod_{i=1}^{n} A_i\right) = \bigcap_{i=1}^{n} \{\omega: X_i(\omega) \in A_i\} \in \mathcal{A}_X$$

Since $\mathcal{R}^n$ is the smallest σ algebra containing all n products of one-dimensional Borel sets, it follows that for every $B \in \mathcal{R}^n$,

$X^{-1}(B) \in \mathcal{A}_X$ so that $\mathcal{A}_X \supset \{X^{-1}(B), B \in \mathcal{R}^n\}$

Conversely, consider the collection $\{X^{-1}(B), B \in \mathcal{R}^n\}$. It is a σ algebra, and every X_i is clearly measurable with respect to $\{X^{-1}(B), B \in \mathcal{R}^n\}$. Hence, $\mathcal{A}_X \subset \{X^{-1}(B), B \in \mathcal{R}^n\}$, and our assertion is proved.

CHAPTER 2

1. For any real number a,

$$\begin{aligned} \mathcal{P}(|X_t - X_s| > \epsilon) &\geq \mathcal{P}(X_t > a + \epsilon, X_s < a - \epsilon) \\ &= [1 - P_t(a+\epsilon)]P_s(a-\epsilon) \\ &\xrightarrow[s\to t]{} [1 - P_t(a+\epsilon)]P_t(a-\epsilon) \end{aligned}$$

and continuity in probability means that

$$[1 - P_t(a^+)]P_t(a) = 0$$

for all t and a. Therefore, at all continuity points P_t is either 0 or 1. It follows that P_t is a function with a single jump of size 1, say, at $f(t)$, and that for each t, $X_t(\omega) = f(t)$ with probability 1. Because the X process is continuous in probability, the function f must be continuous.

2. (a) $$\mathcal{P}(\{\omega: X_t(\omega) = 0 \text{ for at least one } t \text{ in } T_n\} = \mathcal{P}\left(\bigcup_{t\in T_n} \{\omega: X(\omega) = -t\}\right) = \sum_{t\in T_n} \mathcal{P}(X = -t) = 0$$

(b) $$\mathcal{P}(X + t = 0 \text{ for at least one } t \text{ in } [0,1]) = \mathcal{P}(X \in [-1,0]) = \int_{-1}^{0} \frac{1}{\sqrt{2\pi}} e^{-x^2}\, dx > 0$$

3. $$P_t(x) = \mathcal{P}(\{\omega: X_t(\omega) < x\}) = \mathcal{P}(\{\omega: t\omega < x\})$$
$$= \text{Lebesgue measure of } \left[0, \frac{x}{t}\right) \cap [0,1]$$
$$= \min\left(1, \frac{x}{t}\right)$$

$$P_{t,s}(x_1,x_2) = \mathcal{P}\left(\left\{\omega: \omega < \frac{x_1}{t}, \omega < \frac{x_2}{s}\right\}\right)$$
$$= \text{Lebesgue measure of } \left[0, \frac{x_1}{t}\right) \cap \left[0, \frac{x_2}{s}\right) \cap [0,1]$$
$$= \min\left(1, \frac{x_1}{t}, \frac{x_2}{s}\right).$$

4. (a) $$\mu(t) = EX_t = \int_0^1 \omega t\, d\omega = \tfrac{1}{2}t$$

$$R(t,s) = \int_0^1 (\omega t - \tfrac{1}{2}t)(\omega s - \tfrac{1}{2}s)\, d\omega = ts \int_0^1 (\omega - \tfrac{1}{2})^2\, d\omega$$
$$= \frac{ts}{3}\tfrac{1}{4} = \frac{ts}{12}$$

(b) We note that
$$X_t = Z \cos\theta \cos 2\pi t - Z \sin\theta \sin 2\pi t$$
$$= A \cos 2\pi t - B \sin 2\pi t$$

where $A = Z\cos\theta$ and $B = Z\sin\theta$. If we denote the joint density function of A and B by $\tilde{p}$, then

$$\tilde{p}(z\cos\theta, z\sin\theta) \begin{vmatrix} \cos\theta & -z\sin\theta \\ \sin\theta & z\cos\theta \end{vmatrix} = \frac{1}{2\pi} z e^{-\frac{1}{2}z^2}$$

or

$$\tilde{p}(z\cos\theta, z\sin\theta) = \frac{1}{2\pi} e^{-\frac{1}{2}z^2}$$

and

$$\tilde{p}(a,b) = \frac{1}{2\pi} e^{-\frac{1}{2}(a^2+b^2)}$$

It now follows that every linear combination

$$\sum_{i=1}^{n} \alpha_i X_{t_i} = \left(\sum_{i=1}^{n} \alpha_i \cos 2\pi t_i\right) A - \left(\sum_{i=1}^{n} \alpha_i \sin 2\pi t_i\right) B$$

is a Gaussian random variable. By definition, $\{X_t, -\infty < t < \infty\}$ is a Gaussian process.

8. Let $\mathcal{B}_{xs}$ denote the smallest algebra (not σ algebra) such that for every $\tau \leq s$, all sets of the form $\{\omega: X_\tau(\omega) < a\}$ are in $\mathcal{B}_{xs}$. It is clear that $\mathcal{A}_{xs}$ is generated by $\mathcal{B}_{xs}$. Now, every set A in $\mathcal{B}_{xs}$ depends on only a finite collection $X_{t_1}, X_{t_2}, \ldots, X_{t_n}, t_i \leq s$. Therefore, for every $A \in \mathcal{B}_{xs}$

$$EI_A X_t = E\{E[I_A X_t | X_{t_1}, X_{t_2}, \ldots, X_{t_n}, X_s]\}$$

Writing $X_t = X_t^+ - X_t^-$, where both X_t^+ and X_t^- are nonnegative, we have

$$EI_A X_t^+ = EI_A X_s^+ \qquad A \in \mathcal{B}_{xs}$$
$$EI_A X_t^- = EI_A X_s^- \qquad A \in \mathcal{B}_{xs}$$

Each of these four terms defines a finite measure on $\mathcal{B}_{xs}$ which has a unique extension to $\mathcal{A}_{xs}$. It follows that

$$EI_A X_t^+ = EI_A X_s^+ \qquad A \in \mathcal{A}_{xs}$$

for otherwise, we would have two different extensions of the same measure. Similarly,

$$EI_A X_t^- = EI_A X_s^- \qquad A \in \mathcal{A}_{xs}$$

and

$$EI_A X_t = EI_A X_s \qquad A \in \mathcal{A}_{xs}$$

Since X_s is $\mathcal{A}_{xs}$ measurable, we have

$$E^{\mathcal{A}_{xs}} X_t = X_s$$

with probability 1.

11. Suppose that $X_t = f(t) W_{g(t)/f(t)}$, then

$$EX_t X_s = f(t) f(s) \min\left(\frac{g(t)}{f(t)}, \frac{g(s)}{f(s)}\right)$$

Because $g(t)/f(t)$ is nondecreasing, we have

$$\begin{aligned} EX_t X_s &= f(t) f(s) \frac{g(\min(t,s))}{f(\min(t,s))} \\ &= f(\max(t,s)) g(\min(t,s)) \\ &= e^{-|t-s|} = e^{-[\max(t,s) - \min(t,s)]} \end{aligned}$$

It follows that $f(t)g(t) = 1$ so that $g(t) = 1/f(t)$ and

$$\begin{aligned} f(\max(t,s)) g(\min(t,s)) &= \frac{f(\max(t,s))}{f(\min(t,s))} \\ &= e^{-\max(t,s)} e^{-\min(t,s)} \end{aligned}$$

Hence, we can take $f(t) = ke^{-t}$ where k is any nonzero constant. Thus,

$X_t = ke^{-t}W_{[(1/k)e^t]^2}$

12. (*a*) Since $\{X_t, \ -\infty < t < \infty\}$ is Markov, the Chapman-Kolmogorov equation yields

$$\mathcal{P}(X_{t+s} = x_i | X_0 = x_j) = \sum_{k=1}^{n} \mathcal{P}(X_{t+s} = x_i | X_s = x_k)\mathcal{P}(X_s = x_k | X_0 = x_j)$$

or, equivalently,

$$p_{ij}(t+s) = \sum_{k=1}^{n} p_{ik}(t)p_{kj}(s) \qquad t,s > 0$$

In matrix form, this can be rewritten as

$$\mathbf{p}(t+s) = \mathbf{p}(t)\mathbf{p}(s)$$

so that

$$\lim_{t\downarrow 0} \frac{1}{t} [\mathbf{p}(t+s) - \mathbf{p}(s)] = \lim_{t\downarrow 0} \frac{1}{t} [\mathbf{p}(t) - \mathbf{I}]\mathbf{p}(s) = \mathbf{A}\mathbf{p}(s)$$

Hence,

$$\dot{\mathbf{p}}(s) = \mathbf{A}\mathbf{p}(s) \qquad s > 0$$

the unique solution of which corresponding to $\mathbf{p}(0) = \mathbf{I}$ is $\mathbf{p}(s) = e^{s\mathbf{A}}$.

$$(b) \ \sum_{j=1}^{n} \mathcal{P}(X_{t+\tau} = x_i | X_t = x_j)\mathcal{P}(X_t = x_j) = \mathcal{P}(X_{t+\tau} = x_i)$$

$$\sum_{i=1}^{n} \mathcal{P}(X_{t+\tau} = x_i | X_t = x_j) = 1$$

Hence, $\mathbf{p}(\tau)\mathbf{q} = \mathbf{q}$ and $\mathbf{p}^T(\tau)\mathbf{1} = \mathbf{1}$. $\left(\mathbf{1} = \begin{bmatrix} 1 \\ 1 \end{bmatrix}\right)$

If $\mathbf{q} = \begin{bmatrix} \frac{1}{2} \\ \frac{1}{2} \end{bmatrix}$, then we have $\mathbf{p}(\tau)\mathbf{1} = \mathbf{1}$ and $\mathbf{p}^T(\tau)\mathbf{1} = \mathbf{1}$, and $\mathbf{p}(\tau)$ must have the form

$$\mathbf{p}(\tau) = \begin{bmatrix} f(\tau) & 1 - f(\tau) \\ 1 - f(\tau) & f(\tau) \end{bmatrix}$$

and $\dot{\mathbf{p}}(0) = \dot{f}(0) \begin{bmatrix} 1 & -1 \\ -1 & 1 \end{bmatrix} = \mathbf{A}$. Because every entry in $p(\tau)$ is nonnegative, $\dot{f}(0)$ must be less than or equal to zero. Setting $\dot{f}(0) = -\lambda$, we have from part (*a*)

$$\mathbf{p}(\tau) = e^{\tau\mathbf{A}} = \exp\left(-\lambda\tau \begin{bmatrix} 1 & -1 \\ -1 & 1 \end{bmatrix}\right) = \begin{bmatrix} \dfrac{1 + e^{-\lambda\tau}}{2} & \dfrac{1 - e^{-\lambda\tau}}{2} \\ \dfrac{1 - e^{-\lambda\tau}}{2} & \dfrac{1 + e^{-\lambda\tau}}{2} \end{bmatrix}$$

14. Because $\{X_t, -\infty < t < \infty\}$ is stationary, its covariance function depends only on the time difference. Set

$$\rho(t-s) = \frac{E[(X_t - EX_t)(X_s - EX_s)]}{E(X_t - EX_t)^2}$$

then from Solution 2.13, ρ must satisfy

$$\rho(t+s) = \rho(t)\rho(s) \qquad t, s \geq 0$$

It follows that for any positive integers m,n, we have

$$\rho\left(\frac{m}{n}\right) = \rho^m\left(\frac{1}{n}\right)$$

$$\rho(1) = \rho^n\left(\frac{1}{n}\right)$$

so that $\rho\left(\frac{m}{n}\right) = \rho^{m/n}(1) = e^{(m/n)\ln\rho(1)}$. By continuity we must have

$$\rho(t) = e^{t\ln\rho(1)} \qquad t \geq 0$$

and by symmetry

$$\begin{aligned}\rho(t) &= e^{|t|\ln\rho(1)} \\ &= e^{-\lambda|t|}\end{aligned}$$

where we have set $-\lambda = \ln\rho(1)$.

15. (*a*) Consider the characteristic function

$$\begin{aligned}E\exp\left(i\sum_{k=1}^{n} u_k X_{t_k+\tau}\right) &= E\left\{\frac{1}{2\pi}\int_0^{2\pi}\exp\left[iA\sum_{k=1}^{n} u_k\cos(2\pi t_k + 2\pi\tau + \theta)\right]d\theta\right\} \\ &= E\left\{\frac{1}{2\pi}\int_{2\pi\tau}^{2\pi(\tau+1)}\exp\left[iA\sum_{k=1}^{n} u_k\cos(2\pi t_k + \psi)\right]d\psi\right\} \\ &= E\left\{\frac{1}{2\pi}\int_0^{2\pi}\exp\left[iA\sum_{k=1}^{n} u_k\cos(2\pi t_k + \psi)\right]d\psi\right\} \\ &= E\exp\left(i\sum_{k=1}^{n} u_k X_{t_k}\right)\end{aligned}$$

(*b*) $EX_t = EA\left(\frac{1}{2\pi}\int_0^{2\pi}\cos(2\pi t + \theta)\,d\theta\right) = 0$

$$M_T(\omega) = \frac{1}{2T}\int_{-T}^{T} X_t(\omega)\,dt = A(\omega)\frac{1}{2T}\int_{-T}^{T}\cos[2\pi t + \theta(\omega)]\,dt \xrightarrow[T\to\infty]{} 0$$

(c) Let $Y = A \cos \theta$ and $Z = A \sin \theta$, then

$$X_t = Y \cos 2\pi t - Z \sin 2\pi t$$

and $Y = X_0$, $Z = -X_{\frac{1}{4}}$, so that $\{X_t, t \in (-\infty, \infty)\}$ is a Gaussian process if and only if Y and Z are jointly Gaussian. Since

$$EYZ = EA^2 \frac{1}{2\pi} \int_0^{2\pi} \sin \theta \cos \theta \, d\theta = 0$$

$$EY^2 = EZ^2 = \tfrac{1}{2} EA^2 = \sigma^2$$

Y and Z are jointly Gaussian if and only if

$$p_{YZ}(y,z) = \frac{1}{2\pi\sigma^2} \exp \left[-\frac{1}{2\sigma^2} (y^2 + z^2) \right]$$

Hence, by the transformation rule for random variables (see, e.g., Exercise 1.5), we have

$$\frac{1}{2\pi} p_A(r) = p_{YZ}(r \cos \theta, r \sin \theta) \begin{vmatrix} \cos \theta & -r \sin \theta \\ \sin \theta & r \cos \theta \end{vmatrix}$$

$$= \frac{r}{2\pi\sigma^2} \exp \left(-\frac{r^2}{2\sigma^2} \right)$$

and

$$p_A(r) = \frac{r}{\sigma^2} \exp \left(-\frac{1}{2} \frac{r^2}{\sigma^2} \right) \qquad r \geq 0$$

$\{X_t \;\; -\infty < t < \infty\}$ is not Markov because for $t \geq \frac{1}{4}$

$$E(X_t | X_0 = y, X_{\frac{1}{4}} = z) = y \cos 2\pi t - z \sin 2\pi t$$

which depends on y, contrary to the Markov property.

16. (a) Since X_t, Y_t are independent and Markov

$$\begin{aligned} E(Z_4{}^2 | X_1, X_2, X_3, Y_1, Y_2, Y_3) &= E(X_4{}^2 | X_3) + E(Y_4{}^2 | Y_3) \\ &= E[(X_4 - X_3)^2 + 2X_3(X_4 - X_3) + X_3{}^2 | X_3] \\ &\quad + E[(Y_4 - Y_3)^2 + 2Y_3(Y_4 - Y_3) + Y_3{}^2 | Y_3] \\ &= 1 + X_3{}^2 + 1 + Y_3{}^2 \end{aligned}$$

Therefore, $E(Z_4{}^2 | \text{observed data}) = 7$.

Now, by the Schwarz inequality

$$\begin{aligned} E(Z_4 | \text{data}) &\leq \sqrt{E(Z_4{}^2 | \text{data})} \\ &= \sqrt{7} \end{aligned}$$

On the other hand, the Schwarz inequality applied to summations yields

$$(xx_0 + yy_0)^2 \leq (x^2 + y^2)(x_0{}^2 + y_0{}^2)$$

so that

$$\sqrt{x^2 + y^2} \geq \frac{|xx_0 + yy_0|}{\sqrt{x_0^2 + y_0^2}} \geq \frac{(xx_0 + yy_0)}{\sqrt{x_0^2 + y_0^2}}$$
$$= \sqrt{x_0^2 + y_0^2} + \frac{x_0}{\sqrt{x_0^2 + y_0^2}}(x - x_0) + \frac{y_0}{\sqrt{x_0^2 + y_0^2}}(y - y_0)$$

It follows that

$$E(\sqrt{X_4^2 + Y_4^2}\,|X_3, Y_3) \geq \sqrt{X_3^2 + Y_3^2} + \frac{X_3}{\sqrt{X_3^2 + Y_3^2}} E[(X_4 - X_3)|X_3]$$
$$+ \frac{Y_3}{\sqrt{X_3^2 + Y_3^2}} E[(Y_4 - Y_3)|Y_3]$$
$$= \sqrt{X_3^2 + Y_3^2}$$

Hence, $E(Z_4|\text{data}) \geq \sqrt{1 + (2)^2} = \sqrt{5}$.

(*b*) Introduce $\{\Theta_t, t \geq 0\}$ so that

$$X_t = Z_t \cos \Theta_t \qquad Y_t = Z_t \sin \Theta_t$$

For $t_1 < t_2 < \cdots < t_n$

$$E[f(Z_{t_n})|Z_{t_1}, Z_{t_2}, \ldots, Z_{t_{n-1}}] = E\{E[f(Z_{t_n})|Z_{t_j}, \Theta_{t_j}, j = 1, \ldots, n-1\}$$
$$= E\{E[f(Z_{t_n})|Z_{t_{n-1}}, \Theta_{t_{n-1}}]|Z_{t_1}, \ldots, Z_{t_{n-1}}\}$$

Now,

$$E[f(Z_{t_n})|Z_{t_{n-1}}, \Theta_{t_{n-1}}] = E[f(\sqrt{X_{t_n}^2 + Y_{t_n}^2})|X_{t_{n-1}}, Y_{t_{n-1}}]$$

and

$$E[f(Z_{t_n})|Z_{t_{n-1}} = r_0, \Theta_{t_{n-1}} = \theta_0] = E[f(\sqrt{X_{t_n}^2 + Y_{t_n}^2}\,|$$
$$X_{t_{n-1}} = r_0 \cos \theta_0, Y_{t_{n-1}} = r_0 \sin \theta_0]$$
$$= \int_0^\infty dr \int_0^{2\pi} d\theta \frac{rf(r)}{2\pi(t_n - t_{n-1})}$$
$$\exp\left\{-\frac{1}{2(t_n - t_{n-1})}[r^2 + r_0^2 - 2rr_0 \cos(\theta - \theta_0)]\right\}$$

Because $\cos \theta$ is periodic with period 2π, a change in variable yields

$$E[f(Z_{t_n})|Z_{t_{n-1}} = r_0, \Theta_{t_{n-1}} = \theta_0] = \int_0^\infty dr \int_0^{2\pi} d\theta' \frac{rf(r)}{2\pi(t_n - t_{n-1})}$$
$$\exp\left[-\frac{1}{2(t_n - t_{n-1})}(r^2 + r_0^2 - 2rr_0 \cos \theta')\right]$$

which is independent of θ_0. It follows that

$$E[f(Z_{t_n})|Z_{t_{n-1}}, \Theta_{t_{n-1}}] = E[f(Z_{t_n})|Z_{t_{n-1}}]$$

and

$$E[f(Z_{t_n})|Z_{t_1}, \ldots, Z_{t_{n-1}}] = E[f(Z_{t_n})|Z_{t_{n-1}}]$$

so that $\{Z_t, t \geq 0\}$ is Markov (see Proposition 4.6.4 for an easy means of proving this fact).

CHAPTER 3

1. (*a*) Compute

$$\begin{aligned} S(\nu) &= \int_{-\infty}^{\infty} R(\tau)\,d\tau = \int_{-1}^{1} (1 - |\tau|)e^{-i2\pi\nu\tau}\,d\tau \\ &= 2\int_0^1 (1-\tau)\cos 2\pi\nu\tau\,d\tau \\ &= \frac{2}{2\pi\nu}\int_0^1 \sin 2\pi\nu\tau\,d\tau = \frac{2(1-\cos 2\pi\nu)}{(2\pi\nu)^2} \\ &= \left(\frac{\sin \pi\nu}{\pi\nu}\right)^2 \geq 0 \end{aligned}$$

It follows from one-half of Bochner's theorem that R is nonnegative definite.

(*b*) Repeat the same procedure as in (*a*).

(*c*) If $R(t,s) = e^{|t-s|}$ then $R(t,t) = 1$, and it violates (1.8), viz.

$$|R(t,s)| \leq \sqrt{R(t,t)R(s,s)}$$

(*d*) R is continuous at all diagonal points (t,t), but not continuous everywhere, so it cannot be nonnegative definite.

3. (*a*) Since R is continuous and periodic, we can write it in a Fourier series as

$$R(\tau) = \sum_{n=-\infty}^{\infty} R_n e^{in(2\pi/T)t}$$

where $R_n \geq 0$. It follows that

$$\begin{aligned} EZ_m\bar{Z}_n &= \iint_0^T R(t-s)e^{-i(2\pi/T)(mt-ns)}\,dt\,ds \\ &= \sum_{k=-\infty}^{\infty} R_n\delta_{mk}\delta_{nk} = R_n\delta_{mn} \end{aligned}$$

(*b*)
$$\begin{aligned} E\left|X_t - \sum_{n=-N}^{N} Z_n e^{in(2\pi/T)t}\right|^2 &= R(0) - \sum_{n=-N}^{N} E|Z_n|^2 \\ &= R(0) - \sum_{n=-N}^{N} R_n \xrightarrow[N\to\infty]{} 0 \end{aligned}$$

(*c*) We can write

$$R(\tau) = \sum_{n=-\infty}^{\infty} R_n e^{in(2\pi/T)\tau}$$

where $R_n = 1/2(1+n^2)$, $n \neq 0$ and $R_0 = 1$. Since the family $\{e^{in(2\pi/T)t}, 0 \leq t \leq T, n = 0, \pm 1, \ldots\}$ is orthogonal, we can clearly take $\varphi_n(t) = (1/\sqrt{T})e^{in(2\pi/T)t}$ to be the orthonormal eigenfunction. The eigenvalues are $\lambda_n = R_nT$.

4. First, we write

$$\lambda\varphi(t) = \int_0^t (1 - t + s)\varphi(s)\,ds + \int_t^{\frac{1}{2}} (1 - s + t)\varphi(s)\,ds$$

Differentiating it once, we get

$$\lambda\varphi'(t) = -\int_0^t \varphi(s)\,ds + \int_t^{\frac{1}{2}} \varphi(s)\,ds$$

Differentiating once more, we find

$$\lambda\varphi''(t) = -2\varphi(t)$$

as was suggested by the hint. The second of the above equations yields the boundary condition $-\varphi'(0) = \varphi'(\frac{1}{2})$. The first and second equations yield $\varphi(0) + \varphi(\frac{1}{2}) = \frac{3}{2}\varphi'(0)$.

From the equation $\lambda\varphi''(t) = -2\varphi(t)$, we get

$$\varphi(t) = A \cos\sqrt{\frac{2}{\lambda}}\,t + B \sin\sqrt{\frac{2}{\lambda}}\,t$$

Applying the condition $-\varphi'(0) = \varphi'(\frac{1}{2})$ yields

$$\varphi(t) = C \cos\sqrt{\frac{2}{\lambda}}\,(t - \tfrac{1}{4})$$

Applying the second boundary condition yields the transcendental equation

$$\tfrac{3}{4}\sqrt{2/\lambda} = \frac{\cos\frac{1}{4}\sqrt{2/\lambda}}{\sin\frac{1}{4}\sqrt{2/\lambda}} = \cot\tfrac{1}{4}\sqrt{2/\lambda}$$

which is to be solved for the eigenvalues λ. Finally, for normalization we choose $C = 2$ so that

$$\varphi(t) = 2\cos\sqrt{\frac{2}{\lambda}}\,(t - \tfrac{1}{4})$$

6. Since the W process is a Brownian motion, we can write

$$W_{\tau(t)} = \sum_{n=0}^{\infty} \sqrt{\lambda_n}\,\varphi_n(\tau(t))Z_n \qquad 0 \le t \le T$$

where λ_n and φ_n are given (4.32) and (4.33), respectively, and $\{Z_n\}$ are Gaussian and orthonormal. Hence, the desired expansion is obtained by setting

$$\alpha_n(t) = f(t)\sqrt{\lambda_n}\,\varphi_n(t)$$

Suppose that we define $\tau^{-1}(t) = \min\{s: \tau(s) = t\}$. Then,

$$X_{\tau^{-1}(t)} = f(\tau^{-1}(t))W_t \qquad 0 \le t \le \tau(T)$$

Since a Brownian motion is q.m. continuous, the Hilbert space $\mathcal{H}_W$ generated by $\{W_t, 0 \le t \le \tau(T)\}$ is spanned by $\{W_t, t \in S\}$ where S is any dense subset of $[0,\tau(T)]$. It follows that every $Z_n \in \mathcal{H}_X$ if and only if there exists a dense subset S of $[0,\tau(T)]$ such that

$$f(\tau^{-1}(t)) \neq 0 \qquad \text{for every } t \in S$$

7. $R(\tau) = \frac{1}{8}e^{-|\tau|}(3 \cos \tau + \sin |\tau|)$

$$S(\nu) = \int_{-\infty}^{\infty} e^{-i2\pi\nu\tau}R(\tau)\,d\tau = 2\int_0^{\infty} \cos 2\pi\nu\tau R(\tau)\,d\tau$$

$$= \frac{1}{4}\int_0^{\infty} e^{-\tau} \cos 2\pi\nu\tau(3 \cos \tau + \sin \tau)\,d\tau$$

$$= \frac{1}{8}\int_0^{\infty} e^{-\tau}[3 \cos (1 + 2\pi\nu)\tau + 3 \cos (1 - 2\pi\nu\tau) + \sin (1 + 2\pi\nu)\tau + \sin (1 - 2\pi\nu)\tau]\,d\tau$$

$$= \frac{1}{8}\left[\frac{3}{1 + (1 + 2\pi\nu)^2} + \frac{3}{1 + (1 - 2\pi\nu)^2} + \frac{1 + 2\pi\nu}{1 + (1 + 2\pi\nu)^2} + \frac{1 - 2\pi\nu}{1 + (1 - 2\pi\nu)^2}\right]$$

$$= \frac{1}{8}\frac{16 + 4(2\pi\nu)^2}{4 + (2\pi\nu)^4} = \frac{1}{2}\frac{4 + (2\pi\nu)^2}{4 + (2\pi\nu)^4}$$

8. Let $\rho(t - s) = EY_t\bar{X}_s$. Then if $Y_t = \int_{-\infty}^{\infty} e^{i2\pi\nu t}H(\nu)\,d\hat{X}_\nu$

$$\rho(t - s) = \int_{-\infty}^{\infty} e^{i2\pi\nu(t-s)}H(\nu)S(\nu)\,d\nu$$

Therefore,

$$H(\nu) = \frac{1}{S(\nu)}\int_{-\infty}^{\infty} e^{-i2\pi\nu\tau}\rho(\tau)\,d\tau$$

$$= \frac{1}{S(\nu)}\frac{1}{4 + (2\pi\nu)^4}$$

$$= \frac{2}{4 + (2\pi\nu)^2}$$

The covariance funct on of the Y process is given by

$$R_y(\tau) = \int_{-\infty}^{\infty} e^{i2\pi\nu\tau}|H(\nu)|^2\,d\nu$$

$$= \int_{-\infty}^{\infty} \frac{4}{[4 + (2\pi\nu)^2]^2} e^{i2\pi\nu\tau}\,d\nu$$

$$= \frac{1}{8}e^{-2|\tau|}(1 + 2|\tau|)$$

. $E(e^{i2\pi Wt}X_t)\overline{(e^{i2\pi Ws}X_s)} = e^{i2\pi W(t-s)}EX_t\bar{X}_s$

$$= e^{i2\pi W(t-s)}R_x(t - s)$$

$\cos 2\pi WtX_t$ is not wide-sense stationary.

2. $EY_t\bar{Y}_s = \int_{-\infty}^{\infty} |\eta(\nu)|^2 e^{i\varphi(\nu)(t-s)}\,dF(\nu)$

1. (*a*) More generally, $Y_t = \int_{-\infty}^{\infty} H(\nu)e^{i2\pi\nu t}\,d\hat{X}_\nu$ is real valued i

$$H(\nu) = \bar{H}(-\nu)$$

To prove this, first assume $H \in L^2(-\infty, \infty)$, then H i the Fourier transform of a real-valued function h and

$$Y_t = \int_{-\infty}^{\infty} h(t-s)X_s\,ds$$

If $H \notin L^2(-\infty, \infty)$ we truncate $H(\nu)$ to $\nu = \pm n$. It follows then that Y_t is the q.m. limit of a sequence of real-valued processes.

(*b*) Since $EX_tX_s = e^{-|t-s|}$,

$$E|d\hat{X}_\nu|^2 = \frac{2}{1+(2\pi\nu)^2}\,d\nu$$

It follows that

$$\begin{aligned} E\tilde{X}_t\tilde{X}_s = E\tilde{X}_t\bar{\tilde{X}}_s &= \int_{-\infty}^{\infty} |-i \operatorname{sgn} \nu|^2 e^{i2\pi\nu(t-s)} \frac{2}{1+(2\pi\nu)^2}\,d\nu \\ &= e^{-|t-s|} \end{aligned}$$

$$\begin{aligned} E\tilde{X}_tX_s = E\tilde{X}_t\bar{X}_s &= \int_{-\infty}^{\infty} (-i \operatorname{sgn} \nu) e^{i2\pi\nu(t-s)} \frac{2}{1+(2\pi\nu)^2}\,d\nu \\ &= 4\int_0^\infty \frac{\sin 2\pi\nu(t-s)}{1+(2\pi\nu)^2}\,d\nu \end{aligned}$$

(*c*) $$\begin{aligned} Z_t &= \int_{-\infty}^{\infty} (\cos 2\pi\nu_0 t - i \operatorname{sgn} \nu \sin 2\pi\nu_0 t) e^{i2\pi\nu t}\,d\hat{X}_\nu \\ &= \int_{-\infty}^{\infty} e^{-i \operatorname{sgn} \nu (2\pi_0)t + i2\pi\nu t}\,d\hat{X}_\nu \end{aligned}$$

14. (*a*) Check the three cases $\min(t,s) > 0$, $\max(t,s) < 0$, and $\max(t,s) > 0 > \min(t,s)$.

(*b*) $$\begin{aligned} E\zeta_{nt}\bar{\zeta}_{ns} &= n^2(Z_{t+1/n} - Z_t)(\overline{Z_{s+1/n} - Z_s}) \\ &= \frac{n^2}{2}\Bigg(\left|t+\frac{1}{n}\right| + \left|s+\frac{1}{n}\right| - |t-s| + |t| + |s| - |t-s| - |t| \\ &\qquad - \left|s+\frac{1}{n}\right| + \left|t-s-\frac{1}{n}\right| - \left|t+\frac{1}{n}\right| - |s| + \left|t+\frac{1}{n}-s\right|\Bigg) \\ &= \frac{n^2}{2}\left(\left|t-s-\frac{1}{n}\right| + \left|t-s+\frac{1}{n}\right| - 2|t-s|\right) \\ &= \begin{cases} n(1-n|t-s|) & 0 \le |t-s| \le \dfrac{1}{n} \\ 0 & \text{elsewhere} \end{cases} \end{aligned}$$

$$\begin{aligned} S_n(\nu) &= \int_{-1/n}^{1/n} n(1-n\tau)e^{-i2\pi\nu\tau}\,d\tau = 2n\int_0^{1/n} (1-n\tau)\cos 2\pi\nu\tau\,d\tau \\ &= \frac{2n^2}{2\pi\nu}\int_0^{1/n} \sin 2\pi\nu\tau\,d\tau = \frac{2n^2}{(2\pi\nu)^2}\left(1-\cos\frac{2\pi\nu}{n}\right) \\ &= \frac{4n^2}{(2\pi\nu)^2}\sin^2\frac{\pi\nu}{n} = \left(\frac{\sin \pi\nu/n}{\pi\nu/n}\right)^2 \end{aligned}$$

(*c*) $$\begin{aligned} E\left[\int_{-\infty}^{\infty} \zeta_{nt}f(t)\,dt\right]\left[\overline{\int_{-\infty}^{\infty} \zeta_{nt}g(t)\,dt}\right] &= \iint_{-\infty}^{\infty} f(t)\bar{g}(s)R_\zeta^{(n)}(t-s)\,ds \\ &= \int_{-\infty}^{\infty} \hat{f}(\nu)\bar{\hat{g}}(\nu)\left(\frac{\sin \pi\nu/n}{\pi\nu/n}\right)^2 d\nu \xrightarrow[n\to\infty]{\text{(dominated convergence)}} \int_{-\infty}^{\infty} \hat{f}(\nu)\bar{\hat{g}}(\nu)\,d\nu \\ &= \int_{-\infty}^{\infty} f(t)\bar{g}(t)\,dt \end{aligned}$$

15. Set $Z_{nt} = \int_0^t \zeta_{ns}\, ds$. Then, we have

$$
\begin{aligned}
Z_{nt} &= n \int_0^t (Z_{s+1/n} - Z_s)\, ds \\
&= n \left(\int_{1/n}^{t+1/n} Z_s\, ds - \int_0^t Z_s\, ds \right) \\
&= n \left(\int_t^{t+1/n} Z_s\, ds - \int_0^{1/n} Z_s\, ds \right)
\end{aligned}
$$

Therefore,

$$
\begin{aligned}
E|Z_t - Z_{nt}|^2 &= n^2 E \left| \int_t^{t+1/n} (Z_t - Z_s)\, ds + \int_0^{1/n} Z_s\, ds \right| \\
&\le 2n^2 \left[E \left| \int_t^{t+1/n} (Z_t - Z_s)\, ds \right|^2 + E \left| \int_0^{1/n} Z_s\, ds \right|^2 \right] \\
&\le 2n^2 \left[\frac{1}{n} \int_t^{t+1/n} (s - t)\, ds + \frac{1}{n} \int_0^{1/n} s\, ds \right] \\
&= \frac{2}{n} \xrightarrow[n\to\infty]{} 0 \qquad \text{uniformly in } t
\end{aligned}
$$

It follows that

$$
\int_a^b \dot{f}(t) Z_{nt}\, dt + \int_a^b (t)\zeta_{nt}\, dt = f(b)Z_{nb} - (a)Z_{na}
$$

We have already shown that $Z_{nt} \xrightarrow[n\to\infty]{\text{q.m.}} Z_t$ uniformly in t, and by definition,

$$
\int_a^b f(t)\, dZ_t = \lim_{n\to\infty} \text{in q.m.} \int_a^b f(t)\zeta_{nt}\, dt
$$

Hence, the desired result follows.

16. Starting from $Y_t = Y_a - \int_a^t Y_s\, ds + Z_t - Z_a$, we set $U_t = \int_a^t Y_s\, ds$ and write

$$
\dot{U}_t + U_t = Y_a + Z_t - Z_a
$$

Hence, we have

$$
U_t = e^{-t} \left[\int_a^t Y_a e^s\, ds + \int_a^t (Z_s - Z_a) e^s\, ds \right]
$$

and

$$
Y_t = \dot{U}_t = -U_t + Y_a + Z_t - Z_a
$$

If we replace $\int_a^t Z_s e^s\, ds$ by $-\int_0^t e^s\, dZ_s + Z_t e^t - Z_a e^a$, we get

$$
Y_t = Y_a e^{-(t-a)} + \int_a^t e^{-(t-s)}\, dZ_s
$$

which was to be shown.

17. (*a*) Define $X(I_{cd}) = X_d - X_c$ as required. If $f = \sum_{i=1}^n \alpha_i f_i$ and each f_i is an indicator function of a half-open interval, call f a step function and set $X(f) = \sum_{i=1}^n \alpha_i X(f_i)$.

If f and g are step functions, then

$$EX(f)\bar{X}(g) = \int_a^b f(t)\bar{g}(t)\, dt + \iint_a^b \rho(t,s)f(t)\bar{g}(s)\, ds$$

It follows that

$$\int_a^b |f(t)|^2\, dt \le E|X(f)|^2 \le K \int_a^b |f(t)|^2\, dt$$

By taking q.m. limits, $X(f)$ can be defined for any f which is the L^2 limit of a sequence of step functions. But the set of step functions is dense in $L^2(a,b)$, hence our definition is complete.

(*b*) $$E\bar{X}(f)X(g) = \int_a^b f(t)\bar{g}(t)\, dt + \iint_a^b \rho(t,s)f(t)\bar{g}(s)\, dt\, ds$$

(*c*) If Y is a finite sum $Y = \sum_{i=0}^{n} \alpha_i X_{t_i}$, then $Y = \int_a^b \eta(t)\, dX_t + kX_a$

where η is a step function. Every $Y \in \mathcal{H}_X$ is the q.m. limit of a sequence of finite sums. Since $X(\eta_n)$ converges in q.m. if and only if η_n is a L^2-convergent sequence, the result is proved.

18. We use the condition

$$E(A - \hat{A}_t)X_s = 0 \qquad 0 \le s \le t$$

and find

$$s = EX_s \int_0^t h(t,\tau)\, dX_\tau = s \int_0^t h(t,\tau)\, d\tau + \int_0^s h(t,\tau)\, d\tau \qquad 0 \le s \le t$$

Hence, $\int_0^t h(t,\tau)\, d\tau = t/(1+t)$ and

$$\int_0^s h(t,\tau)\, d\tau = \frac{s}{1+t}$$

$$h(t,s) = \frac{1}{1+t} \qquad 0 \le s \le t$$

Therefore, $\hat{A}_t = [1/(1+t)]X_t$.

19. (*a*) We use the condition

$$E(Y - \hat{Y}_t)\bar{X}_s = 0 \qquad s \le t$$

and find

$$\int_0^t h(t,\tau) \frac{\partial}{\partial \tau} (EX_\tau \bar{X}_s)\, d\tau + EX_0\bar{X}_s = \cos 2\pi W s \qquad s \le t$$

Since $EX_t\bar{X}_s = \cos 2\pi Wt \cos 2\pi Ws + e^{-\nu_0|t-s|}$, we have

$$\cos 2\pi Ws \left[2\pi W \int_0^t h(t,\tau) \sin 2\pi W\tau \, d\tau \right] = e^{-\nu_0 s} - \nu_0 \int_0^t h(t,\tau) \operatorname{sgn}(\tau - s) e^{-\nu_0|\tau - s|} \, d\tau \qquad s \le t \quad (*)$$

Differentiating both sides twice with respect to s yields

$$-(2\pi W)^2 \cos 2\pi Ws \left[2\pi W \int_0^t h(t,\tau) \sin 2\pi W\tau \, d\tau \right] = \nu_0^2 e^{-\nu_0^2} - \nu_0 \int_0^t h(t,\tau) \operatorname{sgn}(\tau - s) e^{-\nu_0|\tau-s|} \, d\tau + 2\nu_0 \frac{\partial h(t,s)}{\partial s}$$

or

$$2\nu_0 \frac{\partial h(t,s)}{\partial s} + \left\{ [\nu_0^2 + (2\pi W)^2](2\pi W) \int_0^t h(t,\tau) \sin 2\pi W\tau \, d\tau \right\} \cos 2\pi Ws = 0$$

That means that $h(t,s)$ must have the form

$$h(t,s) = g(t) + f(t) \sin 2\pi Ws$$

and

$$f(t) = -\frac{1}{2\nu_0} [\nu_0^2 + (2\pi W)^2] \int_0^t h(t,\tau) \sin 2\pi W\tau \, d\tau$$

which yields

$$f(t) \left\{ 1 + \frac{1}{2\nu_0} [\nu_0^2 + (2\pi W)^2] \right\} \int_0^t \sin^2 2\pi W\tau \, d\tau + \left\{ \frac{1}{2\nu_0} [\nu_0^2 + (2\pi W)^2] \int_0^t \sin 2\pi W\tau \, d\tau \right\} g(t) = 0 \quad (**)$$

From $(*)$ we find (upon setting $s = 0$)

$$2\pi W \int_0^t h(t,\tau) \sin 2\pi W\tau \, d\tau = 1 - \nu_0 \int_0^t h(t,\tau) e^{-\nu_0\tau} \, d\tau$$

which yields

$$f(t) \left[\nu_0 \int_0^t e^{-\nu_0\tau} \sin 2\pi W\tau \, d\tau - \frac{4\pi\nu_0 W}{\nu_0^2 + (2\pi W)^2} \right] + \left(\nu_0 \int_0^t e^{-\nu_0\tau} \, d\tau \right) g(t) = 1 \quad (***)$$

Equations $(**)$ and $(***)$ suffice to determine $f(t)$ and $g(t)$, which in turn determine $h(t,s)$ completely.

(*b*) We begin by noting that $\int_{-\infty}^{\infty} e^{-\nu_0|\tau|} e^{i2\pi\nu\tau} \, d\tau = 2\nu_0/(|2\pi\nu|^2 + \nu_0^2)$ so that $\{N_t, -\infty < t < \infty\}$ can be viewed as a white noise filtered by transfer function $\sqrt{2\nu_0}/(\nu_0 + i2\pi\nu)$ or

$$\dot{N}_t + \nu_0 N_t = \sqrt{2\nu_0}\, \zeta_t$$

where ζ_t is a standard white-noise process. It follows that

$$\begin{aligned} \dot{X}_t &= (-2\pi W \sin 2\pi Wt) Y + \dot{N}_t \\ &= (-2\pi W \sin 2\pi Wt) Y - \nu_0 (X_t - \cos 2\pi WtY) + \sqrt{2\nu_0}\, \zeta_t \end{aligned}$$

Let $Z_t = e^{\nu_0 t}X_t$. Then

$$\begin{aligned}\dot{Z}_t &= Ye^{\nu_0 t}(\nu_0 \cos 2\pi Wt - 2\pi W \sin 2\pi Wt) + \sqrt{2\nu_0}\, e^{\nu_0 t}\zeta_t \\ \dot{Y} &= 0\end{aligned}$$

which is in the standard form (9.32) and (9.34), with Z_t replacing X_t. Hence, we have $A(t) = 0 = F(t)$, $B(t) = \sqrt{2\nu_0}\, e^{\nu_0 t}$ and

$$H(t) = e^{\nu_0 t}(\nu_0 \cos 2\pi Wt - 2\pi W \sin 2\pi Wt)$$

The estimator $\hat{Y}_t$ must satisfy

$$\dot{\hat{Y}}_t = k(t)[\dot{Z}_t - H(t)\hat{Y}_t]$$

Equations (9.39) and (9.40) now become

$$\begin{aligned}\dot{\Sigma}(t) &= +B^2(t)k^2(t) - 2H(t)\Sigma(t)k(t) \\ K(t)B^2(t) &= \Sigma(t)H(t)\end{aligned}$$

Continuing, we get

$$\dot{\Sigma}(t) = -\frac{H^2(t)}{B^2(t)}\Sigma^2(t)$$

which can be transformed into a linear equation by setting

$$\frac{\dot{u}(t)}{u(t)} = \frac{H^2(t)}{B^2(t)}\Sigma(t)$$

we then get

$$\begin{aligned}\ddot{u}(t) &= \dot{u}(t)\frac{d}{dt}\frac{H^2(t)}{B^2(t)} \\ \dot{u}(t) &= \dot{u}(0)\exp\left[\frac{H^2(t)}{B^2(t)} - \frac{H^2(0)}{B^2(0)}\right]\end{aligned}$$

It is now clear that $K(t)$ can be found explicitly.

20. We begin by factoring $S_x(\nu)$ into the form

$$S_x(\nu) = \frac{1}{[(i2\pi\nu) + 1 + i][(i2\pi\nu) + 1 - i][(i2\pi\nu) - 1 + i][(i2\pi\nu) - 1 - i]}$$

It is clear that an $\hat{h}$ satisfying (9.10) can be taken to be

$$\hat{h}(\nu) = \frac{1}{[(i2\pi\nu) + 1 + i][(i2\pi\nu) + 1 - i]}$$

corresponding to

$$\begin{aligned}h(t) &= e^{-t}\sin t \qquad & t \geq 0 \\ &= 0 & t < 0\end{aligned}$$

The function $g(t)$ can be found by using (9.24) as

$$\begin{aligned} g(t) &= \int_{-\infty}^{\infty} e^{i2\pi\nu t} \left[\frac{S_{xy}(\nu)}{\overline{\hat{h}(\nu)}}\right] \\ &= \int_{-\infty}^{\infty} e^{i2\pi\nu t} e^{i2\pi\nu\alpha} \hat{h}(\nu)\, d\nu \\ &= e^{-(t+\alpha)} \sin(t+\alpha) \qquad t \geq -\alpha \\ &= 0 \qquad t \leq -\alpha \end{aligned}$$

Now, $\hat{\gamma}(\nu)$ is given by

$$\begin{aligned} \hat{\gamma}(\nu) &= \int_0^{\infty} g(t) e^{-i2\pi\nu t}\, dt \\ &= e^{-\alpha} \int_0^{\infty} e^{-t} \frac{e^{i(t+\alpha)} - e^{-i(t+\alpha)}}{2i} e^{-i2\pi\nu t}\, dt \\ &= \frac{e^{-\alpha}}{2i} \left[\frac{e^{i\alpha}}{2\pi i\nu + 1 - i} - \frac{e^{-i\alpha}}{2\pi i\nu + 1 + i}\right] \\ &= e^{-\alpha} \left[\frac{\cos\alpha + \sin\alpha(2\pi i\nu + 1)}{(2\pi i\nu + 1 - i)(2\pi i\nu + 1 + i)}\right] \end{aligned}$$

From (9.29) we get

$$\begin{aligned} \hat{E}(X_{t+\alpha}|\mathcal{H}_X{}^t) &= \int_{-\infty}^{\infty} e^{-\alpha} [\cos\alpha + \sin\alpha(2\pi i\nu + 1)] e^{i2\pi\nu t}\, d\hat{X}_\nu \\ &= (\cos\alpha + \sin\alpha) e^{-\alpha} X_t + e^{-\alpha} \sin\alpha \dot{X}_t \end{aligned}$$

21. Here, $S_y(\nu)$ has the form

$$S_y(\nu) = \frac{1}{1 + (2\pi\nu)^2} + \frac{1}{4 + (2\pi\nu)^4} = \frac{5 + (2\pi\nu)^2 + (2\pi\nu)^4}{[1 + (2\pi\nu)^2][4 + (2\pi\nu)^4]} = |\hat{h}(\nu)|^2$$

The function $\hat{h}(\nu)$ has the form

$$\hat{h}(\nu) = \frac{(2\pi i\nu + z_1)(2\pi i\nu + z_2)}{(2\pi i\nu + p_1)(2\pi i\nu + p_2)(2\pi i\nu + p_3)}$$

where we take $p_1 = 1 + i$, $p_2 = 1 - i$, $p_3 = 1$, and z_1 and z_2 have positive real parts. For this problem

$$S_{xy}(\nu) = S_x(\nu) = \frac{1}{4 + (2\pi\nu)^4}$$

Therefore,

$$\left[\frac{S_{xy}(\nu)}{\overline{\hat{h}(\nu)}}\right] = \frac{-2\pi i\nu + \bar{p}_3}{(-2\pi i\nu + \bar{z}_1)(-2\pi i\nu + \bar{z}_2)(2\pi i\nu + p_1)(2\pi i\nu + p_2)}$$

Since $g(\cdot)$ is the inverse transform of $[S_{xy}(\nu)/\overline{\hat{h}(\nu)}]$ and only $g(t)$ for $t \geq 0$ contributes to $\hat{\gamma}(\nu)$, if we write

$$\left[\frac{\overline{S_{xy}(\nu)}}{\hat{h}(\nu)}\right] = \frac{a_1}{2\pi i\nu + p_1} + \frac{a_2}{2\pi i\nu + p_2} + \frac{a_3}{-2\pi i\nu + \bar{z}_1} + \frac{a_4}{-2\pi i\nu + \bar{z}_2}$$

then, $\hat{\gamma}(\nu) = \dfrac{a_1}{2\pi i\nu + p_1} + \dfrac{a_2}{2\pi i\nu + p_2}\cdot$ Finally,

$$H(\nu) = \frac{\hat{\gamma}(\nu)}{\hat{h}(\nu)} = \frac{[(a_1p_2 + p_1a_2) + (a_1 + a_2)2\pi i\nu](2\pi i\nu + p_3)}{(2\pi i\nu + z_1)(2\pi i\nu + z_2)}$$

Using calculus of residues, we get

$$a_1 = \frac{p_1 + \bar{p}_3}{(p_1 + \bar{z}_1)(p_1 + \bar{z}_2)(p_1 + p_2)}$$

$$a_2 = \frac{p_2 + \bar{p}_3}{(p_2 + \bar{z}_1)(p_2 + \bar{z}_2)(-p_2 + p_1)}$$

Now, it only remains to compute z_1 and z_2 which can be shown to be given by $z_1, z_2 = (5)^{\frac{1}{4}} \exp\left(\pm i/2 \tan^{-1} \sqrt{19}\right)$

22. For this case (9.40) becomes

$$K(t) = \begin{bmatrix} 1 \\ 0 \end{bmatrix} + \mathbf{\Sigma}(t) \begin{bmatrix} 0 \\ 1 \end{bmatrix}$$

and (9.53) becomes

$$\dot{\mathbf{\Sigma}}(t) = \begin{bmatrix} 1 & 0 \\ 0 & 0 \end{bmatrix} - \begin{bmatrix} 1 & 0 \\ 0 & 0 \end{bmatrix} - \mathbf{\Sigma}(t) \begin{bmatrix} 0 & 0 \\ 1 & 0 \end{bmatrix} - \begin{bmatrix} 0 & 1 \\ 0 & 0 \end{bmatrix} \mathbf{\Sigma}(t) + \mathbf{\Sigma}(t) \begin{bmatrix} 0 & 0 \\ 0 & 1 \end{bmatrix} \mathbf{\Sigma}(t) + \begin{bmatrix} 0 & -1 \\ 1 & 0 \end{bmatrix} \mathbf{\Sigma}(t) + \mathbf{\Sigma}(t) \begin{bmatrix} 0 & 1 \\ -1 & 0 \end{bmatrix}$$

Since $\mathbf{\Sigma}(0) = 0$, $\mathbf{\Sigma}(t) \equiv 0$, $t \geq 0$ is a solution. If the solution to (9.53) is unique, as can be shown, then $\mathbf{\Sigma}(t) \equiv 0$ is the only solution, and we have $K(t) = \begin{bmatrix} 1 \\ 0 \end{bmatrix}\cdot$ We can verify by writing

$$\dot{\mathbf{Y}}_t = \mathbf{F}\mathbf{Y}_t + \begin{bmatrix} 1 \\ 0 \end{bmatrix} \zeta_t$$
$$\dot{X}_t = \mathbf{H}\mathbf{Y}_t + \zeta_t$$

Hence,

$$\dot{\mathbf{Y}}_t = \mathbf{F}\mathbf{Y}_t + \begin{bmatrix} 1 \\ 0 \end{bmatrix} (\dot{X}_t - \mathbf{H}\mathbf{Y}_t)$$

23. Let $Y_{2t} = Y_t$ and $Y_{1t} = \dot{Y}_t$. Then

$$\begin{bmatrix} \dot{Y}_{1t} \\ \dot{Y}_{2t} \end{bmatrix} = \begin{bmatrix} 0 & 1 \\ 1 & 0 \end{bmatrix} \begin{bmatrix} Y_{1t} \\ Y_{2t} \end{bmatrix} + \begin{bmatrix} 1 & 0 \\ 0 & 0 \end{bmatrix} \begin{bmatrix} \eta_t \\ \xi_t \end{bmatrix}$$
$$\dot{X}_t = [0 \quad 1] \begin{bmatrix} Y_{1t} \\ Y_{2t} \end{bmatrix} + [0 \quad 1] \begin{bmatrix} \eta_t \\ \xi_t \end{bmatrix}$$

24. Since $e^{-|t-s|} = \int_{-\infty}^{\infty} e^{i2\pi\nu t}2/[1 + (2\pi\nu)^2]\, d\nu$, $(1/\sqrt{2})\left(\frac{d}{dt}\xi_t + \xi_t\right) = \zeta_t$ is a standard white noise. Therefore,

$$\left(\frac{d}{dt} + 1\right)\dot{X}_t = \left(\frac{d}{dt} + 1\right)Y_t + \sqrt{2}\,\zeta_t$$

Let $Z_t = [(d/dt) + 1]X_t$, then we have

$$\dot{Z}_t = [1 \quad 1]\mathbf{Y}_t + [0 \quad \sqrt{2}]\begin{bmatrix}\eta \\ \zeta_t\end{bmatrix}$$

which is in the form of (9.34).

CHAPTER 4

1. Without loss of generality we can assume $a = 0$ and $b = 1$ and set

$$W_t^{(n)} = W_{k/n} \qquad \begin{array}{l} k/n \le t < k + 1/n \\ k = 0, 1, \ldots, n-1 \end{array}$$

Because $E\varphi_t\bar{\varphi}_s$ is clearly continuous on $[0,1]^2$, we have

$$\sum_{k=0}^{n-1} \varphi_{k/n}(W_{k+1/n} - W_{k/n}) \xrightarrow[n\to\infty]{} \int_0^1 \varphi_t\, dW_t$$

On the other hand

$$\begin{aligned}
\varphi_1 W_1 - \varphi_0 W_0 - \int_0^1 \dot{\varphi}_t W_t^{(n)}\, dt &= \varphi_1 W_1 - \varphi_0 W_0 - \sum_{k=0}^{n-1} W_{k/n} \int_{k/n}^{k+1/n} \dot{\varphi}_t\, dt \\
&= \varphi_1 W_1 - \varphi_0 W_0 - \sum_{k=0}^{n-1} W_{k/n}(\varphi_{(k+1)/n} - \varphi_{k/n}) \\
&= \sum_{k=1}^{n} \varphi_{k/n}(W_{k/n} - W_{(k-1)/n}) \\
&= \sum_{k=0}^{n-1} \varphi_{k/n}(W_{(k+1)/n} - W_{k/n}) \\
&\qquad + \sum_{k=1}^{n} (\varphi_{k/n} - \varphi_{(k-1)/n})(W_{k/n} - W_{(k-1)/n})
\end{aligned}$$

The first term goes to $\int_0^1 \varphi_t\, dW_t$ as $n \to \infty$, while

$$\left|\sum_{k=1}^{n} (\varphi_{k/n} - \varphi_{k-1/n})(W_{k/n} - W_{(k-1)/n})\right| \le \sqrt{\sum_{k=1}^{n} (\varphi_{k/n} - \varphi_{(k-1)/n})^2 \sum_{k=1}^{n} (W_{k/n} - W_{(k-1)/n})^2}$$

$\xrightarrow[n\to\infty]{\text{a.s.}} 0$. Since $\int_0^1 \dot{\varphi}_t W_t^{(n)}\, dt \xrightarrow[n\to\infty]{\text{a.s.}} \int_0^1 \dot{\varphi}_t W_t\, dt$, we have

$$\varphi_1 W_1 - \varphi_0 W_0 - \int_0^1 \dot{\varphi}_t W_t\, dt = \int_0^1 \varphi_t\, dW_t$$

which was to be proved.

2. Set $X_t = \ln Z_t$ then

$$dX_t = \frac{1}{Z_t}\,dZ_t - \frac{1}{2}\frac{1}{Z_t^2}\,Z_t^2\varphi_t^2\,dt = \varphi_t\,dW_t - \tfrac{1}{2}\varphi_t^2\,dt$$

Since $X_0 = \ln Z_0 = 0$ we have

$$Z_t = e^{X_t} = \exp\left(\int_0^t \varphi_s\,dW_s - \frac{1}{2}\int_0^t \varphi_s^2\,ds\right)$$

3. Define $Y_t = \ln X_t - \ln X_0$. Then, Ito's differentiation rule yields

$$\begin{aligned} dY_t &= \frac{1}{X_t}\,dX_t - \frac{1}{2}\frac{1}{X_t^2}\,\sigma^2(t)X_t^2\,dt \\ &= m(t)\,dt + \sigma(t)\,dW_t - \frac{\sigma^2(t)}{2}\,dt \end{aligned}$$

Since $Y_0 = 0$, we have

$$Y_t = \int_0^t \sigma(s)\,dW_s + \int_0^t \left[m(s) - \frac{\sigma^2(s)}{2}\right] ds$$

or

$$X_t = X_0 e^{Y_t} = \exp\left\{\int_0^t \sigma(s)\,dW_s + \int_0^t \left[m(s) - \frac{\sigma^2(s)}{2}\right] ds\right\}$$

4. We first write

$$X_t = X_0 + \int_0^t m(X_s,s)\,ds + \int_0^t \sigma(X_s,s)\,dWs$$

By the Cauchy-Schwarz inequality, we have

$$\Big(\sum_{k=1}^N a_k\Big)^2 \le N \sum_{k=1}^N a_k^2$$

It follows that

$$\begin{aligned} EX_t^2 &\le 3\left\{EX_0^2 + \int_0^t E\sigma^2(X_s,s)\,ds + E\left[\int_0^t m(X_s,s)\,ds\right]^2\right\} \\ &\le 3\left[EX_0^2 + K\int_0^t E(1+X_s^2)\,ds + Kt\int_0^t E(1+X_s^2)\,ds\right] \\ &= 3\left[EX_0^2 + K(1+t)\int_0^t E(1+X_s^2)\,ds\right] \end{aligned}$$

If we set $f(t) = \int_0^t E(1+X_s^2)\,ds$, then

$$\frac{d}{dt}f(t) - 3K(1+t)f(t) \le 1 + 3EX_0^2$$

or

$$\frac{d}{dt}\left[e^{-3K(t+t^2/2)}f(t)\right] \le (1+3EX_0^2)e^{-3K(t+t^2/2)}$$

Since $f(0) = 0$, we have

$$f(t) \leq (1 + 3EX_0^2)e^{3K(t+t^2/2)} \int_0^\infty e^{-3K(t+t^2/2)}\, dt = Be^{3K(t+t^2/2)}$$

and

$$EX_t^2 < f'(t) \leq (1 + 3EX_0^2) + 3K(1 + t)f(t) \leq Ae^{3Kt^2}$$

5. We merely have to note that

$$\varphi(W_t) - \varphi(0) = \int_0^t \varphi'(W_s)\, dW_s + \frac{1}{2}\int_0^t \varphi''(W_s)\, ds$$

Therefore, $X_t = \varphi(0) + \int_0^t \varphi'(W_s)\, dW_s$. Since φ'' is bounded (say $|\varphi''| \leq B$), we have $|\varphi'(x)| \leq |\varphi'(0)| + B|x|$. Hence, $\int_0^t E|\varphi'(W_s)|^2\, ds < \infty$, and $\int_0^t \varphi'(W_s)\, dW_s$ is a martingale and so is X_t.

7. First, let $Z_t = \sum_{k=1}^n W_{kt}^2$. Then,

$$(Z_t - Z_s) = \sum_{k=1}^n (W_{kt} - W_{ks})^2 + 2\sum_{k=1}^n W_{ks}(W_{kt} - W_{ks})$$

Therefore,

$$E[(Z_t - Z_s)|W_{k\tau}, \tau \leq s, k = 1, \ldots, n] = \sum_{k=1}^n E(W_{kt} - W_{ks})^2 = n(t - s)$$

and

$$E[(Z_t - Z_s)^2|W_{k\tau}, \tau \leq s, k = 1, \ldots, n] = \sum_{k=1}^n E(W_{kt} - W_{ks})^4$$
$$+ \sum_{k\neq l} E(W_{kt} - W_{ks})^2 E(W_{lt} - W_{ls})^2 + 4\sum_{k=1}^n W_{ks}^2 E(W_{kt} - W_{ks})^2$$
$$+ 4\sum_{k\neq l} W_{ks}W_{ls}E[(W_{kt} - W_{ks})(W_{lt} - W_{ls})] - 3n(t - s)^2 + n(n - 1)(t - s)^2$$
$$+ 4(t - s)Z_s$$

It follows that

$$\lim_{\Delta\downarrow 0} E[(Z_{t+\Delta} - Z_t)|Z_\tau, \tau \leq t] = n$$
$$\lim_{\Delta\downarrow 0} E[(Z_{t+\Delta} - Z_t)^2|Z_t, \tau \leq t] = 4Z_t$$

From Proposition 6.4 we know that Z is Markov and satisfies

$$dZ_t = n\, dt + 2\sqrt{Z_t}\, dW_t$$

for some Brownian motion W. Since $X_t = \sqrt{Z_t}$, we have

$$dX_t = \frac{1}{2}\frac{1}{\sqrt{Z_t}}\,dZ_t - \frac{1}{8}Z_t^{-\frac{3}{2}}(4Z_t)\,dt = \frac{1}{2X_t}(n\,dt + 2X_t\,dW_t) - \frac{1}{2X_t}\,dt$$
$$= \left(\frac{n}{2} - \frac{1}{2}\right)\frac{1}{X_t}\,dt + dW_t$$

We note that neither the stochastic differential equation for Z nor the one for X satisfies the conditions of Proposition 4.1. However, since Z is Markov and $X_t = \sqrt{Z_t}$, the process X is also Markov.

8. (*a*) For $\mu = \mu(X_0,t)$ the white-noise equation is equivalent to a stochastic differential equation.

$$dX_t = \mu(X_0,t)[b + \sigma\,dW_t]$$

or

$$X_t = X_0 + b\int_0^t \mu(X_0,s)\,ds + \sigma\int_0^t \mu(X_0,s)\,dW_s$$

Therefore,

$$E(X_T^2|X_0 = x) = \left[x + b\int_0^T \mu(x,s)\,ds\right]^2 + \sigma^2\int_0^T \mu^2(x,s)\,ds$$

Suppose we denote the above functional of μ by $F(\mu)$. We get a minimizing μ by setting

$$\left.\frac{\partial F(\mu + \epsilon\,\delta\mu)}{\partial\epsilon}\right|_{\epsilon=0} = 0$$

and get

$$\int_0^T \left\{2\sigma^2\mu(x,s) + 2b\left[x + b\int_0^T \mu(x,\tau)\,d\tau\right]\right\}\delta\mu(x,s)\,ds = 0$$

for all $\delta\mu$. This yields

$$\mu(x,s) = \mu(x) = \frac{bx}{(\sigma^2 + b^2T)} \qquad 0 \le s \le T$$

(*b*) For $\mu_t = \alpha(t)X_t$, the equivalent stochastic differential equation is now

$$dX_t = b\alpha(t)X_t\,dt + \sigma\alpha(t)X_t\,dW_t + \tfrac{1}{2}\sigma^2\alpha^2(t)X_t\,dt$$
$$= X_t[m(t)\,dt + \sigma\alpha(t)\,dW_t]$$

We know from Solution 4.3 that the solution is given by

$$X_t = X_0\exp\left\{\int_0^t \sigma\alpha(s)\,dW_s + \int_0^t\left[m(s) - \frac{\sigma^2\alpha^2(s)}{2}\right]ds\right\}$$
$$= X_0\exp\left[\int_0^t \sigma\alpha(s)\,dW_s + \int_0^t b\alpha(s)\,ds\right]$$

Now since $\int_0^T \sigma\alpha(s)\,dW_s$ is a Gaussian random variable with zero mean and variance $\Sigma^2 = \sigma^2\int_0^T \alpha^2(s)\,ds$, we have

$$E\exp\left[2\int_0^T \sigma\alpha(s)\,dW_s\right] = \exp\left[2\sigma^2\int_0^T \alpha^2(s)\,ds\right]$$

Therefore,

$$E(X_T^2|X_0 = x) = x^2 \exp\left[2\int_0^T b\alpha(s)\,ds + 2\sigma^2\int_0^T \alpha^2(s)\,ds\right]$$
$$= x^2 \exp\left\{2\int_0^T \left[\sigma\alpha(s) + \frac{1}{2\sigma}b\right]^2 ds - \frac{1}{2}\left(\frac{b}{\sigma}\right)^2 T\right\}$$
$$\geq x^2 e^{-\frac{1}{2}(b/\sigma)^2\, T}$$

so that $\alpha(s) = b/2\sigma^2$, $0 \leq s \leq T$, is the minimizing function.

9. We use the standard technique of separation of variables. Consider the equation

$$\frac{\partial f(x,t)}{\partial t} = \frac{1}{2}\frac{\partial^2}{\partial x^2}[\sigma^2(x)f(x,t)] - \frac{\partial}{\partial x}[m(x)f(x,t)]$$

If $f(x,t)$ is a product $f(x,t) = g(t)W(x)\varphi(x)$, then we have

$$W(x)\varphi(x)\frac{dg(t)}{dt} = g(t)\left(\frac{d}{dx}\left\{\frac{1}{2}\frac{d}{dx}[\sigma^2(x)W(x)\varphi(x)] - m(x)W(x)\varphi(x)\right\}\right)$$
$$= g(t)\frac{1}{2}\frac{d}{dx}\left[\sigma^2(x)W(x)\frac{d\varphi(x)}{dx}\right]$$

Therefore,

$$\frac{1}{g(t)}\frac{dg(t)}{dt} = \frac{1}{W(x)\varphi(x)}\frac{1}{2}\frac{d}{dx}\left[\sigma^2(x)W(x)\frac{d\varphi(x)}{dx}\right]$$

The two sides, being functions of different variables, must be constants. Hence, if $f(x,t)$ is a product, then it must have the form

$$f_\lambda(x,t) = e^{-\lambda t}\varphi_\lambda(x)$$

when φ_λ satisfies the Sturm-Liouville equation

$$\frac{1}{2}\frac{d}{dx}\left[\sigma^2(x)W(x)\frac{d\varphi(x)}{dx}\right] + \lambda W(x)\varphi(x) = 0$$

Under rather general conditions, it can be shown that every solution $f(x,t)$ can be represented as a linear combination of products. Since $p(x,t|x_0,t_0)$ is a function of $t - t_0$, x, and x_0, it must have the assumed form.

10. The Sturm-Liouville equation in this case is

$$\tfrac{1}{2}f''(x) + \lambda f(x) = 0$$

Let $\lambda = \frac{1}{2}\nu^2$, then $f_\nu(x) = e^{i\nu x}$ are the bounded solutions. Now,

$$\int_{-\infty}^{\infty} e^{-i\nu x}\frac{1}{\sqrt{2\pi t}}\exp\left(-\frac{1}{2}\frac{x^2}{t}\right)dx = e^{-\frac{1}{2}\nu^2 t}$$

By the inversion formula of the Fourier integral, we get

$$\frac{1}{\sqrt{2\pi t}}\exp\left(-\frac{1}{2}\frac{x^2}{t}\right) = \frac{1}{2\pi}\int_{-\infty}^{\infty} e^{-\frac{1}{2}\nu^2 t}e^{i\nu x}\,d\nu$$

and

$$\frac{1}{\sqrt{2\pi(t-t_0)}} \exp\left[-\frac{1}{2}\frac{(x-x_0)^2}{(t-t_0)}\right] = \frac{1}{2\pi}\int_{-\infty}^{\infty} e^{i\nu(x-x_0)}e^{-\frac{1}{2}\nu^2(t-t_0)}\,d\nu$$

11. The Fokker-Planck equation for this case has the form

$$\frac{\partial p}{\partial t} = \tfrac{1}{2}\beta(t)\frac{\partial^2 p}{\partial x^2} - \alpha(t)\frac{\partial}{\partial x}(xp)$$

Assuming p to have the form

$$p(x,t|x_0,t_0) = \frac{1}{\sqrt{2\pi a^2}}\exp\left[-\frac{1}{2a^2}(x-bx_0)^2\right]$$

we get

$$\frac{1}{p}\frac{\partial p}{\partial t} = -\frac{1}{2}\frac{1}{a^2}\frac{d}{dt}(a^2) + \frac{1}{2a^4}(x-bx_0)^2\frac{da^2}{dt} + \frac{x_0}{a^2}(x-bx_0)\frac{db}{dt}$$

$$\frac{1}{p}\frac{\partial p}{\partial x} = -\frac{1}{a^2}(x-bx_0)$$

$$\frac{1}{p}\frac{\partial^2 p}{\partial x^2} = \frac{1}{a^4}(x-bx_0)^2 - \frac{1}{a^2}$$

Upon substituting into the Fokker-Planck equati n, we get

$$\left[\frac{(x-bx_0)^2}{2a^4} - \frac{1}{2a^2}\right]\frac{da^2}{dt} + \frac{x_0}{a^2}(x-bx_0)\frac{db}{dt} = \frac{\beta}{2}\left[\frac{(x-bx_0)^2}{a^4} - \frac{1}{a^2}\right] + \frac{\alpha(t)x}{a^2}(x-bx_0) - \alpha(t)$$

$$= \left[\frac{(x-bx_0)^2}{a^2} - 1\right]\left(\frac{\beta}{2a^2} + \alpha\right) + \frac{\alpha}{a^2}bx_0(x-bx_0)$$

Therefore,

$$\frac{da^2}{dt} = 2a^2\left(\frac{\beta}{2a^2} + \alpha\right) = \beta + 2\alpha a^2$$

$$\frac{db}{dt} = a^2\frac{\alpha}{a^2}b = \alpha b$$

which are to be solved with $a^2(t_0,t_0) = 0$, $b(t_0,t_0) = 1$.

12. (*a*) We assume $\frac{\partial p}{\partial t} \xrightarrow[t\to\infty]{} 0$ so that

$$\frac{1}{2}\frac{d^2p(x)}{dx^2} + \frac{d}{dx}[\operatorname{sgn} x\,p(x)] = 0$$

or

$$\frac{dp(x)}{dx} + 2\operatorname{sgn} x\,p(x) = \text{constant}$$

Since $p'(x),\ p(x) \xrightarrow[|x|\to\infty]{} 0$, this constant mu t be zero. Therefore,

$$p(x) = e^{-2|x|}$$

which is already normalized.

(*b*) Consider the Sturm-Liouville equation

$$\frac{1}{2}\frac{d}{dx}\left[p(x)\frac{d\varphi(x)}{dx}\right] + \lambda p(x)\varphi(x) = 0$$

This can be written as

$$\frac{d^2\varphi(x)}{dx^2} - 2\,\mathrm{sgn}\,x\,\frac{d\varphi(x)}{dx} + \lambda\varphi(x) = 0$$

The solutions satisfying, (1) $d\varphi/dx$ continuous at $x = 0$, and (2) $p\varphi^2$ bounded, are of the form

$$\varphi(x) = A_\nu e^{|x|}\sin\nu x \qquad 0 < \nu < \infty$$

where $\nu^2 = \lambda - 1$. It is convenient to choose A_ν to obtain the normalization

$$\int_{-\infty}^{\infty} \varphi_\nu(x)\varphi_{\nu'}(x)p(x)\,dx = \delta(\nu - \nu')$$

It can be shown that A_ν should be $\sqrt{2/\pi}$. Hence,

$$\varphi(\nu,x) = \sqrt{\frac{2}{\pi}}\,e^{|x|}\sin\nu x$$

The transition-density function must have the form

$$p(x,t|x_0,t_0) = p(x)\int_0^\infty e^{-(\nu^2+1)(t-t_0)}\varphi(\nu,x)\psi(\nu,x_0)\,d\nu$$

Because $p(x,t_0|x_0,t_0) = \delta(x - x_0)$, we have

$$\psi(\nu,x_0) = \int_{-\infty}^{\infty}\varphi(\nu,x)\delta(x - x_0)\,dx = \varphi(\nu,x_0)$$

Therefore,

$$\begin{aligned} p(x,t|x_0,t_0) &= e^{-2|x|}e^{|x|+|x_0|}\frac{2}{\pi}\int_0^\infty e^{-(1+\nu^2)(t-t_0)}\sin\nu x\sin\nu x_0\,d\nu \\ &= e^{-(|x|-|x_0|)}e^{-(t-t_0)}\frac{1}{\pi}\int_0^\infty e^{-\nu^2(t-t_0)}\left[\cos\nu(x - x_0) - \cos\nu(x + x_0)\right]d\nu \\ &= e^{-(|x|-|x_0|)-(t-t_0)}\frac{1}{\sqrt{2\pi}}\left\{\exp\left[-\frac{(x - x_0)^2}{2(t - t_0)}\right] - \exp\left[-\frac{(x + x_0)^2}{2(t - t_0)}\right]\right\} \end{aligned}$$

CHAPTER 5

1. $(H_tf)(x) = E_x(f(X_t))$

$$= \tfrac{1}{2}(1 + e^{-t})f(x) + \tfrac{1}{2}(1 - e^{-t})f(-x) \qquad x = 1, -1$$

If we set $\mathbf{f} = \begin{bmatrix} f(1) \\ f(-1) \end{bmatrix}$, then the operator H_t has a representation as a matrix, and we have

$$\mathbf{H}_t\mathbf{f} = \begin{bmatrix} \frac{1}{2}(1 + e^{-t}) & \frac{1}{2}(1 - e^{-t}) \\ \frac{1}{2}(1 - e^{-t}) & \frac{1}{2}(1 + e^{-t}) \end{bmatrix} \mathbf{f}$$

The generator A also has a matrix representation and can be found by

$$\mathbf{A} = \lim_{t \downarrow 0} \frac{1}{t} (\mathbf{H}_t - \mathbf{I}) = \lim_{t \downarrow 0} \frac{1}{2t} (1 - e^{-t}) \begin{bmatrix} -1 & 1 \\ 1 & -1 \end{bmatrix}$$
$$= \frac{1}{2} \begin{bmatrix} -1 & 1 \\ 1 & -1 \end{bmatrix}$$

We can verify that $\mathbf{H}_t = e^{t\mathbf{A}}$ in many ways. For example, we note that

$$\mathbf{A}^2 = \frac{1}{2} \begin{bmatrix} 1 & -1 \\ -1 & 1 \end{bmatrix} = -\mathbf{A}$$

so that for $n = 1, 2, \ldots,$

$$\mathbf{A}^n = (-1)^{n-1}\mathbf{A} = \frac{(-1)^n}{2} \begin{bmatrix} 1 & -1 \\ -1 & 1 \end{bmatrix}$$

Therefore,

$$e^{t\mathbf{A}} = \sum_{n=0}^{\infty} \frac{t^n}{n!} \mathbf{A}^n = \mathbf{I} + \frac{1}{2} \begin{bmatrix} 1 & -1 \\ -1 & 1 \end{bmatrix} \sum_{n=1}^{\infty} \frac{(-t)^n}{n!}$$
$$= \begin{bmatrix} 1 & 0 \\ 0 & 1 \end{bmatrix} + \frac{1}{2} \begin{bmatrix} 1 & -1 \\ -1 & 1 \end{bmatrix} (e^{-t} - 1)$$
$$= \begin{bmatrix} \frac{1}{2}(1 + e^{-t}) & \frac{1}{2}(1 - e^{-t}) \\ \frac{1}{2}(1 - e^{-t}) & \frac{1}{2}(1 + e^{-t}) \end{bmatrix} = \mathbf{H}_t$$

2. Imitating the example at the end of Sec. 5.2, we find

$$\mathfrak{D}_A = \{f: f \text{ bounded continuous on } [0, \infty), f'' \text{ bounded continuous on } (0, \infty)\}$$

and

$$(Af)(a) = \frac{1}{2} \frac{d^2 f(a)}{da^2} \qquad a > 0$$
$$= 0 \qquad a = 0$$

3. (*a*) With f as defined, we have

$$u_\lambda(a) = E_a \int_0^\infty e^{-\lambda t} f(X_t)\, dt = \int_0^\infty \text{prob } (X_t > c | X_0 = a) e^{-\lambda t}\, dt$$
$$= \int_0^\infty \left[\int_c^\infty \frac{1}{\sqrt{2\pi t}} \exp \left[-\frac{1}{2t} (x - a)^2 \right] dx \right] e^{-\lambda t}\, dt$$

Since $\int_0^\infty \frac{1}{\sqrt{2\pi t}} \exp \left[-\frac{1}{2t} (x - a)^2 \right] e^{-\lambda t}\, dt = \exp (-\sqrt{2\lambda}|x - a|)/\sqrt{2\lambda}$, we find

$$u_\lambda(a) = \int_c^\infty \frac{\exp (-\sqrt{2\lambda}\, |x - a|)}{\sqrt{2\lambda}}\, dx$$

$$= \begin{cases} \dfrac{1}{2\lambda} \exp\left[-\sqrt{2\lambda}\,(c-a)\right] & a \le c \\ \dfrac{1}{\lambda} - \dfrac{1}{2\lambda} \exp\left[-\sqrt{2\lambda}\,(a-c)\right] & a \ge c \end{cases}$$

Equation (3.13) now yields (with $\tau = \tau_c$)

$$u_\lambda(a) = u_\lambda(c) E_a e^{-\lambda \tau_c}$$

Since the Brownian motion X_t starts from 0 at $t = 0$, we get

$$E_0 e^{-\lambda\tau_c} = \frac{u_\lambda(0)}{u_\lambda(c)} = \exp\left(-\sqrt{2\lambda}\,c\right)$$

(b) Assume that τ_c has a density function $q(s)$, $0 \le s \le \infty$ so that

$$\mathcal{P}(\tau_c \le t) = \int_0^t q(s)\,ds$$

Taking the Laplace transform, we get

$$\int_0^\infty e^{-\lambda t} \int_0^t q(s)\,ds\,dt = \frac{1}{\lambda}\int_0^\infty e^{-\lambda t} q(t)\,dt = \frac{1}{\lambda} E e^{-\lambda\tau_c} = \frac{1}{\lambda}\exp\left(-\sqrt{2\lambda}\,c\right)$$

On the other hand,

$$u_\lambda(0) = \int_0^\infty e^{-\lambda t}\mathcal{P}(X_t \ge c)\,dt = \frac{1}{2\lambda}\exp\left(-\sqrt{2\lambda}\,c\right)$$

Because of the uniqueness of Laplace transforms

$$2\mathcal{P}(X_t \ge c) = \mathcal{P}(\tau_c \le t)$$

4. Let $\tau_0 = \min\{t : t > 0, X_t = 0\}$. Then,

$$\mathcal{P}(X_t \ge 0, 0 \le t \le 1) = \int_0^\infty \frac{1}{\sqrt{2\pi}} e^{-a^2/2}\, \mathcal{P}(\tau_0 \ge 1 | X_0 = a)\,da$$

Now, let $p(x,t|x_0,s)$ denote the transition density and define

$$\begin{aligned} f(x) &= 0 \qquad x \ge 0 \\ &= 1 \qquad x < 0 \end{aligned}$$

Then, (3.12) yields

$$u_\lambda(a) = \int_0^\infty e^{-\lambda t}\mathcal{P}(X_t < 0 | X_0 = a)\,dt = \int_{-\infty}^0 \left[\int_0^\infty e^{-\lambda t} p(x,t|a,0)\,dt\right] dx$$

Using τ_0 in (3.13), we get

$$u_\lambda(a) = u_\lambda(0) E_a e^{-\lambda\tau_0}$$

Let $q_a(\cdot)$ denote the density of τ_0 given $X_0 = a$. Then we have

$$\int_0^\infty q_a(s) e^{-\lambda s}\,ds = \frac{u_\lambda(a)}{u_\lambda(0)} = 2\lambda \int_0^\infty e^{-\lambda t}\mathcal{P}(X_t < 0|X_0 = a)\,dt$$

For $a > 0$, we get

$$q_a(t) = 2\frac{\partial}{\partial t}\mathcal{P}(X_t < 0 | X_0 = a)$$

Therefore, for $a > 0$,

$$\mathcal{P}(\tau_0 \geq 1 | X_0 = a) = \int_1^\infty q_a(t)\, dt = 2\{\tfrac{1}{2} - \mathcal{P}(X_1 < 0 | X_0 = a)\}$$

Finally,

$$\mathcal{P}(X_t \geq 0,\, 0 \leq t \leq 1) = \int_0^\infty \frac{1}{\sqrt{2\pi}} e^{-a^2/2}[1 - 2\mathcal{P}(X_1 < 0 | X_0 = a)]\, da$$

$$= \tfrac{1}{2} - 2\int_0^\infty \frac{1}{\sqrt{2\pi}} e^{-a^2/2} \mathcal{P}(X_1 < 0 | X_0 = a)\, da$$

The joint density function for X_0 and X_1 has the form

$$p(x,t;x_0,0) = \frac{1}{2\pi\sqrt{1-\rho^2}} \exp\left[-\frac{1}{2(1-\rho^2)}(x^2 - 2\rho x x_0 + x_0^2)\right]$$

where $\rho = e^{-1}$. Therefore,

$$\int_0^\infty \frac{1}{\sqrt{2\pi}} e^{-a^2/2} \mathcal{P}(X_1 \leq 0 | X_0 = a)\, da$$

$$= \int_0^\infty da \int_{-\infty}^0 dx \frac{1}{2\pi\sqrt{1-\rho^2}} \exp\left[-\frac{1}{2(1-\rho^2)}(x^2 - 2\rho x a + a^2)\right]$$

(Letting $a = r\cos\theta$, $x = r\sin\theta$, we get)

$$= \int_0^\infty r\, dr \int_{-\pi/2}^0 d\theta \frac{1}{2\pi\sqrt{1-\rho^2}} \exp\left[-\frac{r^2}{2(1-\rho^2)}(1 - \rho\sin 2\theta)\right]$$

$$= \frac{\sqrt{1-\rho^2}}{2\pi} \int_{-\pi/2}^0 \frac{1}{1 - \rho\sin 2\theta}\, d\theta$$

$$= \frac{1}{4} - \frac{1}{2\pi}\sin^{-1}\rho$$

Thus, we finally arrive at the expression

$$\mathcal{P}(X_t \geq 0,\, 0 \leq t \leq 1) = \frac{1}{\pi}\sin^{-1}(e^{-1})$$

5. See Solution 4.7 where we determined that

$$m(x) = 2 \qquad \sigma^2(x) = 4x$$

6. By (5.7) we can take

$$u(x) = \int_c^x \exp\left[-\int_0^y \frac{2m(z)}{\sigma^2(z)}\, dz\right] dy$$

Since $2m(z)/\sigma^2(z) = 1/x$, it is convenient to take $c = 1$ and find

$$u(x) = \int_1^x e^{-\ln y}\, dy = \ln x$$

Since $u(0) = -\infty$, S is open at 0.

7. We rewrite (2.12) for $H_t g \in \mathfrak{D}_A$ as

$$\frac{d}{dt} H_t g = A H_t g$$

If we set $f_\lambda = \int_0^\infty e^{-\lambda t} H_t g \, dt$, then

$$\int_0^\infty e^{-\lambda t} \frac{d}{dt} H_t g \, dt = -g + \lambda \int_0^\infty e^{-\lambda t} H_t g \, dt = -g + \lambda f_\lambda$$

Using (2.12), we get

$$\int_0^\infty e^{-\lambda t} \frac{d}{dt} H_t g \, dt = A \int_0^\infty e^{-\lambda t} H_t g \, dt = A f_\lambda$$

Hence,

$$A f_\lambda = \lambda f_\lambda - g$$

8. First, we write

$$\begin{aligned} h(x, t + \delta) &= E_x \left\{ f(X_{t+\delta}) \exp\left[-\int_0^{t+\delta} k(X_s)\, ds \right] \right\} \\ &= E_x \left(E \left\{ f(X_{t+\delta}) \exp\left[-\int_0^{t+\delta} k(X_s)\, ds \right] \middle| X_\tau, 0 \le \tau \le \delta \right\} \right) \end{aligned}$$

Because X is Markov and has a stationary transition function,

$$\begin{aligned} h(x, t + \delta) &= E_x \left(\exp\left[-\int_0^{\delta} k(X_s)\, ds \right] E \left\{ f(X_{t+\delta}) \exp\left[-\int_0^{t} k(X_{s+\delta})\, ds \right] \middle| X_\delta \right\} \right) \\ &= E_x \left\{ \exp\left[-\int_0^{\delta} k(X_s)\, ds \right] h(X_\delta, t) \right\} \\ &= h(x,t) + h'(x,t) E_x(X_\delta - x) + \tfrac{1}{2} h''(x,t) E_x(X_\delta - x)^2 \\ &\qquad - k(x) h(x,t)\, \delta + o(\delta) \end{aligned}$$

Therefore,

$$\frac{\partial h(x,t)}{\partial t} = m(x) h'(x,t) + \tfrac{1}{2}\sigma^2(x) h''(x,t) - k(x) h(x,t)$$

9. Setting $k(x) = \beta(1 + \operatorname{sgn} x/2)$ and $f(x) = 1$ in Kac's theorem, we get

$$u_\lambda(x) = E_x \int_0^\infty e^{-\lambda t} \exp\left(-\beta \int_0^t \frac{1 + \operatorname{sgn} X_s}{2}\, ds \right) dt$$

and u_λ satisfies

$$\frac{1}{2} \frac{d^2 u_\lambda(x)}{dx^2} - \beta \frac{1 + \operatorname{sgn} x}{2} u_\lambda(x) - \lambda u_\lambda(x) = -1$$

Therefore,

$$\frac{1}{2} \frac{d^2 u_\lambda(x)}{dx^2} - (\beta + \lambda) u_\lambda(x) = -1 \qquad x > 0$$

$$\frac{1}{2} \frac{d^2 u_\lambda(x)}{dx^2} - \lambda u_\lambda(x) = -1 \qquad x < 0$$

and u_λ must have the form

$$u_\lambda(x) = A \exp\left[-\sqrt{2(\beta+\lambda)}\,x\right] + \frac{1}{\lambda+\beta} \qquad x > 0$$

$$= B \exp\left(\sqrt{2\lambda}\,x\right) + \frac{1}{\lambda} \qquad x < 0$$

Continuity of u_λ and u'_λ at 0 yields

$$A = \frac{1}{\sqrt{\lambda+\beta}}\frac{1}{\sqrt{\lambda}} - \frac{1}{\lambda+\beta}$$

$$B = \frac{1}{\sqrt{\lambda+\beta}}\frac{1}{\sqrt{\lambda}} - \frac{1}{\lambda}$$

Since the Brownian motion s arts at 0, we only need $u_\lambda(0)$ which is given by

$$u_\lambda(0) = \frac{1}{\sqrt{\lambda(\lambda+\beta)}}$$

which is the double Laplace transform of the distribution of

$$\tau(t) = \int_0^t \left(\frac{1+\operatorname{sgn} X_s}{2}\right) ds$$

If we define $q(s,t)$ as the density of $\tau(t)$, that is, $q(s,t)\,ds = \mathcal{P}(\tau(t) \in ds)$, then

$$u_\lambda(0) = \frac{1}{\sqrt{\lambda(\lambda+\beta)}} = \iint_0^\infty e^{-(\lambda t - \beta s)} q(s,t)\,ds\,dt$$

Inverting the Laplace transform once (wi h respect to β) yields

$$\int_0^\infty e^{-\lambda t} q(s,t)\,dt = \frac{1}{\sqrt{\pi s}}\frac{1}{\sqrt{\lambda}} e^{-\lambda s}$$

Inverting once again yields

$$q(s,t) = \frac{1}{\pi}\frac{1}{\sqrt{s}}\frac{1}{\sqrt{t-s}} \qquad t > s$$

$$= 0 \qquad t < s$$

Finally

$$\mathcal{P}(\tau(1) \le t) = \int_0^t q(s,t)\,ds = \frac{1}{\pi}\int_0^t \frac{1}{\sqrt{s(1-s)}}\,ds = \frac{2}{\pi}\sin^{-1}\sqrt{t}$$

10. Define

$$h(x,t) = E_x \exp\left(-\beta \int_0^t X_s^2\,ds\right)$$

and consider the equation

$$\frac{\partial h}{\partial t} = \frac{1}{2}\frac{\partial^2 h}{\partial x^2} - \rho x^2 h \qquad h(x,0) = 1$$

We attempt a solution of the form

$$h(x,t) = A(t)e^{-\alpha(t)x^2} \qquad \begin{array}{l}\alpha(0) = 0\\ A(0) = 1\end{array}$$

Substituting the trial solution in the differential equation yields

$$\frac{\dot{A}(t)}{A(t)} - x^2\alpha(t) = 2\alpha^2(t)x^2 - \alpha(t) - \beta x^2$$

Equating like terms, we get

$$\dot{\alpha} + 2\alpha^2(t) = \beta$$

$$\frac{\dot{A}(t)}{A(t)} = -\alpha(t)$$

If we let $\alpha(t) = \dot{v}(t)/2v(t)$, then

$$\dot{\alpha}(t) + 2\alpha^2(t) = \frac{\ddot{v}(t)}{2v(t)} = \beta$$

so that

$$\ddot{v}(t) = 2\beta v(t)$$

With the initial condition $\alpha(0) = 0$, we get

$$\alpha(t) = \sqrt{2\beta}\tanh\sqrt{2\beta}\,t$$

and

$$\begin{aligned} A(t) &= \exp\left(-\int_0^t \sqrt{2\beta}\tanh\sqrt{2\beta}\,s\,ds\right) \\ &= \exp\,(-\ln\cosh\sqrt{2\beta}\,t) \\ &= \frac{1}{\cosh\sqrt{2\beta}\,t} \end{aligned}$$

Therefore,

$$h(x,t) = \frac{1}{\cosh\sqrt{2\beta}\,t}\exp\,[-\sqrt{2\beta}\,(\tanh\sqrt{2\beta}\,t)x^2]$$

and

$$h(0,t) = \frac{1}{\cosh\sqrt{2\beta}\,t}$$

and the density function for $Z = \int_0^1 X_t^2\,dt$ can be found by inverting $h(0,1)$, that is,

$$p_Z(z) = \frac{1}{2\pi i}\int_{C-i\infty}^{C+i\infty} e^{\beta z}\frac{1}{\cosh\sqrt{2\beta}}\,d\beta$$

11.

$\sigma^2(x)$	$m(x)$	$u(x)$	Closed or open	Regularity
1	-1	$\frac{e^{2x}-1}{2}$	Closed	Regular
1	$-x$	$\int_0^x e^{y^2}\,dy$	Closed	Regular
x	$-x$	$\frac{e^{2y}-1}{2}$	Closed	Regular
x^2	$1-\frac{3}{2}x$	$\int_0^x y^3e^{-2/y}\,dy$	Open	

CHAPTER 6

1. Let $A \in \mathcal{B}$. Then $M_2(A) = \int_A (dM_2{}^{\mathcal{B}}/dM_1{}^{\mathcal{B}})\, dM_1$. On the other hand, by the definition of conditional expectation, we also have

$$M_2(A) = \int_A \frac{dM_2}{dM_1}\, dM_1 = \int_A E_1{}^{\mathcal{B}} \frac{dM_2}{dM_1}\, dM_1$$

2. (a) Heuristically, we only need to verify that

$$E(\dot{X}_t + X_\tau)(\dot{X}_s + X_s) = \delta(t - s)$$

or

$$\frac{\partial^2 R(t,s)}{\partial t\, \partial s} + \frac{\partial R(t,s)}{\partial s} + \frac{\partial R(t,s)}{\partial t} + R(t,s) = \delta(t - s)$$

Since

$$\frac{\partial^2}{\partial t\, \partial s} \frac{1}{2} e^{-|t-s|} = \delta(t - s) - \tfrac{1}{2} e^{-|t-s|}$$

$$\frac{\partial R(t,s)}{\partial s} + \frac{\partial R(t,s)}{\partial t} = 0$$

the verification is done. Of course, we can also verify by direct computation and find

$$E\left(X_t - X_0 + \int_0^t X_\tau\, d\tau\right)\left(X_s - X_0 + \int_0^t X_\tau\, d\tau\right) = \min\,(t,s)$$

However, this procedure is somewhat tedious.

(b) Consider the distribution of $X_{k/2^n}$, $k = 0, 1, \ldots, 2^n$, and let $p_0{}^{(n)}(\mathbf{x})$ and $p^{(n)}(\mathbf{x})$ denote the density functions under $\mathcal{P}_0$ and $\mathcal{P}$, respectively. Then

$$p_0{}^{(n)}(\mathbf{x}) = \frac{1}{\sqrt{2\pi}} e^{-\frac{1}{2}x_0{}^2} \prod_{k=1}^{2^n} \frac{1}{\sqrt{2\pi 2^{-n}}} \exp\left[-\frac{1}{2(2^{-n})}(x_k - x_{k-1})^2\right]$$

and

$$p^{(n)}(\mathbf{x}) \cong \frac{1}{\sqrt{2\pi}} e^{-\frac{1}{2}x_0{}^2} \prod_{k=1}^{2^n} \frac{1}{\sqrt{2\pi 2^{-n}}} \exp\left[-\frac{1}{2(2^{-n})}(x_k - x_{k-1} + 2^{-n}x_{k-1})\right]$$

Therefore,

$$\frac{p^{(n)}(\mathbf{x})}{p_0{}^{(n)}(\mathbf{x})} \cong \exp\left[-\sum_{k=1}^{2^n} x_{k-1}(x_k - x_{k-1}) - \tfrac{1}{2}2^{-n} \sum_{k=1}^{2^n} x_{k-1}\right]$$

and

$$\ln \frac{p^{(n)}(X_{k/2^n}, k = 0, 1, \ldots, 2^n)}{p_0{}^{(n)}(X_{k/2^n}, k = 0, 1, \ldots, 2^n)}$$

$$= -\sum_{k=0}^{2^n-1} X_{k/2^n}(X_{k+1/2^n} - X_{k/2^n}) - \frac{1}{2}\sum_{k=0}^{2^n-1} \frac{1}{2^n} X_{k/2^n}$$

$$\xrightarrow[n\to 0]{} -\int_0^1 X_t\, dX_t - \frac{1}{2}\int_0^1 X_t\, dt$$

3. We use the Ito differentiation formula and write

$$dX_t = e^{W_t}\,dW_t + \tfrac{1}{2}e^{W_t}\,dt = X_t\,dW_t + \tfrac{1}{2}X_t\,dt$$

Therefore, the quadratic variation is $\int_0^1 X_t^2\,dt$

4. (*a*) Since $E^{\alpha_t}\,dX_t = \varphi_t\,dt$, we write

$$EX_s\varphi_t = \frac{\partial}{\partial t}EX_0X_t = \frac{\partial}{\partial t}R(t,s) \qquad 0 < s < t$$

Since $\varphi_t = \int_0^t h(t,\tau)\,dX_\tau$, we get

$$\int_0^t h(t,\tau)\frac{\partial}{\partial \tau}R(s,\tau)\,d\tau = \frac{\partial}{\partial t}R(t,s) \qquad 0 < s < t$$

We now use (4.2) for $R(t,s)$ and get

$$\frac{\partial}{\partial t}R(t,s) = -\int_0^s K(t,v)\,dv \qquad s < t$$

$$\frac{\partial}{\partial t}R(s,\tau) = 1 - \int_0^s K(\tau,v)\,dv \qquad \tau < s$$

$$= -\int_0^s K(\tau,v)\,dv \qquad \tau > s$$

Therefore,

$$\int_0^s h(t,\tau)\,d\tau - \int_0^t d\tau \int_0^s dv\,h(t,\tau)K(\tau,v) = -\int_0^s K(t,v)\,dv \qquad 0 < s < t$$

Differentiating with respect to s yields

$$h(t,s) - \int_0^t h(t,\tau)K(\tau,s)\,d\tau = -K(t,s) \qquad 0 < s < t$$

(*b*) We attempt a solution of the form $h(t,s) - g(t)f(s)$ and find

$$g(t)f(s)\left[1 - \int_0^t f^2(\tau)\,d\tau\right] = -f(t)f(s)$$

Therefore,

$$g(t) = -\frac{f(t)}{1 - \int_0^t f^2(\tau)\,d\tau}$$

5. (*a*) $dX_t = X_t\,dW_t + X_t\,dt + \tfrac{1}{2}X_t\,dt \qquad (\mathcal{P})$
$dX_t = X_t\,dV_t - X_t\,dt + \tfrac{1}{2}X_t\,dt \qquad (\mathcal{P}_0)$

The quadratic variation on [0,1] is $\int_0^1 X_t^2\,dt$ in either case.

(*b*) and (*c*) Let $Y_t = \ln X_t + t$. Then under $\mathcal{P}_0$, Y_t is a Brownian motion and under $\mathcal{P}$, $Y_t - 2t$ is a Brownian motion. It follows that $\mathcal{P} \equiv \mathcal{P}_0$. Here, we can identify $\varphi_t = 2$ in (4.24), and identify Y in place of X, and get

$$\frac{d\mathcal{P}}{d\mathcal{P}_0} = \exp\left(\int_0^1 2dY_t - \frac{1}{2}\int_0^1 4dt\right) = e^{2Y_1 - 2} = e^{2\ln X_1} = X_1^2$$

6. It suffices to prove that

$$E_0^{\mathcal{A}_t} \exp\left(\int_t^1 \mathbf{Z}_s^T\, d\mathbf{X}_s - \frac{1}{2}\int_t^1 \mathbf{Z}_s^T\mathbf{Z}_s\, ds\right) = 1$$

Since $\{X_s - X_t,\ t \le s \le 1\}$ is a Brownian motion independent of $\{Z_s,\ 0 \le s \le t\}$, we can write

$$E\left[\exp\left(\int_t^1 \mathbf{Z}_s^T\, d\mathbf{X}_s\right) \middle|\ \mathbf{Z}_s,\ 0 \le s \le 1,\ \mathbf{X}_s,\ 0 \le s \le t\right] = \frac{1}{\sqrt{2\pi\sigma^2}}\int_{-\infty}^{\infty} e^y \exp\left(-\frac{1}{2\sigma^2}y^2\right) dy$$

where

$$\sigma^2 = E\left[\left(\int_t^1 \mathbf{Z}_s^T\, d\mathbf{X}_s\right)^2 \middle|\ \mathbf{Z}_s\ \ 0 \le s \le 1,\ \mathbf{X}_s, 0 \le s \le t\right] = \int_t^1 \mathbf{Z}_s^T\mathbf{Z}_s\, ds$$

Since $1/\sqrt{2\pi\sigma^2}\int_{-\infty}^{\infty} e^y \exp\,[-(1/2\sigma^2)y^2]\, dy = e^{\frac{1}{2}\sigma^2}$, we get

$$E\left[\exp\left(\int_t^1 \mathbf{Z}_s^T\, d\mathbf{X}_s\right) \middle|\ \mathbf{Z}_s,\ 0 \le s \le 1,\ \mathbf{X}_s,\ 0 \le s \le t\right] = \exp\left(\frac{1}{2}\int_t^1 \mathbf{Z}_s^T\mathbf{Z}_s\, ds\right)$$

and (5.8) follows. Note: Equation (5.8) is true under quite general conditions even without the assumption of independence between Z and X. The above method of proof, however, will no longer be valid.

7. Equation (5.15) is given by

$$\Lambda_t = 1 + \int_0^t \Lambda_s\mathbf{Z}_s^T\, d\mathbf{X}_s$$

therefore

$$E_0^{\mathcal{A}_{xt}}\Lambda_t = 1 + \int_0^t E_0^{xt}(\Lambda_s\mathbf{Z}_s^T)\, d\mathbf{X}_s$$

Because $\Lambda_s\mathbf{Z}_s$ is $\mathcal{A}_s$ measurable and $\{\mathbf{X}_\tau,\ s \le \tau \le t\}$ is independent of $\mathcal{A}_s$, we have

$$E_0^{\mathcal{A}_{xt}}(\Lambda_s\mathbf{Z}_s) = E_0^{\mathcal{A}_{xs}}(\Lambda_s\mathbf{Z}_s)$$

Using (5.11), we get

$$E_0^{\mathcal{A}_{xs}}(\Lambda_s\mathbf{Z}_s) = (E^{\mathcal{A}_{xs}}\mathbf{Z}_s)E_0^{\mathcal{A}_{xs}}\Lambda_s = \hat{\mathbf{Z}}_s E_0^{\mathcal{A}_{xs}}\Lambda_s$$

Setting $E_0^{\mathcal{A}_{xt}}\Lambda_t = L_t$, we find

$$L_t = 1 + \int_0^t L_s\hat{\mathbf{Z}}_s^T\, d\mathbf{X}_s$$

or

$$dL_t = L_t\hat{\mathbf{Z}}_t^T\, d\mathbf{X}_t$$

If we let $Y_t = \ln L_t$ and use Ito's differentiation rule, we get

$$dY_t = \frac{1}{L_t}\, dL_t - \frac{1}{2}\frac{1}{L_t^2}\, L_t^2\hat{\mathbf{Z}}_t^T\hat{\mathbf{Z}}_t\, dt = \hat{\mathbf{Z}}_t^T\, d\mathbf{X}_t - \tfrac{1}{2}\hat{\mathbf{Z}}_t^T\hat{\mathbf{Z}}_t\, dt$$

which gives

$$Y_t = \int_0^t \hat{\mathbf{Z}}_s{}^T \, d\mathbf{X}_s - \frac{1}{2}\int_0^t \hat{\mathbf{Z}}_s{}^T \hat{\mathbf{Z}}_s \, ds$$

and

$$L_t = \exp\left(\int_0^t \hat{\mathbf{Z}}_s \, d\mathbf{X}_s - \frac{1}{2}\int_0^t \hat{\mathbf{Z}}_s{}^T \hat{\mathbf{Z}}_s \, ds\right)$$

8. (a) $\frac{1}{2}\sigma^2(t) \dfrac{\partial^2 p}{\partial z^2} + \frac{1}{2}\beta^2(t) \dfrac{\partial^2 p}{\partial \hat{z}^2} - \dfrac{\partial}{\partial z}(m(t)zp) - \dfrac{\partial}{\partial \hat{z}}[(\alpha(t)\hat{z} + \beta(t)z)p] = \dfrac{\partial p}{\partial t}$

(b) The requirement means that $p(z,\hat{z},t)$ must have the form

$$p(z,\hat{z},t) = \frac{1}{\sqrt{2\pi a(t)}} \exp\left[-\frac{1}{2a(t)}\hat{z}^2\right] \frac{1}{\sqrt{2\pi b(t)}} \exp\left[-\frac{1}{2b(t)}(z-\hat{z})^2\right]$$

If we substitute the desired form of p into the partial differential equation and collect terms, we get the following equations:

$$\frac{\dot{a}}{2a} = \frac{\beta^2}{2a} + (\alpha + \beta)$$

$$\frac{\dot{b}}{2b} = \left(\frac{\sigma^2}{2b} + \frac{\beta^2}{2b} + m - \beta\right)$$

$$\frac{m}{b} + \frac{\beta}{a} = \frac{\beta^2}{ab} + \frac{\alpha + \beta}{b}$$

The first two equations can be solved for a and b with $a(0) = b(0) = 0$. Upon substituting the results in the third equation, we get a relationship relating α and β to m and σ^2.

(c) We note that only β and not α affects $E(Z_t - \hat{Z}_t)^2 = b(t)$. A standard application of calculus of variation yields the equation

$$\dot{\beta}(t) - 2m(t)\,\beta(t) + \beta^2(t) = \sigma^2(t) \qquad \beta(0) = 0$$

for the minimizing β. The function α can then be found from the previously obtained relationship.

9. The function $\beta(\cdot)$ must satisfy

$$\dot{\beta}(t) - 2\beta(t) + \beta^2(t) = 1 \qquad \beta(0) = 0$$

or

$$\frac{d}{dt}[\beta(t) - 1] + [\beta(t) - 1]^2 = 2$$

Setting $\dot{u}(t)/u(t) = \beta(t) - 1$, we get

$$\frac{\ddot{u}(t)}{u(t)} = 2$$

With the initial condition $u(0)/u(0) = \beta(0) - 1 = -1$, we get

$$\frac{\dot{u}(t)}{u(t)} = \sqrt{2}\,\frac{\sqrt{2}\sinh\sqrt{2}\,t - \cosh\sqrt{2}\,t}{\sqrt{2}\cosh\sqrt{2}\,t - \sinh\sqrt{2}}$$

$$\beta(t) = 1 + \frac{\dot{u}(t)}{u(t)} = \frac{\sinh \sqrt{2}\, t}{\sqrt{2} \cosh \sqrt{2}\, t - \sinh \sqrt{2}\, t}$$

Now, we can solve the equation

$$\frac{\dot{b}(t)}{2b(t)} = \frac{\sigma^2(t)}{2b(t)} + \frac{\beta^2(t)}{2b(t)} + m(t) - \beta(t) \qquad b(0) = 0$$

and get

$$\begin{aligned} b(t) &= \int_0^t [1 + \beta^2(s)] \exp\left\{-\int_s^t [1 + \beta(\tau)]\, d\tau\right\} ds \\ &= \int_0^t [1 + \beta^2(t-u)] \exp\left\{-\int_{t-u}^t [1 + \beta(\tau)]\, d\tau\right\} du \\ &= \int_0^t [1 + \beta^2(t-u)] \exp\left\{-\int_0^u [1 + \beta(t-\tau)]\, d\tau\right\} du \\ &\xrightarrow[t\to\infty]{} \int_0^\infty (1 + \beta^2) e^{-(1+\beta)u}\, du = \frac{1+\beta}{1+\beta} \end{aligned}$$

where $\beta = \beta(\infty) = 1/(\sqrt{2} - 1) = 1 + \sqrt{2}$. Therefore,

$$b(t) \xrightarrow[t\to\infty]{} \frac{1 + (3 + 2\sqrt{2})}{2 + \sqrt{2}} = 2$$

10. The recursive equation for $\hat{Z}_t$ is

$$d\hat{Z}_t = (-\hat{Z}_t - 2\hat{Z}_t^2 + \hat{Z}_t^3)\, dt + (1 - \hat{Z}_t^2)\, dX_t$$

The observation equation is

$$dX_t = Z_t\, dt + dW_t$$

so that

$$d\hat{Z}_t = [-\hat{Z}_t - 2\hat{Z}_t^2 + \hat{Z}_t^3 + (1 - \hat{Z}_t^2)Z_t]\, dt + (1 - \hat{Z}_t^2)\, dW_t$$

It is clear that given Z, the $\hat{Z}$ process is a diffusion process and its density $p(\zeta,t)$ satisfies

$$\frac{1}{2}\frac{\partial^2}{\partial\zeta^2}[(1 - \zeta^2)p] - \frac{\partial}{\partial\zeta}\{[-\zeta(-\zeta^2 + 2\zeta + 1) + (1 - \zeta^2)Z_t]p\} = \frac{\partial p}{\partial t}$$

Assuming that $\partial p/\partial t \xrightarrow[t\to\infty]{} 0$, we get

$$\frac{d}{d\zeta}[(1 - \zeta^2)p(\zeta)] = [-\zeta(1 + 2\zeta - \zeta^2) + (1 - \zeta^2)Z_t]p(\zeta)$$

$$\ln\left[\frac{(1 - \zeta^2)p(\zeta)}{p(0)}\right] = Z_t\zeta + \ln(1 - \zeta^2) + \frac{\zeta^2}{2}$$

$$p(\zeta) = p(0)e^{\zeta^2/2}e^{\zeta Z_t} \qquad -1 \le \zeta \le 1$$

Averaging over $Z_t = 1$ and -1, we get

$$p(\zeta) = p(0)e^{\zeta^2/2} \cosh \zeta$$

where $p(0) = \left(\int_{-1}^1 e^{\zeta^2/2} \cosh \zeta\, d\zeta\right)^{-1}$.

Index